THE
LIFE OF THE FLY

WITH WHICH ARE INTERSPERSED SOME
CHAPTERS OF AUTOBIOGRAPHY

BY

J. HENRI FABRE

TRANSLATED BY
ALEXANDER TEIXEIRA DE MATTOS
FELLOW OF THE ZOOLOGICAL SOCIETY OF LONDON

Fredonia Books
Amsterdam, The Netherlands

The Life of the Fly

by J. Henri Fabre

ISBN 1-58963-026-2

Reprinted from the 1913 edition

Fredonia Books
Amsterdam, The Netherlands
http://www.FredoniaBooks.com

THE LIFE OF THE FLY

CONTENTS

		PAGE
TRANSLATOR'S NOTE	. . . .	7

CHAPTER

I	THE HARMAS	11
II	THE ANTHRAX	28
III	ANOTHER PROBER (PERFORATOR)	63
IV	LARVAL DIMORPHISM . . .	78
V	HEREDITY	111
VI	MY SCHOOLING	131
VII	THE POND	162
VIII	THE CADDIS-WORM . . .	182
IX	THE GREENBOTTLES . . .	212
X	THE GREY FLESH-FLIES . .	233
XI	THE BUMBLE-BEE FLY . .	252
XII	MATHEMATICAL MEMORIES: NEWTON'S BINOMIAL THEOREM	276
XIII	MATHEMATICAL MEMORIES: MY LITTLE TABLE . .	299
XIV	THE BLUEBOTTLE: THE LAYING	316

Contents

CHAPTER   PAGE

		PAGE
XV	THE BLUEBOTTLE: THE GRUB	341
XVI	A PARASITE OF THE MAGGOT	364
XVII	RECOLLECTIONS OF CHILD-HOOD	387
XVIII	INSECTS AND MUSHROOMS	402
XIX	A MEMORABLE LESSON	426
XX	INDUSTRIAL CHEMISTRY	447

TRANSLATOR'S NOTE

THE present volume contains all the essays on Flies, or Diptera, from the *Souvenirs entomologiques,* to which I have added, in order to make the dimensions uniform with those of the other volumes of the series, the purely autobiographical essays comprised in the *Souvenirs.* These essays, though they have no bearing upon the life of the Fly, are among the most interesting that Henri Fabre has written and will, I am persuaded, make a special appeal to the reader. The chapter entitled *The Caddis-worm* has been included as following directly upon *The Pond.*

Since publishing *The Life of the Spider,* I was much struck by a passage in Dr. Chalmers Mitchell's stimulating work, *The Childhood of Animals,* in which the secretary of the Zoological Society of London says:

'I have attempted to avoid the use of terms familiar only to students of zoology and to refrain from anatomical detail, but at the same time to refrain from the irritating habit

7

of assuming that my readers have no knowledge, no dictionaries and no other books.'

I began to wonder whether I had gone too far in simplifying the terminology of the Fabre essays and in appending explanatory footnotes to the inevitable number of outlandish names of insects. But my doubts vanished when I thought upon Fabre's own words in the first chapter of this book:

'If I write for men of learning, for philosophers . . . I write above all things for the young. I want to make them love the natural history which you make them hate; and that is why, while keeping strictly to the domain of truth, I avoid your scientific prose, which too often, alas, seems borrowed from some Iroquois idiom!'

And I can but apologize if I have been too lavish with my notes to this chapter in particular, which introduces to us, as in a sort of litany, a multitude of the insects studied by the author. For the rest, I have continued my system of references to the earlier Fabre books, whether translated by myself or others.

Of the following essays, *The Harmas* has appeared, under another title, in *The Daily*

Translator's Note

Mail; The Pond, Industrial Chemistry and the two chapters on the Bluebottle in *The English Review;* and *The Harmas, The Pond* and *Industrial Chemistry* in the New York *Bookman.* The others are new to England and America, unless any of them should be issued in newspapers or magazines between this date and the publication of the book.

I wish once more to thank Miss Frances Rodwell for her assistance in the details of my work and in the verification of the many references; and my thanks are also due to Mr. Edward Cahen, who has been good enough to revise the two chemistry chapters for me, and to Mr. W. S. Graff Baker, who has performed the same kindly task towards the two chapters entitled *Mathematical Memories.*

ALEXANDER TEIXEIRA DE MATTOS.
CHELSEA, 8 *July,* 1913.

CHAPTER I

THE HARMAS

THIS is what I wished for, *hoc erat in votis:* a bit of land, oh, not so very large, but fenced in, to avoid the drawbacks of a public way; an abandoned, barren, sun-scorched bit of land, favoured by thistles and by Wasps and Bees. Here, without fear of being troubled by the passers-by, I could consult the Ammophila and the Sphex[1] and engage in that difficult conversation whose questions and answers have experiment for their language; here, without distant expeditions that take up my time, without tiring rambles that strain my nerves, I could contrive my plans of attack, lay my ambushes and watch their effects at every hour of the day. *Hoc erat in votis.* Yes, this was my wish, my dream, always cherished, always vanishing into the mists of the future.

And it is no easy matter to acquire a labo-

[1] Two species of Digger or Hunting Wasps. Cf. *Insect Life*, by J. H. Fabre, translated by the author of *Mademoiselle Mori:* chaps. vi to xii and xvi.—*Translator's Note.*

ratory in the open fields, when harassed by a terrible anxiety about one's daily bread. For forty years have I fought, with steadfast courage, against the paltry plagues of life; and the long-wished-for laboratory has come at last. What it has cost me in perseverance and relentless work I will not try to say. It has come; and, with it—a more serious condition —perhaps a little leisure. I say perhaps, for my leg is still hampered with a few links of the convict's chain.

The wish is realized. It is a little late, O my pretty insects! I greatly fear that the peach is offered to me when I am beginning to have no teeth wherewith to eat it. Yes, it is a little late: the wide horizons of the outset have shrunk into a low and stifling canopy, more and more straitened day by day. Regretting nothing in the past, save those whom I have lost; regretting nothing, not even my first youth; hoping nothing either, I have reached the point at which, worn out by the experience of things, we ask ourselves if life be worth the living.

Amid the ruins that surround me, one strip of wall remains standing, immovable upon its solid base: my passion for scientific truth. Is that enough, O my busy insects, to enable me to add yet a few seemly pages to your history? Will my strength not cheat my good intentions? Why, indeed, did I forsake you so long? Friends have reproached me for it. Ah, tell them, tell those friends, who are yours as well as mine, tell them that it was not forgetfulness on my part, not weariness, nor neglect: I thought of you; I was convinced that the Cerceris[1] cave had more fair secrets to reveal to us, that the chase of the Sphex held fresh surprises in store. But time failed me; I was alone, deserted, struggling against misfortune. Before philosophizing, one had to live. Tell them that; and they will pardon me.

Others again have reproached me with my style, which has not the solemnity, nay, better, the dryness of the schools. They fear lest a page that is read without fatigue should not always be the expression of the truth. Were I to take their word for it, we are profound only on condition of being obscure. Come here, one and all of you—you, the sting-bearers, and you, the wing-cased armour-clads—take up my defence and bear witness in my favour. Tell of the intimate terms on which I live with you, of the patience with which I observe you, of

---

[1] A species of Digger Wasp. Cf. *Insect Life:* chaps vi to xii and xvi.—*Translator's Note.*

the care with which I record your actions. Your evidence is unanimous: yes, my pages, though they bristle not with hollow formulas nor learned smatterings, are the exact narrative of facts observed, neither more nor less; and whoso cares to question you in his turn will obtain the same replies.

And then, my dear insects, if you cannot convince those good people, because you do not carry the weight of tedium, I, in my turn, will say to them:

'You rip up the animal and I study it alive; you turn it into an object of horror and pity, whereas I cause it to be loved; you labour in a torture-chamber and dissecting-room, I make my observations under the blue sky to the song of the Cicadas,[1] you subject cell and protoplasm to chemical tests, I study instinct in its loftiest manifestations; you pry into death, I pry into life. And why should I not complete my thought: the boars have muddied the clear stream; natural history, youth's glorious study, has, by dint of cellular improvements, become a hateful and repulsive thing. Well, if I write for men of learning, for philosophers, who,

[1]The Cicada is the *Cigale,* an insect akin to the Grasshopper and found more particularly in the south of France. Cf. *Social Life in the Insect World,* by J. H. Fabre, translated by Bernard Miall: chaps. i to iv.— *Translator's Note.*

one day, will try to some extent to unravel the tough problem of instinct, I write also, I write above all things for the young. I want to make them love the natural history which you make them hate; and that is why, while keeping strictly to the domain of truth, I avoid your scientific prose, which too often, alas seems borrowed from some Iroquois idiom!'

But this is not my business for the moment: I want to speak of the bit of land long cherished in my plans to form a laboratory of living entomology, the bit of land which I have at last obtained in the solitude of a little village. It is a *harmas*, the name given, in this district,[1] to an untilled, pebbly expanse abandoned to the vegetation of the thyme. It is too poor to repay the work of the plough; but the sheep passes there in spring, when it has chanced to rain and a little grass shoots up.

My harmas, however, because of its modicum of red earth swamped by a huge mass of stones, has received a rough first attempt at cultivation: I am told that vines once grew here. And, in fact, when we dig the ground before planting a few trees, we turn up, here and there, remains of the precious stock, half-car-

[1] The country round Sérignan, in Provence.—*Translator's Note.*

bonized by time. The three-pronged fork, therefore, the only implement of husbandry that can penetrate such a soil as this, has entered here; and I am sorry, for the primitive vegetation has disappeared. No more thyme, no more lavender, no more clumps of kermes-oak, the dwarf oak that forms forests across which we step by lengthening our stride a little. As these plants, especially the first two, might be of use to me by offering the Bees and Wasps a spoil to forage, I am compelled to reinstate them in the ground whence they were driven by the fork.

What abounds without my mediation is the invaders of any soil that is first dug up and then left for a long time to its own resources. We have, in the first rank, the couch-grass, that execrable weed which three years of stubborn warfare have not succeeded in exterminating. Next, in respect of number, come the centauries, grim-looking one and all, bristling with prickles or starry halberds. They are the yellow-flowered centaury, the mountain centaury, the star-thistle and the rough centaury: the first predominates. Here and there, amid their inextricable confusion, stands, like a chandelier with spreading, orange flowers for lights, the fierce Spanish oyster-plant, whose spikes are strong as nails. Above it, towers the Illyrian cotton-thistle, whose straight and solitary stalk soars to a height of three to six feet and ends in large pink tufts. Its armour hardly yields before that of the oyster-plant. Nor must we forget the lesser thistle-tribe, with first of all, the prickly or 'cruel' thistle, which is so well armed that the plant-collector knows not where to grasp it; next, the spear-thistle, with its ample foliage, ending each of its veins with a spear-head; lastly, the black knap-weed, which gathers itself into a spiky knot. In among these, in long lines armed with hooks, the shoots of the blue dewberry creep along the ground. To visit the prickly thicket when the Wasp goes foraging, you must wear boots that come to mid-leg or else resign yourself to a smarting in the calves. As long as the ground retains a few remnants of the vernal rains, this rude vegetation does not lack a certain charm, when the pyramids of the oyster-plant and the slender branches of the cotton-thistle rise above the wide carpet formed by the yellow-flowered centaury saffron heads; but let the droughts of summer come and we see but a desolate waste, which the flame of a match would set ablaze from one end to the other. Such is, or rather was, when I took possession of it, the Eden of bliss where I mean to live henceforth alone with the insect. Forty years of desperate struggle have won it for me.

Eden, I said; and, from the point of view that interests me, the expression is not out of place. This cursed ground, which no one would have had at a gift to sow with a pinch of turnip-seed, is an earthly paradise for the Bees and Wasps. Its mighty growth of thistles and centauries draws them all to me from everywhere around. Never, in my insect-hunting memories, have I seen so large a population at a single spot; all the trades have made it their rallying-point. Here come hunters of every kind of game, builders in clay, weavers of cotton goods, collectors of pieces cut from a leaf or the petals of a flower, architects in pasteboard, plasterers mixing mortar, carpenters boring wood, miners digging underground galleries, workers handling goldbeater's skin and many more.

Who is this one? An Anthidium.[1] She scrapes the cobwebby stalk of the yellow-flowered centaury and gathers a ball of wadding which she carries off proudly in the tips of her mandibles. She will turn it, under ground, into cotton-felt satchels to hold the

[1] A Tailor-bee.—*Translator's Note.*

store of honey and the egg. And these others,
so eager for plunder? They are Megachiles,[1]
carrying under their bellies their black, white
or blood-red reaping-brushes. They will leave
the thistles to visit the neighbouring shrubs and
there cut from the leaves oval pieces which will
be made into a fit receptacle to contain the
harvest. And these, clad in black velvet? They
are Chalicodomæ,[2] who work with cement
and gravel. We could easily find their masonry on the stones in the harmas. And these,
noisily buzzing with a sudden flight? They
are the Anthophoræ,[3] who live in the old walls
and the sunny banks of the neighbourhood.

Now come the Osmiæ. One stacks her cells
in the spiral staircase of an empty snail-shell;
another, attacking the pith of a dry bit of
bramble, obtains for her grubs a cylindrical
lodging and divides it into floors by means of
partition-walls; a third employs the natural
channel of a cut reed; a fourth is a rent-free
tenant of the vacant galleries of some Mason-
bee. Here are the Macroceræ and the Euceræ, whose males are proudly horned; the
Dasypodæ, who carry an ample brush of bris-

[1] Leaf-cutting Bees.—*Translator's Note.*
[2] Mason-bees. Cf. *Insect Life:* chaps. xx to xxii.—*Translator's Note.*
[3] A species of Wild Bees.—*Translator's Note.*

tles on their hind-legs for a reaping implement; the Andrenæ, so manifold in species; the slender-bellied Halicti.[1] I omit a host of others. If I tried to continue this record of the guests of my thistles, it would muster almost the whole of the honey-yielding tribe. A learned entomologist of Bordeaux, Professor Pérez, to whom I submit the naming of my prizes, once asked me if I had any special means of hunting, to send him so many rarities and even novelties. I am not at all an experienced and, still less, a zealous hunter, for the insect interests me much more when engaged in its work than when struck on a pin in a cabinet. The whole secret of my hunting is reduced to my dense nursery of thistles and centauries.

By a most fortunate chance, with this populous family of honey-gatherers was allied the whole hunting tribe. The builders' men had distributed here and there in the harmas great mounds of sand and heaps of stones, with a view to running up some surrounding walls. The work dragged on slowly; and the mate-

[1]Osmiæ, Macroceræ, Euceræ, Dasypodæ, Andrenæ and Haliciti are all different species of Wild Bees. For the Haliciti, cf. *The Life and Love of the Insect*, by J. Henri Fabre, translated by Alexander Teixeira de Mattos: chaps. xv and xvi.—*Translator's Note.*

rials found occupants from the first year. The Mason-bees had chosen the interstices between the stones as a dormitory where to pass the night, in serried groups. The powerful Eyed Lizard, who, when close-pressed, attacks both man and dog, wide-mouthed, had selected a cave wherein to lie in wait for the passing Scarab;[1] the Black-eared Chat, garbed like a Dominican, white-frocked with black wings, sat on the top stone, singing his short rustic lay: his nest, with its sky-blue eggs, must be somewhere in the heap. The little Dominican disappeared with the loads of stones. I regret him: he would have been a charming neighbour. The Eyed Lizard I do not regret at all.

The sand sheltered a different colony. Here, the Bembeces[2] were sweeping the threshold of their burrows, flinging a curve of dust behind them; the Languedocian Sphex was dragging her Ephippigera[3] by the antennæ; a Stizus[4] was storing her preserves of Cicadellæ.[5]

[1] A Dung-beetle also known as the Sacred Beetle. Cf. *Insect Life:* chaps i and ii; and *The Life and Love of the Insect:* chaps. i to iv.—*Translator's Note.*
[2] A species of Digger-wasps. Cf. *Insect Life:* chap. xvi.—*Translator's Note.*
[3] A species of green Grasshopper.—*Translator's Note.*
[4] A species of Hunting Wasp.—*Translator's Note.*
[5] Froghoppers.—*Translator's Note.*

The Life of the Fly

To my sorrow, the masons ended by evicting the sporting tribe; but, should I ever wish to recall it, I have but to renew the mounds of sand: they will soon all be there.

Hunters that have not disappeared, their homes being different, are the Ammophilæ, whom I see fluttering, one in spring, the others in autumn, along the garden-walks and over the lawns, in search of a Caterpillar; the Pompili,[1] who travel alertly, beating their wings and rummaging in every corner in quest of a Spider. The largest of them waylays the Narbonne Lycosa,[2] whose burrow is not infrequent in the harmas. This burrow is a vertical well, with a curb of fescue-grass intertwined with silk. You can see the eyes of the mighty Spider gleam at the bottom of the den like little diamonds, an object of terror to most. What a prey and what dangerous hunting for the Pompilus! And here, on a hot summer afternoon, is the Amazon-ant, who leaves her barrack-rooms in long battalions and marches far afield to hunt for slaves. We

[1] The Pompilus is a species of Digger or Hunting Wasp, known also as the Ringed Calicurgus. Cf. *The Life and Love of the Insect:* chap. xii.—*Translator's Note.*
[2] Known also as the Black-bellied Tarantula. Cf. *The Life and Love of the Insect:* chap. xii; and *The Life of the Spider:* chaps. i and iii to vi.—*Translator's Note.*

will follow her in her raids when we find time. Here again, around a heap of grasses turned to mould, are Scoliæ[1] an inch and a half long, who fly gracefully and dive into the heap, attracted by a rich prey, the grubs of Lamellicorns, Oryctes and Cetoniæ.[2]

What subjects for study! And there are more to come. The house was as utterly deserted as the ground. When man was gone and peace assured, the animal hastily seized on everything. The Warbler took up his abode in the lilac-shrubs; the Greenfinch settled in the thick shelter of the cypresses; the Sparrow carted rags and straw under every slate; the Serin-finch, whose downy nest is no bigger than half an apricot, came and chirped in the plane-tree-tops; the Scops made a habit of uttering his monotonous, piping note here, of an evening; the bird of Pallas Athene, the Owl, came hurrying along to hoot and hiss.

In front of the house is a large pond, fed by the aqueduct that supplies the village-pumps with water. Here, from half a mile and more around, come the Frogs and Toads in the lovers' season. The Natterjack, some-

[1] Large Hunting Wasps. Cf. *The Life and Love of the Insect:* chap. xi.—*Translator's Note.*
[2] Different species of Beetles. The Cetonia is the Rose-chafer.—*Translator's Note.*

times as large as a plate, with a narrow stripe of yellow down his back, makes his appointments here to take his bath; when the evening twilight falls, we see hopping along the edge the Midwife Toad, the male, who carries a cluster of eggs, the size of peppercorns, wrapped round his hindlegs: the genial paterfamilias has brought his precious packet from afar, to leave it in the water and afterwards retire under some flat stone, whence he will emit a sound like a tinkling bell. Lastly, when not croaking amid the foliage, the Tree-frogs indulge in the most graceful dives. And so, in May, as soon as it is dark, the pond becomes a deafening orchestra: it is impossible to talk at table, impossible to sleep. We had to remedy this by means perhaps a little too rigorous. What could we do? He who tries to sleep and cannot needs becomes ruthless.

Bolder still, the Wasp has taken possession of the dwelling-house. On my door-sill, in a soil of rubbish, nestles the White-banded Sphex: when I go indoors, I must be careful not to damage her burrows, not to tread upon the miner absorbed in her work. It is quite a quarter of a century since I last saw the saucy Cricket-hunter. When I made her acquaintance, I used to visit her at a few miles' distance: each time, it meant an expedition under the blazing August sun. To-day, I find her at my door; we are intimate neighbours. The embrasure of the closed window provides an apartment of a mild temperature for the Pelopæus.[1] The earth-built nest is fixed against the freestone wall. To enter her home, the Spider-huntress uses a little hole left open by accident in the shutters. On the mouldings of the Venetian blinds, a few stray Mason-bees build their group of cells; inside the outer shutters, left ajar, a Eumenes[2] constructs her little earthen dome, surmounted by a short, bell-mouthed neck. The common Wasp and the Polistes[3] are my dinner-guests: they visit my table to see if the grapes served are as ripe as they look.

Here, surely—and the list is far from complete—is a company both numerous and select, whose conversation will not fail to charm my solitude, if I succeed in drawing it out. My dear beasts of former days, my old friends, and others, more recent acquaintances, all are here, hunting, foraging, building in close proximity. Besides, should we wish to vary the

[1] A species of Mason-wasp.—*Translator's Note.*
[2] A species of Mason-wasp.—*Translator's Note.*
[3] A species of Solitary Wasp.—*Translator's Note.*

scene of observation, the mountain[1] is but a few hundred steps away, with its tangle of arbutus, rock-roses and arborescent heather; with its sandy spaces dear to the Bembeces; with its marly slopes exploited by different Wasps and Bees. And that is why, foreseeing these riches, I have abandoned the town for the village and come to Sérignan to weed my turnips and water my lettuces.

Laboratories are being founded, at great expense, on our Atlantic and Mediterranean coasts, where people cut up small sea-animals, of but meagre interest to us; they spend a fortune on powerful microscopes, delicate dissecting-instruments, engines of capture, boats, fishing-crews, aquariums, to find out how the yolk of an Annelid's[2] egg is constructed, a question whereof I have never yet been able to grasp the full importance; and they scorn the little land-animal, which lives in constant touch with us, which provides universal psychology with documents of inestimable value, which too often threatens the public wealth by destroying our crops. When shall we have an entomological laboratory for the study not of

[1]Mont Ventoux, an outlying summit of the Alps, 6,270 feet high. Cf. *Insect Life:* chap. xiii.—*Translator's Note.*
[2]A red-blooded Worm.—*Translator's Note.*

the dead insect, steeped in alcohol, but of the
living insect; a laboratory having for its object
the instinct, the habits, the manner of living,
the work, the struggles, the propagation of
that little world, with which agriculture and
philosophy have most seriously to reckon?

To know thoroughly the history of the de-
stroyer of our vines might perhaps be more important than to know how this or that nerve-
fibre of a Cirriped[1] ends; to establish by experiment the line of demarcation between in-
tellect and instinct; to prove, by comparing
facts in the zoological progression, whether
human reason be an irreducible faculty or not:
all this ought surely to take precedence of the
number of joints in a Crustacean's antenna.
These enormous questions would need an army
of workers; and we have not one. The fashion is all for the Mollusc and the Zoophytes.[2]
The depths of the sea are explored with many
drag-nets; the soil which we tread is consistent-
ly disregarded. While waiting for the fashion
to change, I open my harmas laboratory of
living entomology; and this laboratory shall
not cost the ratepayers one farthing.

[1]Cirripeds are sea-animals with hair-like legs, including the Barnacles and Acorn-shells.—*Translator's Note*.
[2]Zoophytes are plant-like sea-animals, including Star-
fishes, Jelly-fishes, Sea-anemones and Sponges.—*Translator's Note*.

CHAPTER II

THE ANTHRAX

I MADE the acquaintance of the Anthrax in 1855 at Carpentras, at the time when the life-history of the Oil-beetles was causing me to search the tall slopes beloved of the Anthophora-bees.[1] Her curious pupæ, so powerfully equipped to force an outlet for the perfect insect incapable of the least effort, those pupæ armed with a multiple ploughshare at the fore, a trident at the rear and rows of harpoons on the back wherewith to rip open the Osmia-bee's cocoon and break through the hard crust of the hill-side, betokened a field that was worth cultivating. The little that I said about her at the time brought me urgent entreaties: I was asked for a circumstantial chapter on the strange Fly. The stern necessities of life postponed to an ever-retreating future my beloved investigations, so miserably stifled. Thirty years have passed; at last, a little leisure is at hand; and here, in the harmas of my village, with an ardour that has in no wise grown old,

[1] A species of Mason-bees.—*Translator's Note.*

The Anthrax

I have resumed my plans of yore, still alive
like the coal smouldering under the ashes.
The Anthrax has told me her secrets, which
I in my turn am going to divulge. Would
that I could address all those who cheered me
on this path, including first and foremost the
revered Master of the Landes.[1] But the ranks
have thinned, many have been promoted to
another world and their disciple lagging behind them can but record, in memory of those
who are no more, the story of the insect clad
in deepest mourning.

In the course of July, let us give a few sideward knocks to the bracing pebbles and detach
the nests of the Chalicodoma of the Walls[2]
from their supports. Loosened by the shock,
the dome comes off cleanly, all in one piece.
Moreover—and this is a great advantage—
the cells come into view wide open on the base
of the exposed nest, for at this point they have
no other wall than the surface of the pebble.
In this way, without any scraping, which would
be wearisome work for the operator and
dangerous to the inhabitants of the dome, we

[1] Léon Dufour (1780-1865), also described by the author as the "Wizard of the Landes." Cf. *The Life of
the Spider:* chap. i.—*Translator's Note.*
[2] A Mason-bee. Cf. *Insect Life:* chap. xx.—*Translator's
Note.*

have all the cells before our eyes, together
with their contents, consisting of a silky, am-
ber-yellow cocoon, as delicate and translucent
as an onion-peeling. Let us split the dainty
wrapper with the scissors, chamber by chamber, nest by nest. If fortune be at all propitious, as it always is to the persevering,
we shall end by finding that the cocoons har-
bour two larvæ together, one more or less
faded in appearance, the other fresh and
plump. We shall also find some, no less plentiful, in which the withered larva is accompanied by a family of little grubs wriggling
uneasily around it.

Examination at once reveals the tragedy
that is happening under the cover of the cocoon. The flacid and faded larva is the
Mason-bee's. A month ago, in June, having
finished its mess of honey, it wove its silken
sheath for a bedchamber wherein to take the
long sleep which is the prelude to the metamorphosis. Bulging with fat, it is a rich and
defenceless morsel for whoso is able to reach
it. Then, in spite of apparently unsurmount-
able obstacles, the mortar wall and the tent
without an opening, the flesh-eating larvæ appeared in the secret retreat and are now glut-
ting themselves on the sleeper. Three differ-

ent species take part in the carnage, often in the same nest, in adjoining cells. The diversity of shapes informs us of the presence of more than one enemy; the final stage of the creatures will tell us the names and qualities of the three invaders.

Forestalling the secrets of the future for the sake of greater clearness, I will anticipate the actual facts and come at once to the results produced. When it is by itself on the body of the Mason-bee's larva, the murderous grub belongs either to *Anthrax trifasciata*, MEIGEN, or to *Leucospis gigas*, FAB. But, if numerous little worms, often a score and more, swarm around the victim, then it is a Chalcidid's family which we have before us. Each of these ravagers shall have its biography. Let us begin with the Anthrax.

And first the grub, as it is after consuming its victim, when it remains the sole occupant of the Mason-bee's cocoon. It is a naked worm, smooth, legless and blind, of a creamy deadwhite, each segment a perfect ring, very much curved when at rest, but with the tendency to become almost straight when disturbed. Through the diaphanous skin, the lens distinguishes patches of fat, which are the cause of its characteristic colouring. When younger, as a tiny grub a few millimetres long, it is streaked with two different kinds of stains, some white, opaque and of a creamy tint, others translucent and of the palest amber. The former come from adipose masses in course of formation; the second from the nourishing fluid or from the blood which laves those masses.

Including the head, I count thirteen segments. In the middle of the body these segments are well-marked, being separated by a slight groove; but in the fore-part they are difficult to count. The head is small and is soft, like the rest of the body, with no sign of any mouth-parts even under the close scrutiny of the lens. It is a white globule, the size of a tiny pin's head and continued at the back by a pad a little larger, from which it is separated by a scarcely appreciable crease. The whole is a sort of nipple swelling slightly on the upper surface; and its double structure is so difficult to perceive that at first we take it for the animal's head alone, though it includes both the head and the prothorax, or first segment of the thorax.

The mesothorax, or middle segment of the thorax, which is two or three times larger in diameter, is flattened in front and separated from the nipple formed by the prothorax and the head by a deep, narrow, curved fissure. On its front surface are two pale-red stigmata, or respiratory orifices, placed pretty close together. The metathorax, or last segment of the thorax, is a little larger still in diameter and protrudes. These abrupt increases in circumference result in a marked hump, sloping sharply towards the front. The nipple of which the head forms part is set at the bottom of this hump.

After the metathorax, the shape becomes regular and cylindrical, while decreasing slightly in girth in the last two or three segments. Close to the line of separation of the last two rings, I am able to distinguish, not without difficulty, two very small stigmata, just a little darker in colour. They belong to the last segment. In all, four respiratory orifices, two in front and two behind, as is the rule among Flies. The length of the full-sized larva is 15 to 20 millimetres[1] and its breadth 5 to 6.[2]

Remarkable in the first place by the protuberance of its thorax and the smallness of its head, the grub of the Anthrax acquires ex-

[1] .58 to .78 inch.—*Translator's Note.*
[2] .19 to .23 inch.—*Translator's Note.*

ceptional interest by its manner of feeding.
Let us begin by observing that, deprived of
all, even the most rudimentary walking-ap-
paratus, the animal is absolutely incapable of
shifting its position. If I disturb its rest, it
curves and straightens itself in turns by a series
of contractions, it tosses about violently where
it lies, but does not manage to progress. It
fidgets and gets no farther. We shall see later
the magnificent problem raised by this inert-
ness.

For the moment, a most unexpected fact
claims all our attention. I refer to the extreme readiness with which the Anthrax' larva
quits and returns to the Chalicodoma-grub on
which it is feeding. After witnessing flesh-
eating larvæ at hundreds and hundreds of
meals, I suddenly find myself confronted with
a manner of eating that bears no relation to
anything which I have seen before. I feel myself in a world that baffles my old experience.
Let us recall the table-manners of a larva living on prey, the Ammophila's for instance,
when devouring its caterpillar. A hole is
made in the victim's side; and the head and
neck of the nursling dive deep into the
wound, to root luxuriously among the entrails.
There is never a withdrawal from the gnawed

The Anthrax

belly, never a recoil to interrupt the feast and
to take breath awhile. The vivacious animal
always goes forward, chewing, swallowing, digesting, until the caterpillar's skin is emptied
of its contents. Once seated at table, it does
not budge as long as the victuals last. To
tease it with a straw is not always enough to
induce it to withdraw its head outside the
wound; I have to use violence. When removed by force and then left to its own devices, the creature hesitates for a long time,
stretches itself and mouths around, without
trying to open a passage through a new
wound. It needs the attacking-point that has
just been abandoned. If it finds the spot, it
makes its way in and resumes the work of eating; but its future is jeopardized from this
time forward, for the game, now perhaps
tackled at inopportune points, is liable to go
bad.

With the Anthrax' grub, there is none of
this mangling, none of this persistent clinging
to the entrance-wound. I have but to tease
it with the tip of a hair-pencil and forthwith
it retires; and the lens reveals no wound at the
abandoned spot, no such effusion of blood as
there would be if the skin were perforated.
When its sense of security is restored, the grub

once more applies its pimple-head to the fostering larva, at any point, no matter where; and, so long as my curiosity does not prevent it, keeps itself fixed there, without the least effort, or the least perceptible movement that could account for the adhesion. If I repeat the touch with the pencil, I see the same sudden retreat and, soon after, the same contact just as readily renewed.

This facility for gripping, quitting and re-gripping, now here, now there and always without a wound, the part of the victim whence the nourishment is drawn tells us of itself that the mouth of the Anthrax is not armed with mandibular fangs capable of digging into the skin and tearing it. If the flesh were gashed by any such pincers, one or two attempts would be necessary before they could be released or reapplied; besides, each point bitten would display a lesion. Well, there is nothing of the kind: a conscientious examination through the magnifying-glass shows conclusively that the skin is intact; the grub glues its mouth to its prey or withdraws it with an ease that can only be explained by a process of simple contact. This being so, the Anthrax does not chew its food as do the other carnivorous grubs; it does not eat, it inhales.

36

The Anthrax

This method of taking nourishment implies an exceptional apparatus of the mouth, into which it behoves us to enquire before continuing. My most powerful magnifying-glass at last discovers, at the centre of the pimple-head, a small spot of an amber-russet colour; and that is all. For a more exhaustive examination we will employ the microscope. I cut off the strange pimple with the scissors, wash it in a drop of water and place it on the object-slide. The mouth now stands revealed as a round spot which, for hue and for the smallness of its size, may be compared with the front stigmata. It is a small conical crater, with sides of a pale yellowish-red and with faint, more or less concentric lines. At the bottom of this funnel is the opening of the gullet, itself tinted red in front and promptly spreading into a cone at the back. There is not the slightest trace of mandibular fangs, of jaws, of mouth-parts for seizing and grinding. Everything is reduced to the bowl-shaped opening, with a delicate lining of horny texture, as is shown by the amber hue and the concentric streaks. When I look for some term to designate this digestive entrance, of which so far I know no other example, I can find only that of a sucker or cupping-glass. Its attack is a mere kiss, but what a perfidious kiss!

We know the machine; now let us see the working. To facilitate observation, I shifted the new-born Anthrax-grub, together with the Chalicodoma-grub, its wet-nurse, from the natal cell into a glass tube. I was thus able, by employing as many tubes as I wanted, to follow from start to finish, in all its most intimate details, the strange repast which I am going to describe.

The worm is fixed by its sucker to any convenient part of the nurse, plump and fat as butter. It is ready to break off its kiss suddenly, should anything disquiet it, and to resume it as easily when tranquillity is restored. No Lamb enjoys greater liberty with its mother's teat. After three or four days of this contact of the nurse and nurseling, the former, at first replete and endowed with the glossy skin that is a sign of health, begins to assume a withered aspect. Her sides fall in, her fresh colour fades, her skin becomes covered with little folds and gives evidence of an appreciable shrinking in this breast which, instead of milk, yields fat and blood. A week is hardly past before the progress of the exhaustion becomes startlingly rapid. The

The Anthrax

nurse is flabby and wrinkled, as though borne
down by her own weight, like a very slack
object. If I move her from her place, she
flops and sprawls like a half-filled water-bottle
over the new supporting-plane. But the
Anthrax' kiss goes on emptying her: soon she
is but a sort of shrivelled lard-bag, decreasing
from hour to hour, from which the sucker
draws a few last oily drains. At length, between the twelfth and the fifteenth day, all
that remains of the larva of the Mason-bee is
a white granule, hardly as large as a pin's
head.

This granule is the water-bottle drained to
the last drop, is the nurse's breast emptied of
all its contents. I soften the meagre remnant
in water; then, keeping it still immersed, I
blow into it through an extremely attenuated
glass tube. The skin fills out, distends and
resumes the shape of the larva, without there
being an outlet anywhere for the compressed
air. It is intact, therefore; it is free of any
perforation, which would be forthwith revealed under the water by an escape of gas.
And so, under the Anthrax' cupping-glass, the
oily bottle has been drained by a simple transpiration through the membrane; the substance
of the nurse-grub has been transfused into the

body of the nurseling by a process akin to that known in physics as endosmosis.[1] What should we say to a method of being suckled by the mere application of the mouth to a teatless breast? What we see here may be compared with that: without any outlet, the milk of the Chalicodoma-grub passes into the stomach of the Anthrax' larva.

Is it really an instance of endosmosis? Might it not rather be atmospheric pressure that stimulates the flow of nourishing fluids and distils them into the Anthrax' cup-shaped mouth, working, in order to create a vacuum, almost like the suckers of the Cuttlefish? All this is possible, but I shall refrain from deciding, preferring to assign a large share to the unknown in this extraordinary method of nutrition. It ought, I think, to provide physiologists with a field of research in which new views on the hydrodynamics of live fluids might well be gleaned; and this field trenches upon others that would also yield rich harvests. The brief span of my days compels me to set the problem without seeking to solve it.

And the second problem is this: the Chali-

[1] The transmission of a fluid inward through a porous partition which separates it from any fluid of a different density.—*Translator's Note.*

The Anthrax

codoma-grub destined to feed the Anthrax is without a wound of any kind. The mother of the tiny larva is a feeble Fly deprived of whatsoever weapon capable of injuring her offspring's prey. Moreover, she is absolutely powerless to penetrate the Mason-bee's fortress, powerless as a fluff of down against a rock. On this point there is no doubt: the future wet-nurse of the Anthrax has not been paralyzed as are the live provisions collected by the Hunting Wasps; she has received no bite nor scratch nor contusion of any sort; she has experienced nothing out of the common: in short, she is in her normal state. The billeted nurseling arrives, we shall presently see how; he arrives, scarcely visible, almost defying the scrutiny of the lens; and, having made his preparations, he instals himself, he, the atom, upon the monstrous nurse, whom he is to drain to the very husk. And she, not paralyzed by a preliminary vivisection, endowed with all her normal vitality, lets him have his way, lets herself be sucked dry, with the utmost apathy. Not a tremor in her outraged flesh, not a quiver of resistance. No corpse could show greater indifference to the bite which it receives.

Ah, but the maggot has chosen the hour of

attack with traitorous cunning! Had it appeared upon the scene earlier, when the larva was consuming its store of honey, things of a surety would have gone badly with it. The assaulted one, feeling herself bled to death by that ravenous kiss, would have protested with much wriggling of body and grinding of mandibles. The position would have ceased to be tenable and the intruder would have perished. But at this hour all danger has disappeared. Enclosed in its silken tent, the larva is seized with the lethargy that precedes the metamorphosis. Its condition is not death, but neither is it life. It is an intermediary condition; it is almost the latent vitality of grain or egg. Therefore there is no sign of irritation on the larva's part under the needle with which I stir it and still less under the sucker of the Anthrax-grub, which is able to drain the affluent breast in perfect safety.

This lack of resistance, induced by the torpor of the transformation, appears to me necessary, in view of the weakness of the nurseling as it leaves the egg, whenever the mother is herself incapable of depriving the victim of the power of self-defence. And so the nonparalyzed larvæ are attacked during the period of the nymphosis. We shall soon see other instances of this.

Motionless though it be, the Chalicodoma-grub is none the less alive. The primrose tint and the glossy skin are unequivocal signs of health. Were it really dead, it would, in less than twenty-four hours, turn a dirty brown and, soon after, decompose into a fluid putrescence. Now here is the marvellous thing: during the fortnight, roughly, that the Anthrax' meal lasts, the butter-colour of the larva, an unfailing symptom of the presence of life, continues unaltered and does not change into brown, the sign of putrefaction, until hardly anything remains; and even then the brown hue is often absent. As a rule, the look of live flesh is preserved until the final pellet, formed of the skin, the sole residue, makes its appearance. This pellet is white, with not a speck of tainted matter, proving that life persists until the body is reduced to nothing.

We here witness the transfusion of one animal into another, the change of Chalicodoma-substance into Anthrax-substance; and, as long as the transfusion is not complete, as long as the eaten has not disappeared altogether and become the eater, the ruined organism fights against destruction. What manner of life is this, which may be compared with the life of a night-light whose extinction is not accomplished until the last drop of oil has burnt away? How is any creature able to fight against the final tragedy of corruption up to the last moment in which a nucleus of matter remains as the seat of vital energy? The forces of the living creature are here dissipated not through any disturbance of the equilibrium of those forces, but for the want of any point of application for them: the larva dies because materially there is no more of it.

Can we be in the presence of the diffusive life of the plant, a life which persists in a fragment? By no means: the grub is a more delicate organic structure. There is unity between the several parts; and none of them can be jeopardized without involving the ruin of the others. If I myself give the larva a wound, if I bruise it, the whole body very soon turns brown and begins to rot. It dies and decomposes by the mere prick of a needle; it keeps alive, or at least preserves the freshness of the live tissues, so long as it is not entirely emptied by the Anthrax' sucker. A nothing kills it; an atrocious wasting does not.

The Anthrax

No, I fail to understand the problem; and I bequeath it to others.

All that I can see by way of a glimpse—and even then I put forward my suspicions with extreme reserve—all that I am permitted to surmise is reduced to this: the substance of the sleeping larva as yet has no very definite static existence; it is like the raw materials collected for a building; it is waiting for the elaboration that is to make a Bee of it. To mould those shapeless lumps of the future insect, the air, that prime adjuster of living things, circulates among them, passing through a network of ducts. To organize them, to direct the placing of them, the nervous system, the embryo of the animal, distributes its ramifications over them. Nerve and air-duct, therefore, are the essentials; the rest is so much material in reserve for the process of the metamorphosis. As long as that material is not employed, as long as it has not acquired its final equilibrium, it can grow less and less; and life, though languishing, will continue all the same on the express condition that the respiratory organs and the nervous filaments be respected. It is as it were the flame of the lamp, which, whether full or empty, continues to give light so long as the wick is soaked in oil. Nothing but fluids, the plastic materials held in reserve, can be distilled by the Anthrax' sucker through the unpierced skin of the grub; no part of the respiratory and nervous systems passes. As the two essential functions remain unscathed, life goes on until exhaustion is completed. On the other hand, if I myself injure the larva, I disturb the nervous or air-conducting filaments; and the bruised part spreads a taint, followed by putrefaction, all over the body.

I have elsewhere, speaking of the Scolia[1] devouring the Cetonia-grub, enlarged upon this refined art of eating which consists in consuming the prey while killing it only at the last mouthfuls. The Anthrax has the same requirements as his competitors who dine off fresh viands. He needs meat of that day, taken from a single joint that has to last a fortnight without going bad. His method of consuming reaches the highest level of art: he does not cut into his prey, he sips it little by little through his sucker. In this way, any dangerous risk is averted. Whether he imbibe at this spot or at that, even if he abandon the sucking-process and resume it later, by no acci-

[1] A Digger-wasp who feeds her larvæ on the grubs of the Cetonia, or Rose-chafer. Cf. *The Life and Love of the Insect:* chap. xi.—*Translator's Note.*

dent can he ever attack that which it is incumbent upon him to respect lest corruption supervene. The others have a fixed position on the victim, a place at which their mandibles have to bite and enter. If they move away from it, if they miss the appointed path, they imperil their existence. The Anthrax, more highly favoured, puts his mouth where it suits him; he leaves off when he pleases and when he pleases starts again.

Unless I labour under a delusion, I think that I see the necessity for this privilege. The egg of the carnivorous burrower is firmly fixed on the victim at a point which varies considerably, it is true, according to the nature of the prey, but which is uniform for the same species of prey; moreover—and this is an important condition—the point of adhesion of that egg is always the head, whereas the egg of a Bee, of the Osmia, for instance, is fixed to the mess of honey by the hinder-end. When hatched, the new-born Wasp-grub has not to choose for itself, at its risk and peril, the suitable point at which to take the first cut in the quarry without fear of killing it too quickly: all that it need do is to bite at the spot where it has just been born. The mother, with her unfailing instinct, has already made the dangerous choice; she has stuck her egg on the propitious spot and, by the very act of doing so, marked out the course for the inexperienced grub to follow. The tact of ripe age here guides the young larva's behaviour at table.

The conditions are very different in the Anthrax' case. The egg is not placed upon the victuals, it is not even laid in the Mason-bee's cell. This is the natural consequence of the mother's feeble frame and of her lack of any instrument, such as a probe or auger, capable of piercing the mortar wall. It is for the newly-hatched grub to make its own way into the dwelling. It enters, finds itself in the presence of ample provisions, the larva of the Mason-bee. Free of its actions, it is at liberty to attack the prey where it chooses; or rather the attacking-point will be decided at haphazard by the first contact of the mouth in quest of food. Grant this mouth a set of carving-tools, jaws and mandibles; in short, suppose the grub of the Fly to possess a manner of eating similar to that of the other carnivorous larvæ; and the nursling is at once threatened with a speedy death. He will split open his nurse's belly, he will dig without any rule to guide him, he will bite at random, essentials as well as accessories; and, from one

The Anthrax

day to the next, he will set up gangrene in the violated mass, even as I myself do when I give it a wound. For the lack of an attacking-point prescribed for him at birth, he will perish on the damaged provisions. His freedom of action will have killed him.

Certainly, liberty is a noble attribute, even in an insignificant grub; but it also has its dangers everywhere. The Anthrax escapes the peril only on the condition of being, so to speak, muzzled. His mouth is not a fierce forceps that tears asunder; it is a sucker that exhausts but does not wound. Thus restrained by this safety-appliance, which changes the bite into a kiss, the grub has fresh victuals until it has finished growing, although it knows nothing of the rules of methodical consumption at a fixed point and in a predetermined direction.

The considerations which I have set forth seem to me strictly logical: the Anthrax, owing to the very fact that he is free to take his nourishment where he pleases on the body of the fostering larva, must, for his own protection, be made incapable of opening his victim's body. I am so utterly convinced of this harmonious relation between the eater and the eaten that I do not hesitate to set it up as a

principle. I will therefore say this: whenever
the egg of any kind of insect is not fastened
to the larva destined for its food, the young
grub, free to select the attacking-point and to
change it at will, is as it were muzzled and consumes its provisions by a sort of suction, without inflicting any appreciable wound. This re-
striction is essential to the maintenance of the
victuals in good condition. My principle is
already supported by examples many and various, whose depositions are all to the same
effect. The witnesses include, after the Anthrax, the Leucospis[1] and his rivals, whose
evidence we shall hear presently; the *Ephialtes
mediator*,[2] who feeds, in the dry brambles, on
the larva of the Black Psen;[3] the Myodites,
that strange, Fly-shaped Beetle whose grub
consumes the larva of the Cockchafer. All—
Flies, Ichneumon-flies and Beetles—scrupulously spare their foster-mother; they are careful not to tear her skin, so that the vessel may
keep its liquid good to the last.

The wholesomeness of the victuals is not

[1]The Leucospis is a parasitic insect that forms the subject of the chapter immediately following in the *Souvenirs entomologiques,* but, being a Chalcidid, is not included
in the present volume.—*Translator's Note.*
[2]A genus of Ichneumon-fly.—*Translator's Note.*
[3]A species of Digger-wasp.—*Translator's Note.*

the only condition imposed: I find a second, which is no less essential. The substance of the fostering larva must be sufficiently fluid to ooze through the unbroken skin under the action of the sucker. Well, the necessary fluidity is realized as the time of the metamorphosis draws near. When they wished Medea to restore Pelias to the vigour of youth, his daughters cut the old king's body to pieces and boiled it in a cauldron, for there can be no new existence without a prior dissolution. We must pull down before we can rebuild; the analysis of death is the first step towards the synthesis of life. The substance of the grub that is to be transformed into a Bee begins, therefore, by disintegrating and dissolving into a fluid broth. The materials of the future insect are obtained by a general recasting. Even as the founder puts his old bronzes into the melting-pot in order afterwards to cast them in a mould whence the metal will issue in a different shape, so life liquefies the grub, a mere digesting-machine, now thrown aside, and out of its running matter produces the perfect insect, Bee, Butterfly or Beetle, the final manifestation of the living creature.

Let us open a Chalicodoma-grub under the

microscope, during the period of torpor. Its contents consists almost entirely of a liquid broth, in which swim numberless oily globules and a fine dust of uric acid, a sort of offthrow of the oxidized tissues. A flowing thing, shapeless and nameless, is all that the animal is, if we add abundant ramified air-ducts, some nervous filaments and, under the skin, a thin layer of muscular fibres. A condition of this kind accounts for a fatty transpiration through the skin when the Anthrax' sucker is at work. At any other time, when the larva is in the active period or else when the insect has reached the perfect stage, the firmness of the tissues would resist the transfusion and the suckling of the Anthrax would become a diffi-cult matter, or even impossible. In point of fact, I find the grub of the Fly established, in the vast majority of cases, on the sleeping larva and sometimes, but rarely, on the pupa. Never do I see it on the vigorous larva eating its honey; and hardly ever on the insect brought to perfection, as we find it enclosed in its cell all through the autumn and winter. And we can say the same of the other grub-eaters that drain their victims without wound-ing them: all are engaged in their death-deal-ing work during the period of torpor, when the tissues are fluidified. They empty their patient, who has become a bag of running grease with a diffused life; but not one, among those I know, reaches the Anthrax' perfection in the art of extraction.

Nor can any be compared with the Anthrax as regards the means brought into play in order to leave the cell. These others, when they become perfect insects, have implements for sapping and demolishing, stout mandibles, capable of digging the ground, of pulling down clay partition-walls and even of reducing the Mason-bee's tough cement to powder. The Anthrax, in her final form, has nothing like this. Her mouth is a short, soft proboscis, good at most for soberly licking the sugary exudations of the flowers; her slim legs are so feeble that to move a grain of sand were an excessive task for them, enough to strain every joint; her great, stiff wings, which must remain full-spread, do not allow her to slip through a narrow passage; her delicate suit of downy velvet, from which you take the bloom by merely breathing on it, could not withstand the rough contact of the gallery of a mine. Unable herself to enter the Mason-bee's cell to lay her egg, she cannot leave it either, when the time comes to free herself and appear in broad daylight in her wedding-dress. The larva, on its side, is powerless to prepare the way for the coming flight. That buttery little cylinder, owning no tools but a sucker so flimsy that it barely arrives at substance and so small that it is almost a geometrical point, is even weaker than the adult insect, which at least flies and walks. The Mason-bee's cell represents to it a granite cave. How to get out? The problem would be insoluble to those two incapables, if nothing else played its part.

Among insects, the nymph, or pupa, the transition-stage between the larval and the adult form, is generally a striking picture of every weakness of a budding organism. A sort of mummy tight-bound in swaddling-clothes, motionless and impassive, it awaits the resurrection. Its tender tissues flow in every direction; its limbs, transparent as crystal, are held fixed in their place, along the side, lest a movement should disturb the exquisite delicacy of the work in course of accomplishment. Even so, to secure his recovery, is a broken-boned patient held captive in the surgeon's bandages. Absolute stillness is necessary in both cases, lest they be crippled or even die.

Well, here, by a strange inversion that confuses all our views on life, a Cyclopean task

The Anthrax

is laid upon the nymph of the Anthrax. It is
the nymph that has to toil, to strive, to exhaust itself in efforts to burst the wall and
open the way out. To the embryo falls the
desperate duty, which shows no mercy to the
nascent flesh; to the adult insect the joy of
resting in the sun. This transposition of functions has as its result a well-sinker's equipment
in the nymph, an eccentric, complicated equipment which nothing suggested in the larva and
which nothing recalls in the perfect insect.
The set of tools includes an assortment of
ploughshares, gimlets, hooks and spears and
of other implements that are not found in our
trades nor named in our dictionaries. Let us
do our best to describe the strange piercing-
gear.

In a fortnight at most, the Anthrax has
consumed the Chalicodoma-grub, whereof
naught remains but the skin, gathered into a
white granule. By the time that July is nearly
over, it becomes rare to find any nurselings
left upon their nurses. From this period
until the following May, nothing fresh happens. The Anthrax retains its larval shape
without any appreciable change and lies motionless in the Mason-bee's cocoon, beside the
pellet remains. When the fine days of May

arrive, the grub shrivels and casts its skin and the nymph appears, fully clad in a stout, reddish, horny hide.

The head is round and large, separated from the thorax by a strangulated furrow, crowned on top and in front with a sort of diadem of six hard, sharp, black spikes, arranged in a semicircle whose concave side faces downward. These spikes decrease slightly in length from the summit to the ends of the arch. Taken together, they suggest the radial crowns which we see the Roman emperors of the Decadence wear on the medals. This six-fold ploughshare is the chief excavating-tool. Lower down, on the median line, the instrument is finished off with a separate group of two small black spikes, placed close together.

The thorax is smooth, the wing-cases large, folded under the body like a scarf and coming almost to the middle of the abdomen. This has nine segments, of which four, starting with the second, are armed, on the back, down the middle, with a belt of little horny arches, pale-brown in colour, drawn up parallel to one another, set in the skin by their convex surfaces and finishing at both ends with a hard, black point. Altogether, the belt thus forms a double row of little thorns, with a hollow in between. I count about twenty-five twin-toothed arches to one segment, which gives a total of two hundred spikes for the four rings thus armed.

The use of this rasp, or grater, is obvious: it gives the nymph a purchase on the wall of its gallery as the work proceeds. Thus anchored on a host of points, the stern pioneer is able to hit the obstacle harder with its diadem of awls. Moreover, to make it more difficult for the instrument to recoil, long, stiff bristles, pointing backwards, are scattered here and there among the climbing-belts. There are some besides on the other segments, both on the ventral and the dorsal surface. On the flanks, they are thicker and arranged as it were in clusters.

The sixth segment carries a similar belt, but a much less powerful one, consisting of a single row of unassuming thorns. The belt is weaker still on the seventh segment; lastly, on the eighth, it is reduced to a mere rough brown shading. Commencing with the sixth, the rings decrease in width and the abdomen ends in a cone, the extremity of which, formed of the ninth segment, constitutes a weapon of a new kind. It is a sheaf of eight brown spikes.

The last two exceed the others in length and stand out from the group in a double terminal ploughshare.

There is a round air-hole in front, on either side of the thorax, and similar stigmata on the flanks of each of the first seven abdominal segments. When at rest, the nymph is curved into a bow. When about to act, it suddenly unbends and straightens itself. It measures 15 to 20 millimetres[1] long and 4 to 5 millimetres[2] across.

Such is the strange perforating-machine that is to prepare an outlet for the feeble Anthrax through the Mason-bee's cement. The structural details, so difficult to explain in words, may be summed up as follows: in front, on the forehead, a diadem of spikes, the ramming- and digging-tool; behind, a many-bladed ploughshare which fits into a socket and allows the pupa to slacken suddenly in readiness for an attack on the barrier which has to be demolished; on the back, four climbing-belts, or graters, which keep the animal in position by biting on the walls of the tunnel with their hundreds of teeth; and, all over the body,

[1] .58 to .78 inch.—*Translator's Note.*
[2] .15 to .19 inch.—*Translator's Note.*

long, stiff bristles, pointing backwards, to prevent falls or recoils.

A similar structure exists in the other species of Anthrax with slight variations of detail. I will confine myself to one instance, that of *Anthrax sinuata*, who thrives at the cost of *Osmia tricornis*. Her nymph differs from that of *Anthrax trifasciata*, the Anthrax of the Mason-bee, in possessing less powerful armour. Its four climbing-belts consist of only fifteen to seventeen double-spiked arches, instead of twenty-five; also, the abdominal segments, from the sixth onwards, are supplied merely with stiff bristles, without a trace of horny spikes. If the evolution of the various Anthrax-flies were better known to us, the number of these arches would, I believe, be of great service to entomology in the differentiation of species. I see it remaining constant for any given species, with marked variations between one species and another. But this is not my business: I merely call the attention of the classifiers to this field of study and pass on.

About the end of May, the colouring of the nymph, hitherto a light red, alters greatly and forecasts the coming transformation. The head, the thorax and the scarf formed by the wings become a handsome, shiny black. A

dark band shows on the back of the four segments with their two rows of spikes; three
spots appear on the two next rings; the anal
armour becomes darker. In this manner we
foresee the black livery of the coming insect.
The time has arrived for the pupa to work at
the exit-gallery.

I was anxious to see it in action, not under
natural conditions, which would be impracticable, but in a glass tube in which I confine it
between two thick stoppers of sorghum-pith.
The space thus marked off is about the same
size as the natal cell. The partitions front
and back, although not so stout as the Chalicodoma's masonry, are nevertheless firm enough
not to yield except to prolonged efforts; on
the other hand, the side-walls are smooth and
the toothed belts will not be able to grip
them: a most unfavourable condition for the
worker. No matter: in the space of a single
day, the pupa pierces the front partition, three
quarters of an inch thick. I see it fixing its
double ploughshare against the back partition, arching into a bow and then suddenly releasing itself and striking the plug in front of
it with its barbed forehead. Under the impact
of the spikes, the sorghum slowly crumbles to
pieces. It is slow in coming away; but it

comes away all the same, atom by atom. At long intervals, the method changes. With its crown of awls driven into the pith, the animal frets and fidgets, sways on the pivot of its anal armour. The work of the auger follows that of the pick-axe. Then the blows recommence, interspersed with periods of rest to recover from the fatigue. At last, the hole is made. The pupa slips into it, but does not pass through entirely: the head and thorax appear outside; the abdomen remains held in the gallery.

The glass cell, with its lack of supports at the side, has certainly perplexed my subject, which does not seem to have made use of all its methods. The hole through the sorghum is wide and irregular; it is a clumsy breach and not a gallery. When made through the Mason-bee's walls, it is cylindrical, fairly neat and exactly of the animal's diameter. So I hope that, under natural conditions, the pupa does not give quite so many blows with the pick-axe and prefers to work with the drill.

Narrowness and evenness in the exit-tunnel are necessary to it. It always remains half caught in it and even pretty securely fixed by the graters on its back. Only the head and thorax emerge into the outer air. This is a last precaution for the final deliverance. A fixed support is, in fact, indispensable to the Anthrax for issuing from her horny sheath, unfurling her great wings and extricating her slender legs from their scabbards. All this very delicate work would be endangered by any lack of steadiness.

The pupa, therefore, remains fixed by the graters of its back in the narrow exit-gallery and thus supplies the stable equilibrium essential to the new birth. All is ready. It is time now for the great act. A transversal cleft makes its appearance on the forehead, at the bottom of the perforating diadem; a second, but longitudinal slit divides the skull in two and extends down the thorax. Through this cross-shaped opening, the Anthrax suddenly appears, all moist with the humours of life's laboratory. She steadies herself upon her trembling legs, dries her wings and takes to flight, leaving at the window of the cell her nymphal slough, which keeps intact for a very long period. The sad-coloured Fly has five or six weeks before her, wherein to explore the clay nests amid the thyme and to take her small share of the joys of life. In July, we shall see her once more, busy this time with the entrance into the cell, which is even stranger than the exit.

CHAPTER III

ANOTHER PROBER (PERFORATOR)

WHAT can he be called, this creature whose style and title I dare not inscribe at the head of the chapter? His name is *Monodontomerus cupreus*, SM. Just try it, for fun: *Mo-no-don-to-me-rus*. What a gorgeous mouthful! What an idea it gives one of some beast of the Apocalypse! We think, when we pronounce the word, of the prehistoric monsters: the Mastodon, the Mammoth, the ponderous Megatherium. Well, we are misled by the scientific label: we have to do with a very paltry insect, smaller than the common Gnat.

There are good people like that, only too happy to serve science with resounding appellations that might come from Timbuctoo; they cannot name you a Midge without striking terror into you. O ye wise and revered ones, ye christeners of animals, I am willing, in my study, to make use—but not undue use —of your harsh terminology, with its conglomeration of syllables; but there is a danger of their leaving the sanctum and appearing before the public, which is always ready to show its lack of deference for terms that do not respect its ears. I, wishing to speak like everybody else, so that I may be understood by all, and persuaded that science has no need of this Brobdignagian jargon, make a point of avoiding technical nomenclature when it becomes too barbarous, when it threatens to lumber the page the moment my pen attempts it. And so I abandon *Monodontomerus.*

It is a puny little insect, almost as tiny as the Midges whom we see eddying in a ray of sunshine at the end of autumn. Its dress is golden-bronze; its eyes are coral-red. It carries a naked sword, that is to say, the sheath of its drill stands out slantwise at the tip of its belly, instead of lying in a hollow groove along the back, as it does with the Leucospis. This scabbard holds the latter half of the inoculating-filament, which extends below the animal to the base of the abdomen. In short, its utensil is that of the Leucospis, with this difference, that its lower-half sticks out like a rapier.

This mite that bears a sword upon her rump is yet another persecutor of the Mason-bees and not one of the least formidable. She exploits their nests at the same time as the

Another Prober

Leucospis. I see her, like the Leucospis, slowly explore the ground with her antennæ; I see her, like the Leucospis, bravely drive her dagger into the stone wall. More taken up with her work, less conscious perhaps of danger, she pays no heed to the man who is observing her so closely. Where the Leucospis flies, she does not budge. So great is her assurance that she comes right into my study, to my work-table, and disputes my ownership of the nests whose occupants I am examining. She operates under my lens, she operates just beside my forceps. What risk does she run? What can one do to a thing so very small? She is so certain of her safety that I can take the Mason's nest in my hand, move it, put it down and take it up again without the insect's raising any objection: it continues its work even when my magnifying-glass is placed over it.

One of these heroines has come to inspect a nest of the Chalicodoma of the Walls,[1] most

[1] The author divides Réamur's Mason-bees, roughly, into two species: *Chalicodoma muraria*, or the Mason-bee of the Walls, who builds her nests out of doors; and *C. sicula*, or the Mason-bee of the Sheds, who builds under the inner ceilings of barns. Cf. *Insect Life:* chap. xx. The conclusions in that chapter have, however, since been modified by the author in an essay entitled *Some further Enquiries into Mason-bees*, which has not yet been published in English.—*Translator's Note.*

of whose cells are occupied by the numerous cocoons of a parasite, the Stelis. The contents of these cells, which have been partially ripped up to satisfy my curiosity, are very much exposed to view. The windfall appears to be appreciated, for I see the dwarf ferret about from cell to cell for four days on end, see her choose her cocoon and insert her awl in the most approved fashion. I thus learn that sight, although an indispensable guide in searching, does not decide upon the proper spot for the operation. Here is an insect exploring not the stony exterior of the Mason's dwelling, but the surface of cocoons woven of silk. The explorer has never found herself placed in such circumstances, nor has any of her race before her, every cocoon, under normal conditions, being protected by a surrounding wall. No matter: despite the profound difference in the surfaces, the insect does not waver. Warned by a special sense, an undecipherable riddle to ourselves, it knows that the object of its search lies hidden under this unfamiliar casing. The sense of smell has already been shown[1] to be out of the question; that of sight is now eliminated in its turn.

[1] In the chapter on the Leucopsis aforesaid.—*Translator's Note.*

Another Prober

That she should bore through the cocoons of the Stelis, a parasite of the Mason-bee, does not surprise me at all: I know how indifferent my bold visitor is to the nature of the victuals destined for her family. I have noticed her presence in the homes of Bees differing greatly in size and habits: Anthophoræ, Osmiæ, Chalicodomæ, Anthidia. The Stelis exploited on my table is one victim more; and that is all. The interest does not lie there. The interest lies in the manœuvres of the insect, which I am able to follow under the most favourable conditions.

Bent sharply at right angles, like a couple of broken matches, the antennæ feel the cocoon with their tips alone. The terminal joint is the home of this strange sense which discerns from afar what no eye sees, no scent distinguishes and no ear hears. If the point explored be found suitable, the insect hoists itself on tiptoe so as to give full scope to the play of its mechanism; it brings the tip of the belly a little forward; and the entire ovipositor—inoculating-needle and scabbard—stands perpendicular to the cocoon, in the centre of the quadrilateral described by the four hind-legs, an eminently favourable position for obtaining the maximum effect. For some time, the whole of the awl bears on the cocoon, feeling all round with its point, groping about; then, suddenly, the boring-needle is released from its sheath, which falls back along the body, while the needle strives to make its entrance. The operation is a difficult one. I see the insect make a score of attempts, one after the other, without succeeding in piercing the tough wrapper of the Stelis. Should the instrument not penetrate, it retreats into its sheath and the insect resumes its scrutiny of the cocoon, sounding it point by point with the tips of its antennæ. Then further thrusts are tried until one succeeds.

The eggs are little spindles, white and gleaming like ivory, about two-thirds of a millimetre[1] in length. They have not the long, curved peduncle of the Leucospis' eggs; they are not suspended from the ceiling of the cocoon like these, but are laid without order around the fostering larva. Lastly, in a single cell and with a single mother, there is always more than one laying; and the number of eggs varies considerably in each. The Leucospis, because of her great size, which rivals that of her victim, the Bee, finds in each cell

---

[1]About one-fortieth of an inch.—*Translator's Note.*

provisions enough for one and one alone.
When, therefore, there is more than one set
of eggs in any one cell, this is due to a mistake
on her part and not a premeditated result.
Where the whole ration is required for the
meals of a single grub, she would take good
care not to instal several if she could help it.
Her competitor is not called upon to observe
the same discretion. A Chalicodoma-grub
gives the dwarf the wherewithal to portion a
score of her little ones, who will live in common and in all comfort on what a single son
of the giantess would eat up by himself. The
tiny boring-engineer, therefore, always settles
a numerous family at the same banquet. The
bowl, ample for a dozen or two, is emptied in
perfect harmony.

Curiosity made me count the brood, to see
if the mother was able to estimate the victuals
and to proportion the number of guests to the
sumptuousness of the fare provided. My
notes mention fifty-four larvæ in the cell of a
Masked Anthophora (*Anthophora personata*). No other census attained this figure.
Possibly, two different mothers had laid their
eggs in this crowded habitation. With the
Mason-bee of the Walls, I see the number of
larvæ vary, in different cells, between four

and twenty-six; with the Mason-bee of the Sheds, between five and thirty-six; with the Three-horned Osmia, who supplied me with the largest number of records, between seven and twenty-five; with the Blue Osmia (*Osmia cyanea,* KIRB.), between five and six; with the Stelis (*Stelis nasuta*), between four and twelve.

The first return and the last two seem to point to some relation between the abundance of provisions and the number of consumers. When the mother comes upon the bountiful larva of the Masked Anthophora, she gives it half-a-hundred to feed; with the Stelis and the Blue Osmia, niggardly rations both, she contents herself with half-a-dozen. To introduce into the dining-room only the number of boarders that the bill of fare will allow would certainly be a most deserving performance, especially as the insect is placed under very difficult conditions to judge the contents of the cell. These contents, which lie hidden under the ceiling, are invisible; and the insect can derive its information only from the outside of the nest, which varies in the different species. We should therefore have to admit the existence of a particular power of discrimination, a sort of discernment of the species, which is recognized as large or small from the outward aspect of its house. I refuse to go to this length in my conjectures, not that instinct seems to me incapable of such feats, but because of the particulars obtained from the Three-horned Osmia and the two Mason-bees.

In the cells of these three species, I see the number of larvæ put out to nurse vary in so elastic a fashion that I must abandon all idea of proportionate adjustment. The mother, without troubling unduly whether there be an excess or a dearth of provisions for her family, has filled the cells as her fancy prompted, or rather according to the number of ripe ovules contained in her ovaries at the time of the laying. If food be overplentiful, the brood will be all the better for it and will grow bigger and stronger; if food be scarce, the famished youngsters will not die, but will remain smaller. Indeed, with both the larva and the full-grown insect, I have often observed a difference in size which varies according to the density of the population, the members of a small colony being double the size of their overcrowded neighbours.

The grubs are white, tapering at both ends, sharply segmented and covered all over their bodies with a coat of fine, soft hairs which is invisible except under the lens. The head consists of a little knob much smaller in diameter than the body. In this head, the microscope reveals mandibles consisting of fine spikes of a tawny red, which spread into a wide, colourless base. Deprived of any indentation, incapable of chewing anything between their awl-shaped ends, these two tools serve at best to fix the grub slightly at some point of the fostering larva. Useless for carving, therefore, the mouth is a pure osculatory sucker, which drains the provisions by a process of exudation through the skin. We see here repeated what the Anthrax and the Leucospis have already shown us: the gradual exhaustion of a victim which the parasite consumes without killing it.

It is a curious spectacle even after that of the Anthrax. We have here twenty or thirty starvelings, all with their mouths pressed, as for a kiss, to the body of the plump larva, which, from day to day, fades and shrinks without the least appreciable wound, thus keeping fresh until reduced to a shrivelled slough. If I disturb the gluttonous swarm, all, with a sudden recoil, let go, drop off and flounder around the foster-mother. They are no less prompt in resuming their savage kisses. I need not add that neither at the point where they leave off nor at the point where they recommence is there the faintest trace of liquid. The oily exudation occurs only when the pump is at work. To linger over this strange method of feeding is superfluous after what I have said about the Anthrax.

The appearance of the full-grown insect takes place at the beginning of summer, after nearly a whole year's stay in the invaded dwelling. The large number of inhabitants of one and the same cell led me to think that the work of deliverance ought to present a certain interest. They are all equally anxious to clear the walls of the prison at the earliest possible moment and to come forth into the great festival of the sun: do they all at the same time, in a confused horde, attack the ceiling which has to be pierced? Is the work of deliverance arranged in the general interest? Or is individual selfishness the only rule? These are the questions which observation will answer.

A little in advance of the proper season, I transfer each family into a short glass tube, which will represent the natal cell. A good,

thick cork, quite a centimetre[1] deep, is the obstacle to be pierced for an outlet. Well, instead of the mad haste and the ruinous lack of
organization which I expected to find, my
broods show me in their glass prison an exceedingly well-regulated workshop. One insect, one only, works at perforating the cork.
Patiently, with its mandibles, grain by grain,
it digs a tunnel the width of its body. The
gallery is so narrow that, in order to return
to the tube, the worker has to move backwards. It is a slow process; and it takes
hours and hours to dig the hole, a hard job
for the frail miner.

Should her fatigue become too great, the
excavator leaves the forefront and mingles
with the crowd, to polish and dust herself.
Another, the first neighbour at hand, at once
takes her place and is herself relieved by a
third when her task is done. Others again
take their turn, always one at a time, so much
so that the works are never at a standstill and
never over-crowded. Meanwhile, the multitude keeps out of the way, quietly and patiently. There is no anxiety as to the deliverance. Success will come: of that they are all
convinced. While waiting, one washes her

[1] .39 inch.—*Translator's Note.*

Another Prober

antennæ by passing them through her mouth,
another polishes her wings with her hind-legs,
another frisks about to while away the period
of inaction. Some are making love, a sovran
means of killing time, whether one be born
that day or twenty years ago.

Some, I said, make love. These favoured
ones are rare; they hardly count. Is it through
indifference? No, but the gallants are lacking.
The sexes are very unequally represented in
the population of a cell: the males are in a
wretched minority and sometimes even completely absent. This poverty did not escape
the older observers. Brullé,[1] the only author
whom I am able to consult in my hermitage,
says, literally:

'The males do not appear to be known.'

I, for my part, know them; but, considering
their feeble number, I keep asking myself
what part they play in a harem so disproportionate to their forces. A few figures will
show us what my hesitations are based upon.

In twenty-two Osmia-cocoons (*Osmia tricornis*), the total census of the inmates yields
three hundred and fifty-four, of whom forty-

[1]Gaspard August Brullé (1809-1873), the author of
many works on natural history and one of the founders
of the Société entomologique de France.—*Translator's
Note.*

seven are males and three hundred and seven females. The average number of inmates, therefore, is sixteen individuals; and there are six females at least to one male. This disparity is maintained, in more or less marked proportions, whatever the species of the Bee invaded. In the cocoons of the Mason-bee of the Sheds, I discover the average proportion to be six females to one male; in those of the Mason-bee of the Walls, I find one male to fifteen females.

These facts, which I am unable to state with any greater precision, are enough to give rise to the suspicion that the males, who are even tinier dwarfs than the females and who, moreover, like all insects, are injured by a single act of pairing, must, in most cases, remain strangers to the females. Can the mothers, in fact, dispense with their assistance, without being deprived of offspring on that account? I do not say yes, but I do not say no. The duality of the sexes is a hard problem. Why two sexes? Why not just one? It would have been much simpler and saved a great deal of foolery. Why such a thing as sex, when the tuber of the Jerusalem artichoke can do without it? These are the pregnant questions suggested to me, in the end, by *Monodonto-*

Another Prober

merus cupreus, the insect so infinitesimal in body and so overpowering in name that I had really vowed never to speak of it again by its official designation.

CHAPTER IV

LARVAL DIMORPHISM

IF the reader has paid any attention to the story of the Anthrax, he must have perceived that my narrative is incomplete. The Fox in the fable saw how the Lion's visitors entered his den, but did not see how they went out. With us, it is the converse: we know the way out of the Mason-bee's fortress, but we do not know the way in. To leave the cell of which he has eaten the owner, the Anthrax becomes a perforating-machine, a living tool from which our own industry might take a hint if it required new drills for boring rocks. When the exit-tunnel is opened, this tool splits like a pod bursting in the sun; and from the stout framework there escapes a dainty Fly, a velvety flake, a soft fluff that astounds us by its contrast with the roughness of the depths whence it ascends. On this point, we know pretty well what there is to know. There remains the entrance into the cell, a puzzle that has kept me on the alert for a quarter of a century.

Larval Dimorphism

To begin with, it is evident that the mother cannot lodge her egg in the cell of the Mason-bee, which has been long closed and barricaded with a cement wall by the time that the Anthrax makes her appearance. To penetrate it, she would have to become an excavating-tool once more and resume the cast-off rags which she left behind in the exit-window; she would have to retrace her steps, to be reborn a pupa; and life knows none of these retrogressions. The full-grown insect, if endowed with claws, mandibles and plenty of perseverance, might at a pinch force the mortar casket; but the Fly is not so endowed. Her slender legs would be strained and deformed by merely sweeping away a little dust; her mouth is a sucker for gathering the sugary exudations of the flowers and not the solid pincers needed for the crumbling of cement. There is no auger either, no bore copied from that of the Leucospis, no implement of any kind that can work its way into the thickness of the wall and dispatch the egg to its destination. In short, the mother is absolutely incapable of settling her eggs in the chamber of the Mason-bee.

Can it be the grub that makes its own way into the store-room, that same grub which we have seen draining the Chalicodoma with its

The Life of the Fly

leech-like kisses? Let us call the creature to
mind: a little oily sausage, which stretches
and curls up just where it lies, without being
able to shift its position. Its body is a smooth
cylinder; its mouth simply a circular lip. Not
one ambulatory organ does it possess; not even
hairs, protuberances or wrinkles to enable it to
crawl. The animal is made for digestion and
immobility. Its organization is incompatible
with movement; everything tells us so in the
clearest fashion. No, this grub is even less able
than the mother to make its way unaided into
the Mason's dwelling. And yet the provisions are there; those provisions must be
reached: it is a matter of life or death; to be
or not to be. Then how does the Fly set
about it? It would be vain for me to question probabilities, too often illusory; to obtain
a reply of any value, I have but one resource;
I must attempt the nearly impossible and
watch the Anthrax from the egg onwards.

Although Anthrax-flies are fairly common,
in the sense of there being several different
species, they are not plentiful when it is a case
of wanting a colony populous enough to admit
of continuous observation. I see them, now
here, now there, in the fiercely sun-scorched
places, flitting hither and thither on the old

Larval Dimorphism

walls, the slopes and the sand, sometimes in small platoons, most often singly. I can expect nothing of those vagabonds, who are here to-day and gone to-morrow, for I know nothing of their settlements. To keep a watch on them, one by one, in the blazing heat, is very painful and very unfruitful, as the swift-winged insect has a habit of disappearing one knows not whither just when a prospect of capturing its secret begins to offer. I have wasted many a patient hour at this pursuit, without the least result.

There might be some chance of success with Anthrax-flies whose home was known to us beforehand, especially if insects of the same species formed a pretty numerous colony. The enquiries begun with one would be continued with a second and with more, until a complete verdict was forthcoming. Now, in the course of my long entomological career, I have met with but two species of Anthrax that fulfilled this condition and were to be found regularly: one at Carpentras; the other at Sérignan. The first, *Anthrax sinuata*, FALLEN, lives in the cocoons of *Osmia tricornis*, who herself builds her nest in the old galleries of the Hairy-footed Anthophora; the second, *Anthrax trifasciata*, MEIGEN, exploits the

The Life of the Fly

Chalicodoma of the Sheds. I will consult both.

Once more, here am I, somewhat late in life, at Carpentras, whose rude Gallic name sets the fool smiling and the scholar thinking. Dear little town where I spent my twentieth year and left the first bits of my fleece upon life's bushes, my visit of to-day is a pilgrimage; I have come to lay my eyes once more upon the place which saw the birth of the liveliest impressions of my early days. I bow, in passing, to the old college where I tried my prentice hand as a teacher. Its appearance is unchanged; it still looks like a penitentiary. Those were the views of our mediæval educational system. To the gaiety and activity of boyhood, which were considered unwholesome, it applied the remedy of narrowness, melancholy and gloom. Its houses of instruction were, above all, houses of correction. The freshness of Virgil was interpreted in the stifling atmosphere of a prison. I catch a glimpse of a yard between four high walls, a sort of bear-pit, where the scholars fought for room for their games under the spreading branches of a plane-tree. All around were cells that looked like horse-boxes, without light or air; those were the class-rooms. I speak in the past tense, for doubtless the present day has seen the last of this academic destitution.

Here is the tobacco-shop where, on Wednesday evening, coming out of the college, I would buy on credit the wherewithal to fill my pipe and thus to celebrate on the eve the joys of the morrow, that blessed Thursday[1] which I considered so well-employed in solving hard equations, experimenting with new chemical reagents, collecting and identifying my plants. I would make my timid request, pretending to have come out without my money, for it is hard for a self-respecting man to admit that he is penniless. My candour appears to have inspired some little confidence; and I obtained credit, an unprecedented thing, with the representative of the revenue.[2] Ah, why did not I open a shop and expose for sale some packets of candles, a dozen dried cod, a barrel of sardines and a few cakes of soap! I am no more of a fool nor any less industrious than another; and I should have made my way. But,

[1]Thursday is the weekly half-holiday in the French schools.—*Translator's Note.*
[2]The government in France has the sole control of the tobacco-trade, which forms an important branch of the inland revenue.—*Translator's Note.*

as it was, what could I expect? As an accoucheur of brains, a moulder of intellects, I had no claim even to bread and cheese.

Here is my former habitation, occupied since by droning monks. In the embrasure of that window, sheltered from profane hands, between the closed outer shutters and the panes, I used to keep my chemicals, bought for a few sous cheated out of the weekly budget in the early days of our housekeeping. The bowl of a pipe was my crucible, a sweet-jar my retort, mustard-pots my receptacles for oxides and sulphides. My experiments, harmless or dangerous, were made on a corner of the fire beside the simmering broth.

How I should love to see that room again where I pored over differentials and integrals, where I calmed my poor burning head by gazing at Mont Ventoux, whose summit held in store for my coming expedition[1] those denizens of arctic climes, the saxifrage and the poppy! And to see my familiar friend, the blackboard which I hired at five francs a year from a crusty joiner, that board whose value I paid many times over, though I could never buy it outright, for want of the necessary cash! The

[1]Cf. *Insect Life:* chap. xiii.—*Translator's Note.*

Larval Dimorphism

conic sections which I described on that blackboard, the learned hieroglyphics!

Though all my efforts, which were the more
deserving because I had to work alone, led to
almost nothing in that congenial calling, I
would begin it all over again if I could. I
should love to be conversing for the first time
with Leibnitz[1] and Newton,[2] with Laplace[3]
and Lagrange,[4] with Cuvier[5] and Jussieu,[6]
even if I had afterwards to solve that other
arduous problem: how to procure one's daily
bread. Ah, young men, my successors, what
an easy time you have of it to-day! If you
don't know it, then let me tell you so by means
of these few pages from the life of one of
your elders.

But let us not forget our insects, while list-

[1]Gottfried Wilhelm Baron von Leibnitz (1646-1716),
the discoverer of the differential and integral calculus.—
Translator's Note.
[2]Sir Isaac Newton (1642-1727), discoverer of the law
of gravitation.—*Translator's Note.*
[3]Pierre Simon Marquis de Laplace (1749-1827), author
of *La Mécanique céleste.*—*Translator's Note.*
[4]Joseph Louis Comte Lagrange (1736-1813), author of
La Mécanique analytique.—*Translator's Note.*
[5]Georges Léopold Chrétien Frédéric Dagobert Baron
Cuvier (1769-1832), the founder of the science of comparative anatomy.—*Translator's Note.*
[6]Bernard Jussieu (1699-1777), the most celebrated of a
family of five famous French botanists, consisting of three
brothers, a nephew and a grand-nephew.—*Translator's
Note.*

85

ening to the echoes of illusions and difficulties
roused in my memories by the cupboard-wind-
ow and the hired blackboard. Let us go
back to the sunken roads of the Lègue, which
have become classic, so they say, since the appearance of my notes on the Oil-beetles.[1] Ye
illustrious ravines, with your sun-baked slopes,
if I have contributed a little to your fame, you,
in your turn, have given me many fair hours
of forgetfulness in the happiness of learning.
You, at least, did not lure me with vain hopes;
all that you promised you gave me and often
a hundredfold. You are my promised land,
where I would have sought at the last to pitch
my observer's tent. My wish was not to be
realized. Let me, at least, in passing, greet
my beloved animals of the old days.

I raise my hat to *Cerceris tuberculata,*
whom I see engaged on that slant, storing her
Cleonus.[2] As I saw her then, so I see her
now: the same staggering attempts to hoist
the prey to the mouth of the burrow; the same
brawls between males watching in the brush-

---

[1] The essays on the Oil-beetles have not yet been translated into English. But cf. Chap. XX of the present
volume.—*Translator's Note.*
[2] A large species of Weevil. For the habits of the two
wasps known as *Cerceris bupresticida* and *Cerceris tuberculata,* cf. *Insect Life:* chaps. iii to v.—*Translator's Note.*

wood of the kermes oak. The sight of them sends a younger blood coursing through my veins; I receive as it were the breath of a new spring-time of life. Time presses; let us pass on.

Another bow on this side. I hear buzzing up above, on that ledge, a colony of Sphex-wasps, stabbing their Crickets.[1] We will give them a friendly glance, but no more. My acquaintances here are too numerous; I have not the leisure to renew my former relations with all of them. Without stopping, a wave of the hat to the Philanthi,[2] who send the long avalanches of rubbish streaming down from their nests; and to *Stizus ruficornis*,[3] who stacks her Praying Mantes[4] between two flakes of sandstone; and to the Silky Ammophila,[5] with the red legs, who collects an underground store of Loopers;[6] and to the

---

[1] Cf. *Insect Life:* chaps. vi to xii.—*Translator's Note.*

[2] *Philanthus apivorus*, a Bee-hunting Wasp. Cf. *Social Life in the Insect World:* chap. xiii.—*Translator's Note.*

[3] A Hunting Wasp.—*Translator's Note.*

[4] Predatory insects, akin to the Locusts and Crickets, which, when at rest, adopt an attitude resembling that of prayer. Cf. *Social Life in the Insect World:* chaps. v to vii.—*Translator's Note.*

[5] A Digger-wasp.—*Translator's Note.*

[6] Also known as Measuring-worms, the larvæ or Caterpillars of the Geometrid Moth.—*Translator's Note.*

The Life of the Fly

Tachtyti,[1] devourers of Locusts; and to the Eumenes, builders of clay cupolas on a bough.

Here we are at last. This high, perpendicular rock, facing the south to a length of some hundreds of yards and riddled with holes like a monstrous sponge, is the time-honoured dwelling-place of the Hairy-footed Anthophora and of her rent-free tenant, the Three-horned Osmia. Here also swarm their exterminators: the Sitaris-beetle, the parasite of the Anthophora; the Anthrax-fly, the murderer of the Osmia. Ill-informed as to the proper period, I have come rather late, on the 10th of September. I should have been here a month ago, or even by the end of July, to watch the Fly's operations. My journey threatens to be fruitless: I see but a few rare Anthrax-flies, hovering round the face of the cliff. We will not despair, however, and we will begin by consulting the locality.

The Anthophora's cells contain this Bee in the larval stage. Some of them provide me with the Oil-beetle and the Sitaris, rare finds at one time, to-day of no use to me. Others contain the Melecta[2] in the form of a highly-

[1]Hunting Wasps.—*Translator's Note.*
[2]A Parasitic Bee.—*Translator's Note.*

coloured pupa, or even in that of the full-grown insect. The Osmia, still more precocious, though dating from the same period, shows herself exclusively in the adult form, a bad omen for my investigations, for what the Anthrax demands is the larva and not the perfect insect. The Fly's grub doubles my apprehensions. Its development is complete, the larva on which it feeds is consumed, perhaps several weeks ago. I no longer doubt but that I have come too late to see what happens in the Osmia's cocoons.

Is the game lost? Not yet. My notes contain evidence of Anthrax-flies hatching in the latter half of September. Besides, those whom I now see exploring the rock are not there to take exercise: their preoccupation is the settling of the family. These belated ones cannot tackle the Osmia, who, with her firm, adult flesh, would not suit the nurseling's delicate needs and who, moreover, powerful as she is, would offer resistance. But in autumn a less numerous colony of honey-gatherers takes the place, upon the slope, of the spring colony, from which it differs in species. In particular, I see the Diadem Anthidium[1]

[1] A Clothier-bee, who lines her nest with wool and cotton.—*Translator's Note.*

at work, entering her galleries at one time
with her harvest of pollen-dust and at another
with her little bale of cotton. Might not
these autumnal Bees be themselves exploited
by the Anthrax, the same that selected the
Osmia as her victim a couple of months earlier? This would explain the presence of the
Anthrax-flies whom I now see fussing about.

A little reassured by this conjecture, I take
my stand at the foot of the rock, under a broil-
ing sun; and, for half a day, I follow the
evolutions of my Flies. They flit quietly in
front of the slope, at a few inches from the
earthy covering. They go from one orifice to
the next, but without even penetrating. For
that matter, their big wings, extended crosswise even when at rest, would resist their entrance into a gallery, which is too narrow to
admit those spreading sails. And so they explore the cliff, going to and fro and up and
down, with a flight that is now sudden, now
smooth and slow. From time to time, I see
the Anthrax quickly approach the wall and
lower her abdomen as though to touch the
earth with the end of her ovipositor. This
proceeding takes no longer than the twinkling
of an eye. When it is done, the insect alights
elsewhere and rests. Then it resumes its sober

Larval Dimorphism

flight, its long investigations and its sudden blows with the tip of its belly against the layer of earth. The Bombylii[1] observe similar tactics when soaring at a short height above the ground.

I at once rushed to the spot touched, lens in hand, in the hope of finding the egg which everything told me was laid during that tap of the abdomen. I could distinguish nothing, in spite of the closest attention. It is true that my exhaustion, together with the blinding light and scorching heat, made examination very difficult. Afterwards, when I made the acquaintance of the tiny thing that issues from that egg, my failure no longer surprised me. In the leisure of my study, with my eyes rested and with my most powerful glasses held in a hand no longer shaking with excitement and fatigue, I have the very greatest difficulty in finding the infinitesimal creature, though I know exactly where it lies. Then how could I see the egg, worn out as I was under the sun-baked cliff, how discover the precise spot of a laying performed in a moment by an insect seen only at a distance? In the painful conditions wherein I found myself, failure was inevitable.

[1] Bee-flies.—*Translator's Note.*

The Life of the Fly

Despite my negative attempts, therefore, I remain convinced that the Anthrax-flies strew their eggs one by one, on the spots frequented by those Bees who suit their grubs. Each of their sudden strokes with the tip of the abdomen represents a laying. They take no precaution to place the germ under cover; for that matter, any such precaution would be rendered impossible by the mother's structure. The egg, that delicate object, is laid roughly in the blazing sun, between grains of sand, in some wrinkle of the calcined chalk. That summary installation is sufficient, provided the coveted larva be near at hand. It is for the young grub now to manage as best it can at its own risk and peril.

Though the sunken roads of the Lègue did not tell me all that I wished to know, they at least made it very probable that the coming grub must reach the victualled cell by its own efforts. But the grub which we know, the one that drains the bag of fat which may be a Chalicodoma-larva or an Osmia-larva, cannot move from its place, still less indulge in journeys of discovery through the thickness of a wall and the web of a cocoon. So an imperative necessity presents itself: there must perforce be an initial larva-form, capable of moving and organized for searching, a form under which the grub would attain its end. The Anthrax would thus possess two larval states: one to penetrate to the provisions; the other to consume them. I allow myself to be convinced by the logic of it all; I already see in my mind's eye the wee animal coming out of the egg, endowed with sufficient power of motion not to dread a walk and with sufficient slenderness to glide into the smallest crevices. Once in the presence of the larva on which it is to feed, it doffs its travelling-dress and becomes the obese animal whose duty it is to grow big and fat in immobility. This is all very coherent; it is all deduced like a geometrical proposition. But to the wings of imagination, however smooth their flight, we must prefer the sandals of observed facts, the slow sandals with the leaden soles. Thus shod, I proceed.

Next year, I resume my investigations, this time on the Anthrax of the Chalicodoma, who is my neighbour in the surrounding wastelands and will allow me to repeat my visits daily, morning and evening if need be. Taught by my earlier studies, I now know the exact period of the Bec's hatching and therefore of the Anthrax' laying, which must take place soon after. *Anthrax trifasciata* settles her

The Life of the Fly

family in July, or in August at latest. Every morning, at nine o'clock, when the heat begins to be unendurable and when, to use Favier's[1] expression, an extra log is flung on the bonfire of the sun, I take the field, prepared to come back with my head aching from the glare, provided that I bring home the solution of my puzzle. A man must have the devil in him to leave the shade at this time of the year. And what for, pray? To write the story of a Fly! The greater the heat, the better my chance of success. What causes me to suffer torture fills the insect with delight; what prostrates me braces the Fly. Come along!

The road shimmers like a sheet of molten steel. From the dusty and melancholy olive-trees rises a mighty, throbbing hum, a great *andante* whose executants have the whole sweep of woods for their orchestra. 'Tis the concert of the Cicadæ, whose bellies sway and rustle with increasing frenzy as the temperature rises. The strident scrapings of the Cicada of the Ash, the *Carcan* of the district, lend their rhythm to the one-note symphony of the Common Cicada. This is the moment: come along! And, for five or six weeks, oftenest

[1] An ex-soldier, recurring in many of the essays, the author's gardener and factotum.—*Translator's Note.*

94

in the morning, sometimes in the afternoon, I set myself to explore the flinty plateau.

The Chalicodoma's nests abound, but I cannot see a single Anthrax make a black speck upon their surface. Not one, busy with her laying, settles in front of me. At most, from time to time, I can just see one passing far away, with an impetuous rush. I lose her in the distance; and that is all. It is impossible to be present at the laying of the egg. I know the little that I learnt from the cliffs in the Lègue and nothing more.

As soon as I recognize the difficulty, I hasten to enlist assistants. Shepherds—mere small boys—keep the sheep in these stony meadows, where the flocks graze, to the greater glory of our local mutton, on the camphor-saturated *badafo*, that is to say, spike-lavender. I explain as well as I can the object of my search; I talk to them of a big black Fly and the nests on which she ought to settle, the clay nests so well-known to those who have learnt how to extract the honey with a straw in spring-time and spread it on a crust of bread. They are to watch that Fly and take good note of the nests on which they may see her alight; and, on the same evening, when they bring their flocks back to

the village, they are to tell me the result of
their day's work. On receiving their favourable report, I will go with them, next day, to
continue the observations. They shall be paid
for their trouble, of course. These latter-day
Corydons have not the manners of antiquity:
they reck little of the seven-holed flute ce-
mented with wax, or of the beechen bowl, preferring the coppers that will take them to
the village-inn on Sunday. A reward in
ready money is promised for each nest that
fulfils the desired conditions; and the bargain
is enthusiastically accepted.

There are three of them; and I make a
fourth. Shall we manage it, among us all?
I thought so. By the end of August, however, my last illusions were dispelled. Not
one of us had succeeded in seeing the big
black Fly perching on the dome of the Mason-
bee.

Our failure, it seems to me, can be explained
thus: outside the spacious front of the Anthophora's settlement, the Anthrax is in permanent residence. She visits, on the wing, every
nook and corner, without moving away from
the native cliff, because it would be useless to
go farther. There is board and lodging here,
indefinitely, for all her family. When some

spot is deemed favourable, she hovers round
inspecting it, then comes up suddenly and
strikes it with the tip of her abdomen. The
thing is done, the egg is laid. So I picture it,
at least. Within a radius of a few yards and
in a flight broken by short intervals of rest
in the sun, she carries on her search of likely
places for the laying and dissemination of her
eggs. The insect's assiduous attendance upon
the same slope is caused by the inexhaustible
wealth of the locality exploited.

The Anthrax of the Chalicodoma labours
under very different conditions. Stay-at-home
habits would be detrimental to her. With
her rushing flight, made easy by the long and
powerful spread of her wings, she must travel
far and wide if she would found a colony.
The Bee's nests are not discovered in groups,
but occur singly on their pebbles, scattered
more or less everywhere over acres of ground.
To find a single one is not enough for the
Fly: on account of the many parasites, not all
the cells, by a long way, contain the desired
larva; others, too well protected, would not
allow of access to the provisions. Very many
nests are necessary, perhaps, for the eggs of
one alone; and the finding of them calls for
long journeys.

The Life of the Fly

I therefore picture the Anthrax coming and going in every direction across the stony plain. Her practised eye requires no slackened flight to distinguish the earthen dome which she is seeking. Having found it, she inspects it from above, still on the wing; she taps it once and yet once again with the tip of her ovipositor and forthwith makes off, without having set foot on the ground. Should she take a rest, it will be elsewhere, no matter where, on the soil, on a stone, on a tuft of lavender or thyme. Given these habits—and my observations in the Carpentras roads make them seem exceedingly probable—it is small wonder that the perspicacity of my young shepherds and myself should have come to naught. I was expecting the impossible: the Anthrax does not halt on the Mason-bee's nest to proceed with her laying in a methodical fashion; she merely pays a flying visit.

And so I develop my theory of a primary larval form, differing in every way from the one which I know. The organization of the Anthrax must be such, at the beginning, as to permit of its moving on the surface of the dome where the egg has been dropped so carelessly; the nascent grub must be supplied with tools to pierce the concrete wall and enter the

Larval Dimorphism

Bee's cell through some cranny. The Fly-grub, perhaps dragging the remnants of the egg behind it, must set out in quest of board and lodging almost as soon as it is born. It will succeed under the guidance of instinct, that faculty which waits not to number the days and which is as far-seeing at the moment of hatching as after the trials of a busy life. This primary grub does not seem to me outside the limits of possibility; I see it, if not in the body, at least in its actions, as plainly as though it were really under the lens. It exists, if reason be not a vain and empty guide; I must find it; I shall find it. Never, in the history of my investigations, has the logic of things been more insistent; never has it directed me with greater certainty towards a magnificent biological theory.

While vainly trying to witness the laying of the eggs, I enquire, at the same time, into the contents of the Mason-bee's nests, in quest of the grub just issued from the egg. My own harvest and that of my young shepherds, whose zeal I employ in a task less difficult than the first, procure me heaps of nests, enough to fill baskets and baskets. These are all inspected at leisure, on my work-table, with the excitement which the certainty of an approach-

ing fine discovery never fails to give. The
Mason's cocoons are taken from the cells, inspected without, opened and inspected within.
My lens explores their innermost recesses;
speck by speck, it explores the Chalicodoma's
slumbering larva; it explores the inner walls
of the cells. Nothing, nothing, nothing! For
a fortnight and more, nests were rejected and
heaped up in a corner; my study was crammed
with them. What hecatombs of unfortunate
sleepers removed from their silken bags and
doomed, for the most part, to a wretched end,
despite the care which I took to put them in a
place of safety, where the work of the transformation might be pursued! Curiosity makes
us cruel. I continue to rip up cocoons. And
nothing, nothing! It needed the sturdiest
faith to make me persevere. That faith I
possessed; and well for me that I did.

On the 25th of July—the date deserves to
be recorded—I saw, or rather seemed to see,
something move on the Chalicodoma's larva.
Was it an illusion born of my hopes? Was it
a bit of diaphanous down stirred by my
breath? It was not an illusion, it was not a
bit of down, it was really and truly a grub.
What a moment, followed by what perplexi-
ties! The thing has nothing in common with

the larva of the Anthrax, it suggests rather some microscopic Thread-worm that, by accident, has made its way through the skin of its host and come to enjoy itself outside. I do not reckon my discovery as of much value, because I am so greatly puzzled by the creature's appearance. No matter: we will take a small glass tube and place inside it the Chalicodoma-grub and the mysterious thing wriggling on the surface. Suppose it should be what I am looking for? Who knows?

Once warned of the probable difficulty of seeing the animalcule for which I am hunting, I redouble my attention, so much so that, in a couple of days, I am the owner of half a score of tiny worms similar to the one which caused me such excitement. Each of them is lodged in a glass tube with its Chalicodoma-grub. The infinitesimal thing is so small, so diaphanous, blends to such good purpose with its host that the least fold of skin conceals it from my view. After watching it one day through the lens, I sometimes fail to find it again on the morrow. I think that I have lost it, that it has perished under the weight of the overturned larva and returned to that nothing to which it was so closely akin. Then it moves and I see it again. For a whole fortnight, there was no limit to my perplexity. Was it really the original larva of the Anthrax? Yes, for I at last saw my bantlings transform themselves into the larva previously described and make their first start at draining their victims with kisses. A few moments of satisfaction like those which I then enjoyed make up for many a weary hour.

Let us resume the story of the wee animal, now recognized as the genuine origin of the Anthrax. It is a tiny worm about a millimetre[1] long and almost as slender as a hair. It is very difficult to see because of its transparency. When tucked away in a fold of the skin of its fostering larva, an excessively fine skin, it remains undiscoverable to the lens. But the feeble creature is very active: it tramps over the sides of the rich morsel, walks all round it. It covers the ground pretty quickly, buckling and unbuckling by turns, very much after the manner of the Looper-caterpillar. Its two extremities are its chief points of support. When at a standstill, it moves its front half in every direction, as though to explore the space around it; when walking, it swells out, magnifies its segments and then looks like a bit of knotted string.

[1].039 inch.—*Translator's Note.*

Larval Dimorphism

The microscope shows us thirteen rings, including the head. This head is small, slightly horny, as is proved by its amber colour, and bristles in front with a small number of short, stiff hairs. On each of the three segments of the thorax there are two long hairs, fixed to the lower surface; and there are two similar and still longer hairs at the end of the terminal ring. These four pairs of bristles, three in front and one behind, are the locomotory organs, to which we must add the hairy edge of the head and also the anal button, a sustaining base which might very well work with the aid of a certain stickiness, as happens with the primary larva of the Sitaris.[1] We see, through the transparent skin, two long airtubes running parallel to each other from the first thoracic segment to the last abdominal segment but one. They ought to end in two pairs of breathing-holes which I have not succeeded in distinguishing quite plainly. Those two big respiratory vessels are characteristic of the grubs of Flies. Their mouths correspond exactly with the points at which the two sets of stigmata open in the Anthrax-larva in its second form.

[1] A Parasitic Beetle, noted for the multiplicity of transformations undergone by the grub.—*Translator's Note.*

The Life of the Fly

For a fortnight, the feeble grub remains in the condition which I have described, without growing and very probably also without nourishment. Assiduous though my visits be, I never perceive it taking any refreshment. Besides, what would it eat? In the cocoon invaded there is nothing but the larva of the Mason-bee; and the worm cannot make use of this before acquiring the sucker that comes with the second form. Nevertheless, this life of abstinence is not a life of idleness. The animalcule explores its dish, now here, now elsewhere; it runs all over it with Looper strides; it pries into the neighbourhood by lifting and shaking its head.

I see a need for this long wait under a transitory form that requires no feeding. The egg is laid by the mother on the surface of the nest, somewhere near a suitable cell, I dare say, but still at a distance from the fostering larva, which is protected by a thick rampart. It is for the new-born grub to make its own way to the provisions, not by violence and house-breaking, of which it is incapable, but by patiently slipping through a maze of cracks, first tried, then abandoned, then tried again. It is a very difficult task, even for this most slender worm, for the Bee's masonry is exceedingly compact. There are no chinks due to bad building; no fissures due to the weather; nothing but an apparently impenetrable homogeneity. I see but one weak part and that only in a few nests: it is the line where the dome joins the surface of the stone. An imperfect soldering between two materials of different nature, cement and flint, may leave a breach wide enough to admit besiegers as thin as a hair. Nevertheless, the lens is far from always finding an inlet of this kind on the nests occupied by Anthrax-flies.

And so I am ready to allow that the animalcule wandering in search of its cell has the whole area of the dome at its disposal when selecting an entrance. Where the fine auger of the Leucospis can enter, is there not room enough for the even slimmer Anthrax-grub? True, the Leucospis possesses muscular force and a hard boring-tool. The Anthrax is extremely weak and has nothing but invincible patience. It does at great length of time what the other, furnished with superior implements, accomplishes in three hours. This explains the fortnight spent by the Anthrax under the initial form, the object of which is to overcome the obstacle of the Mason's wall, to pierce through the texture of the cocoon and to reach the victuals.

I even believe that it takes longer. The work is so laborious and the worker so feeble! I cannot tell how long it is since my bantlings attained their object. Perhaps, aided by easy roads, they had reached their fostering larvæ long before the completion of their first baby-hood, the end of which they were spending be-fore my eyes, with no apparent purpose, in exploring their provisions. The time had not yet come for them to change their skins and take their seats at the table. Their fellows must still, for the most part, be wandering through the pores of the masonry; and this was what made my search so vain at the start.

A few facts seem to suggest that the entrance into the cell may be delayed for several months by the difficulty of the passages. There are a few Anthrax-grubs beside the remains of pupæ not far removed from the final meta-morphosis; there are others, but very rarely, on Mason-bees already in the perfect state. These grubs are sickly and appear to be ail-ing; the provisions are too solid and do not lend themselves to the delicate suckling of the worms. Who can these laggards be but ani-malcules that have roamed too long in the

Larval Dimorphism

walls of the nest? Failing to make their entrance at the proper time, they no longer find viands to suit them. The primary larva of the Sitaris continues from the autumn to the following spring. Even so the initial form of the Anthrax might well continue, not in inactivity, but in stubborn attempts to overcome the thick bulwark.

My young worms, when transferred with their provisions into tubes, remained stationary, on the average, for a couple of weeks. At last, I saw them shrink and then rid themselves of their epidermis and become the grub which I was so anxiously expecting as the final reply to all my doubts. It was indeed, from the first, the grub of the Anthrax, the cream-coloured cylinder with the little button of a head, followed by a hump. Applying its cupping-glass to the Mason-bee, the worm, without delay, began its meal, which lasts another fortnight. The reader knows the rest.

Before taking leave of the animalcule, let us devote a few lines to its instinct. It has just awakened to life under the fierce kisses of the sun. The bare stone is its cradle, the rough clay its welcomer, as it makes its entrance into the world, a poor thread of scarce-cohering albumen. But safety lies within; and behold the atom of animated glair embarking on its struggle with the flint. Obstinately, it sounds each pore; it slips in, crawls on, retreats, begins again. The radicle of the germinating seed is no more persevering in its efforts to descend into the cool earth than is the Anthrax-grub in creeping into the lump of mortar. What inspiration urges it towards its food at the bottom of the clod, what compass guides it? What does it know of those depths, of what lies therein or where? Nothing. What does the root know of the earth's fruitfulness? Again nothing. Yet both make for the nourishing spot. Theories are put forward, most learned theories, introducing capillary action, osmosis and cellular imbibition, to explain why the caulicle ascends and the radicle descends. Shall physical or chemical forces explain why the animalcule digs into the hard clay? I bow profoundly, without understanding or even trying to understand. The question is far above our inane means.

The biography of the Anthrax is now complete, save for the details relating to the egg, as yet unknown. In the vast majority of insects subject to metamorphoses, the hatching yields the larval form which will remain un-

Larval Dimorphism

changed until the nymphosis. By virtue of a remarkable variation, revealing a new vein of observation to the entomologist, the Anthrax-flies, in the larval state, assume two successive shapes, differing greatly one from the other, both in structure and in the part which they are called upon to play. I will describe this double stage of the organism by the phrase 'larval dimorphism.' The initial form, that issuing from the egg, I will call 'the primary larva;' the second form shall be 'the secondary larva.' Among the Anthrax-flies, the function of the primary larva is to reach the provisions, on which the mother is unable to lay her egg. It is capable of moving and endowed with ambulatory bristles, which allow the slim creature to glide through the smallest interstices in the wall of a Bee's nest, to slip through the woof of the cocoon and to make its way to the larva intended for its successor's food. When this object is attained, its part is played. Then appears the secondary larva, deprived of any means of progression. Relegated to the inside of the invaded cell, as incapable of leaving it by its own efforts as it was of entering, this one has no mission in life but that of eating. It is a stomach that loads itself, digests and goes on adding to its reserves. Next comes the pupa, armed for the exit even as the primary larva was equipped for entering. When the deliverance is accomplished, the perfect insect appears, busy with its laying. The Anthrax cycle is thus divided into four periods, each of which corresponds with special forms and functions. The primary larva enters the casket containing provisions; the secondary larva consumes these provisions; the pupa brings the insect to light by boring through the enclosing wall; the perfect insect strews its eggs; and the cycle starts afresh.

CHAPTER V

HEREDITY

FACTS which I have set forth elsewhere prove that certain Dung-beetles[1] make an exception to the rule of paternal indifference —a general rule in the insect world—and know something of domestic cooperation. The father works with almost the same zeal as the mother in providing for the settlement of the family. Whence do these favoured ones derive a gift that borders on morality?

One might suggest the cost of installing the youngsters. Once they have to be furnished with a lodging and to be left the wherewithal to live, is it not an advantage, in the interests of the race, that the father should come to the mother's assistance? Work divided between the two will ensure the comfort which solitary work, its strength overtaxed, would deny. This seems excellent reasoning; but it is much more

---

[1] The Lunary Copris and the Bison Oritis, the essay on whom has not yet appeared in English; *Minotaurus typhæus*, for whom see *The Life and Love of the Insect:* chap. x; and the Sisyphus, for whom see *Social Life in the Insect World:* chap. xii.—*Translator's Note.*

The Life of the Fly

often contradicted than confirmed by the facts.
Why is the Sisyphus a hard-working paterfamilias and the Sacred Beetle[1] an idle vagabond? And yet the two pill-rollers practise
the same industry and the same method of
rearing their young. Why does the Lunary
Copris know what his near kinsman, the Spanish Copris,[2] does not? The first assists his
mate, never forsakes her. The second seeks a
divorce at an early stage and leaves the nuptial
roof before the children's rations are massed
and kneaded into shape. Nevertheless, on
both sides, there is the same big outlay on a
cellarful of egg-shaped pills, whose neat rows
call for long and watchful supervision. The
similarity of the produce leads one to believe
in similarity of manners; and this is a mistake.

Let us turn elsewhere, to the Wasps and
Bees, who unquestionably come first in the lay-
ing-up of a heritage for their offspring.
Whether the treasure hoarded for the benefit
of the sons be a pot of honey or a bag of game,
the father never takes the smallest part in the
work. He does not so much as give a sweep
of the broom when it comes to tidying the

[1]See *Insect Life:* chap. i; and *The Life and Love of the
Insect:* chaps. i to iv.—*Translator's Note.*
[2]See *The Life and Love of the Insect:* chap. v.—*Translator's Note.*

outside of the dwelling. To do nothing is his invariable rule. The bringing-up of the family, therefore, however expensive it may be in certain cases, has not given rise to the instinct of paternity. Then where are we to look for a reply?

Let us make the question a wider one. Let us leave the animal, for a moment, and occupy ourselves with man. We have our own instincts, some of which take the name of genius when they attain a degree of might that towers over the plain of mediocrity. We are amazed by the unusual, springing out of flat commonplaces; we are spell-bound by the luminous speck shining in the wonted darkness. We admire; and, failing to understand whence came those glorious harvests in this one or in that, we say of them:

"They have the gift."

A goatherd amuses himself by making combinations with heaps of little pebbles. He becomes an astoundingly quick and accurate reckoner without other aid than a moment's reflection. He terrifies us with the conflict of enormous numbers which blend in an orderly fashion in his mind, but whose mere statement overwhelms us by its inextricable confusion.

The Life of the Fly

This marvellous arithmetical juggler has an instinct, a genius, a gift for figures.

A second, at the age when most of us delight in tops and marbles, leaves the company of his boisterous playmates and listens to the echo of celestial harps singing within him. His head is a cathedral filled with the strains of an imaginary organ. Rich cadences, a secret concert heard by him and him alone, steep him in ecstasy. All hail to that predestined one who, some day, will rouse our noblest emotions with his musical chords. He has an instinct, a genius, a gift for sounds.

A third, a brat who cannot yet eat his bread and jam without smearing his face all over, takes a delight in fashioning clay into little figures that are astonishingly lifelike for all their artless awkwardness. He takes a knife and makes the briar-root grin into all sorts of entertaining masks; he carves boxwood in the semblance of a horse or sheep; he engraves the effigy of his dog on sandstone. Leave him alone; and, if Heaven second his efforts, he may become a famous sculptor. He has an instinct, a gift, a genius for form.

And so with others in every branch of human activity: art and science, industry and commerce, literature and philosophy. We have within us, from the start, that which will distinguish us from the vulgar herd. Now to what do we owe this distinctive character? To some throwback of atavism, men tell us. Heredity, direct in one case, remote in another, hands it down to us, increased or modified by time. Search the records of the family and you will discover the source of the genius, a mere trickle at first, then a stream, then a mighty river.

The darkness that lies behind that word heredity! Metaphysical science has tried to throw a little light upon it and has succeeded only in making unto itself a barbarous jargon, leaving obscurity more obscure than before. As for us, who hunger after lucidity, let us relinquish abstruse theories to whoso delights in them and confine our ambition to observable facts, without pretending to explain the quackery of the plasma. Our method certainly will not reveal to us the origin of instinct; but it will at least show us where it would be waste of time to look for it.

In this sort of research, a subject known through and through, down to its most intimate peculiarities, is indispensable. Where shall we find that subject? There would be a host of them and magnificent ones, if it were

possible to read the sealed pages of others'
lives; but no one can sound an existence outside his own and even then he can think himself lucky if a retentive memory and the habit
of reflection give his soundings the proper accuracy. As none of us is able to project himself into another's skin, we must needs, in
considering this problem, remain inside our
own.

To talk about one's self is hateful, I know.
The reader must have the kindness to excuse
me for the sake of the study in hand. I shall
take the silent Beetle's place in the witness-
box, cross-examining myself in all simplicity
of soul, as I do the animal, and asking myself
whence that one of my instincts which stands
out above the others is derived.

Since Darwin bestowed upon me the title
of 'incomparable observer,' the epithet has
often come back to me, from this side and
from that, without my yet understanding what
particular merit I have shown. It seems to
me so natural, so much within everybody's
scope, so absorbing to interest one's self in
everything that swarms around us! However,
let us pass on and admit that the compliment
is not unfounded.

My hesitation ceases if it is a question of

admitting my curiosity in matters that concern the insect. Yes, I possess the gift, the instinct that impels me to frequent that singular world; yes, I know that I am capable of spending on those studies an amount of precious time which would be better employed in making provision, if possible, for the poverty of old age; yes, I confess that I am an enthusiastic observer of the animal. How was this characteristic propensity, at once the torment and delight of my life, developed? And, to begin with, how much does it owe to heredity?

The common people have no history: persecuted by the present, they cannot think of preserving the memory of the past. And yet what surpassingly instructive records, comforting too and pious, would be the family-papers that should tell us who our forebears were and speak to us of their patient struggles with harsh fate, their stubborn efforts to build up, atom by atom, what we are to-day. No story would come up with that for individual interest. But, by the very force of things, the home is abandoned; and, when the brood has flown, the nest is no longer recognized.

I, a humble journeyman in the toilers' hive, am therefore very poor in family-recollections.

The Life of the Fly

In the second degree of ancestry, my facts become suddenly obscured. I will linger over
them a moment for two reasons: first, to enquire into the influence of heredity; and, secondly, to leave my children yet one more page
concerning them.

I did not know my maternal grandfather.
This venerable ancestor was, I have been
told, a process-server in one of the poorest
parishes of the Rouergue.[1] He used to en-
gross on stamped paper in a primitive spelling. With his well-filled pen-case and inkhorn, he went drawing out deeds up hill and
down dale, from one insolvent wretch to
another more insolvent still. Amid his atmosphere of pettifoggery, this rudimentary
scholar, waging battle on life's acerbities, certainly paid no attention to the insect; at most,
if he met it, he would crush it under foot.
The unknown animal, suspected of evil-doing,
deserved no further enquiry. Grandmother, on
her side, apart from her housekeeping and her
beads, knew still less about anything. She
looked on the alphabet as a set of hieroglyphics only fit to spoil your sight for nothing,

[1] A district of the province of Guienne, having Rodez
for its capital. The author's maternal grandfather, Salgues by name, was the *huissier*, or, as we should say,
sheriff's officer, of Saint Léons.—*Translator's Note.*

unless you were scribbling on paper bearing
the government stamp. Who in the world, in
her day, among the small folk, dreamt of
knowing how to read and write? That luxury
was reserved for the attorney, who himself
made but a sparing use of it. The insect, I
need hardly say, was the least of her cares. If
sometimes, when rinsing her salad at the tap,
she found a Caterpillar on the lettuce-leaves,
with a start of fright she would fling the loathsome thing away, thus cutting short relations reputed dangerous. In short, to both my
maternal grandparents, the insect was a creature of no interest whatever and almost always a repulsive object, which one dared not
touch with the tip of one's finger. Beyond a
doubt, my taste for animals was not derived
from them.

I have more precise information regarding
my grandparents on the father's side,[1] for

[1]Pierre Jean Fabre, son of Pierre Fabre, a peasant proprietor, and of Anne Fages, his wife, and Élisabeth Poujade, daughter of Antoine Poujade and Françoise Azémar, his wife. They were married in 1791. Pierre
Fabre, a labourer, father to Pierre Jean Fabre and grandfather to Antoine Fabre, the father of Jean Henri Casimir
Fabre, our author, was the son of Jean Fabre and of
Françoise Desmazes, his wife, and was married in 1759
to Anne Fages, daughter of Pierre Fages and of Anne
Baumelou, his wife.—*Translator's Note.*

their green old age allowed me to know them both. They were people of the soil, whose quarrel with the alphabet was so great that they had never opened a book in their lives; and they kept a lean farm on the cold granite ridge of the Rouergue table-land. The house, standing alone among the heath and broom, with no neighbour for many a mile around and visited at intervals by the wolves, was to them the hub of the universe. But for a few surrounding villages, whither the calves were driven on fair-days, the rest was only very vaguely known by hearsay. In this wild solitude, the mossy fens, with their quagmires oozing with iridescent pools, supplied the cows, the principal source of wealth, with rich, wet grass. In summer, on the short swards of the slopes, the sheep were penned day and night, protected from beasts of prey by a fence of hurdles propped up with pitchforks. When the grass was cropped close at one spot, the fold was shifted elsewhither. In the centre was the shepherd's rolling hut, a straw cabin. Two watch-dogs, equipped with spiked collars, were answerable for tranquillity if the thieving wolf appeared in the night from out the neighbouring woods.

Padded with a perpetual layer of cow-dung.

in which I sank to my knees, broken up with shimmering puddles of dark-brown liquid manure, the farm-yard also boasted a numerous population. Here the lambs skipped, the geese trumpeted, the fowls scratched the ground and the sow grunted with her swarm of little pigs hanging to her dugs.

The harshness of the climate did not give husbandry the same chances. In a propitious season, they would set fire to a stretch of moorland bristling with gorse and send the swing-plough across the ground enriched with the cinders of the blaze. This yielded a few acres of rye, oats and potatoes. The best corners were kept for hemp, which furnished the distaffs and spindles of the house with the material for linen and was looked upon as grandmother's private crop.

Grandfather, therefore, was, before all, a herdsman versed in matters of cows and sheep, but completely ignorant of aught else. How dumbfoundered he would have been to learn that, in the remote future, one of his family would become enamoured of those insignificant animals to which he had never vouchsafed a glance in his life! Had he guessed that that lunatic was myself, the scapegrace seated at the table by his side, what a smack I should have caught in the neck, what a wrathful look!

"The idea of wasting one's time with that nonsense!" he would have thundered.

For the patriarch was not given to joking. I can still see his serious face, his unclipped head of hair, often brought back behind his ears with a flick of the thumb and spreading its ancient Gallic mane over his shoulders. I see his little three-cornered hat, his small-clothes buckled at the knees, his wooden shoes, stuffed with straw, that echoed as he walked. Ah, no! Once childhood's games were past, it would never have done to rear the Grasshopper and unearth the Dung-beetle from his natural surroundings.

Grandmother, pious soul, used to wear the eccentric head-dress of the Rouergue highlanders: a large disk of black felt, stiff as a plank, adorned in the middle with a crown a finger's-breadth high and hardly wider across than a six-franc piece. A black ribbon fastened under the chin maintained the equilibrium of this elegant, but unsteady circle. Pickles, hemp, chickens, curds and whey, butter; washing the clothes, minding the children, seeing to the meals of the household: say that and you have summed up the strenuous woman's round of ideas. On her left side, the distaff, with its load of flax; in her right hand, the spindle turning under a quick twist of her thumb, moistened at intervals with her tongue: so she went through life, unweariedly, attending to the order and the welfare of the house. I see her in my mind's eye particularly on winter evenings, which were more favourable to family-talk. When the hour came for meals, all of us, big and little, would take our seats round a long table, on a couple of benches, deal planks supported by four rickety legs. Each found his wooden bowl and his tin spoon in front of him. At one end of the table always stood an enormous rye-loaf, the size of a cartwheel, wrapped in a linen cloth with a pleasant smell of washing, and remained until nothing was left of it. With a vigorous stroke, grandfather would cut off enough for the needs of the moment; then he would divide the piece among us with the one knife which he alone was entitled to wield. It was now each one's business to break up his bit with his fingers and to fill his bowl as he pleased.

Next came grandmother's turn. A capacious pot bubbled lustily and sang upon the flames in the hearth, exhaling an appetizing savour of bacon and turnips. Armed with a long metal ladle, grandmother would take from it, for each of us in turn, first the broth, wherein to soak the bread, and next the ration of turnips and bacon, partly fat and partly lean, filling the bowl to the top. At the other end of the table was the pitcher, from which the thirsty were free to drink at will. What appetites we had and what festive meals those were, especially when a cream-cheese, home-made, was there to complete the banquet!

Near us blazed the huge fire-place, in which whole tree-trunks were consumed in the extreme cold weather. From a corner of that monumental, soot-glazed chimney, projected, at a convenient height, a bracket with a slate shelf, which served to light the kitchen when we sat up late. On this we burnt chips of pine-wood, selected among the most translucent, those containing the most resin. They shed over the room a lurid red light, which saved the walnut-oil in the lamp.

When the bowls were emptied and the last crumb of cheese scraped up, grandam went back to her distaff, on a stool by the chimney-corner. We children, boys and girls, squatting on our heels and putting out our hands to the cheerful fire of furze, formed a circle round her and listened to her with eager ears. She told us stories, not greatly varied, it is true, but still wonderful, for the wolf often played a part in them. I should have very much liked to see this wolf, the hero of so many tales that made our flesh creep; but the shepherd always refused to take me into his straw hut, in the middle of the fold, at night. When we had done talking about the horrid wolf, the dragon and the serpent and when the resinous splinters had given out their last gleams, we went to sleep the sweet sleep that toil gives. As the youngest of the household, I had a right to the mattress, a sack stuffed with oat-chaff. The others had to be content with straw.

I owe a great deal to you, dear grandmother: it was in your lap that I found consolation for my first sorrows. You have handed down to me, perhaps, a little of your physical vigour, a little of your love of work; but certainly you were no more accountable than grandfather for my passion for insects.

Nor was either of my own parents. My mother, who was quite illiterate, having known no teacher than the bitter experience of a harassed life, was the exact opposite of what my tastes required for their develop-

ment. My peculiarity must seek its origin
elsewhere: that I will swear. But I do not
find it in my father, either. The excellent
man, who was hard-working and sturdily-
built like grandad, had been to school as a
child. He knew how to write, though he took
the greatest liberties with spelling; he knew
how to read and understood what he read,
provided the reading presented no more
serious literary difficulties than occurred in the
stories in the almanack. He was the first of
his line to allow himself to be tempted by the
town and he lived to regret it. Badly off, having but little outlet for his industry, making[1]
God knows what shifts to pick up a livelihood,
he went through all the disappointments of
the countryman turned townsman. Persecuted by bad luck, borne down by the burden,
for all his energy and good-will, he was far indeed from starting me in entomology. He
had other cares, cares more direct and more
serious. A good cuff or two when he saw me
pinning an insect to a cork was all the encouragement that I received from him. Perhaps
he was right.

The conclusion is positive: there is nothing

[1]The author's father kept a café in more than one small
town in the south of France.—*Translator's Note.*

in heredity to explain my taste for observation. You may say that I do not go far enough back. Well, what should I find beyond the grandparents where my facts come to a stop? I know, partly. I should find even more uncultured ancestors: sons of the soil, ploughmen, sowers of rye, neat-herds; one and all, by the very force of things, of not the least account in the nice matters of observation.

And yet, in me, the observer, the enquirer into things began to take shape almost in infancy. Why should I not describe my first discoveries? They are ingenuous in the extreme, but will serve notwithstanding to tell us something of the way in which tendencies first show themselves. I was five or six years old. That the poor household might have one mouth less to feed, I had been placed in grandmother's care, as I have just been saying. Here, in solitude, my first gleams of intelligence were awakened amidst the geese, the calves and the sheep. Everything before that is impenetrable darkness. My real birth is at that moment when the dawn of personality rises, dispersing the mists of unconsciousness and leaving a lasting memory. I can see myself plainly, clad in a soiled frieze frock flapping against my bare heels; I remember the handkerchief hanging from my waist by a bit of string, a handkerchief often lost and replaced by the back of my sleeve.

There I stand one day, a pensive urchin, with my hands behind my back and my face turned to the sun. The dazzling splendour fascinates me. I am the Moth attracted by the light of the lamp. With what am I enjoying the glorious radiance: with my mouth or my eyes? That is the question put by my budding scientific curiosity. Reader, do not smile: the future observer is already practising and experimenting. I open my mouth wide and close my eyes: the glory disappears. I open my eyes and shut my mouth: the glory reappears. I repeat the performance, with the same result. The question's solved: I have learnt by deduction that I see the sun with my eyes. Oh, what a discovery! That evening, I told the whole house all about it. Grandmother smiled fondly at my simplicity: the others laughed at it. 'Tis the way of the world.

Another find. At nightfall, amidst the neighbouring bushes, a sort of jingle attracted my attention, sounding very faintly and softly through the evening silence. Who is making that noise? Is it a little bird chirping in his nest? We must look into the matter and that quickly. True, there is the wolf, who comes out of the woods at this time, so they tell me. Let's go all the same, but not too far: just there, behind that clump of groom. I stand on the look-out for long, but all in vain. At the faintest sound of movement in the brush-wood, the jingle ceases. I try again next day and the day after. This time, my stubborn watch succeeds. Whoosh! A grab of my hand and I hold the singer. It is not a bird; it is a kind of Grasshopper whose hind-legs my playfellows have taught me to like: a poor recompense for my prolonged ambush. The best part of the business is not the two haunches with the shrimpy flavour, but what I have just learnt. I now know, from personal observation, that the Grasshopper sings. I did not publish my discovery, for fear of the same laughter that greeted my story about the sun.

Oh, what pretty flowers, in a field close to the house! They seem to smile to me with their great violet eyes. Later on, I see, in their place, bunches of big red cherries. I taste them. They are not nice and they have no stones. What can those cherries be? At

the end of the summer, grandfather comes
with a spade and turns my field of observation
topsy-turvy. From under ground there comes,
by the basketful and sackful, a sort of round
root. I know that root; it abounds in the
house; time after time I have cooked it in the
peat-stove. It is the potato. Its violet flower
and its red fruit are pigeon-holed for good and
all in my memory.

With an ever-watchful eye for animals
and plants, the future observer, the little
six-year-old monkey, practised by himself,
all unawares. He went to the flower,
he went to the insect, even as the Large
White Butterfly goes to the cabbage and
the Red Admiral to the thistle. He looked
and enquired, drawn by a curiosity whereof
heredity did not know the secret. He bore
within him the germ of a faculty unknown to
his family; he kept alive a glimmer that was
foreign to the ancestral hearth. What will become of that infinitesimal spark of childish
fancy? It will die out, beyond a doubt, unless education intervene, giving it the fuel of
example, fanning it with the breath of experience. In that case, schooling will explain
what heredity leaves unexplained. This is
what we will examine in the next chapter.

CHAPTER VI

MY SCHOOLING

I AM back in the village, in my father's house. I am now seven years old; and it is high time that I went to school. Nothing could have turned out better: the master is my godfather. What shall I call the room in which I was to become acquainted with the alphabet? It would be difficult to find the exact word, because the room served for every purpose. It was at once a school, a kitchen, a bedroom, a dining-room and, at times, a chickenhouse and a piggery. Palatial schools were not dreamt of in those days; any wretched hovel was thought good enough.

A broad fixed ladder led to the floor above. Under the ladder stood a big bed in a boarded recess. What was there upstairs? I never quite knew. I would see the master sometimes bring down an armful of hay for the ass, sometimes a basket of potatoes which the housewife emptied into the pot in which the little porkers' food was cooked. It must have been a loft of sorts, a storehouse of provisions for man and beast. Those two apartments composed the whole building.

To return to the lower one, the schoolroom: a window faces south, the only window in the house, a low, narrow window whose frame you can touch at the same time with your head and both your shoulders. This sunny aperture is the only lively spot in the dwelling, it overlooks the greater part of the village, which straggles along the slopes of a slanting valley. In the window-recess is the master's little table.

The opposite wall contains a niche in which stands a gleaming copper pail full of water. Here the parched children can relieve their thirst when they please, with a cup left within their reach. At the top of the niche are a few shelves bright with pewter plates, dishes and drinking-vessels, which are taken down from their sanctuary on great occasions only.

More or less everywhere, at any spot which the light touches, are crudely-coloured pictures, pasted on the walls. Here is Our Lady of the Seven Dolours, the disconsolate Mother of God opening her blue cloak to show her heart pierced with seven daggers. Between the sun and moon, which stare at you with their great, round eyes, is the Eternal Father, whose robe swells as though puffed out with the storm. To the right of the window, in the embrasure, is the Wandering Jew. He wears a three-cornered hat, a large, white leather apron, hobnailed shoes and a stout stick. 'Never was such a bearded man seen before or after,' says the legend that surrounds the picture. The draughtsman has not forgotten this detail: the old man's beard spreads in a snowy avalanche over the apron and comes down to his knees. On the left is Geneviève of Brabant, accompanied by the roe, with fierce Golo hiding in the bushes, sword in hand. Above hangs *The Death of Mr. Credit*, slain by defaulters at the door of his inn; and so on and so on, in every variety of subject, at all the unoccupied spots of the four walls.

I was filled with admiration of this picture-gallery, which held one's eyes with its great patches of red, blue, green and yellow. The master, however, had not set up his collection with a view to training our minds and hearts. That was the last and least of the worthy man's ambitions. An artist in his fashion, he had adorned his house according to his taste; and we benefited by the scheme of decoration.

While the gallery of halfpenny pictures

made me happy all the year round, there was
another entertainment which I found particularly attractive in winter, in frosty weather,
when the snow lay long on the ground. Against
the far wall stands the fireplace, as monumental in size as at my grandmother's. Its
arched cornice occupies the whole width of the
room, for the enormous redoubt fulfils more
than one purpose. In the middle is the hearth,
but, on the right and left, are two breast-high
recesses, half wood and half stone. Each of
them is a bed, with a mattress stuffed with
chaff of winnowed corn. Two sliding planks
serve as shutters and close the chest if the
sleeper would be alone. This dormitory,
sheltered under the chimney-mantel, supplies
couches for the favoured ones of the house,
the two boarders. They must lie snug in there
at night, with their shutters closed, when the
north-wind howls at the mouth of the dark
valley and sends the snow awhirl. The rest
is occupied by the hearth and its accessories:
the three-legged stools; the salt-box, hanging
against the wall to keep its contents dry; the
heavy shovel which it takes two hands to wield;
lastly, the bellows similar to those with which
I used to blow out my cheeks in grandfather's
house. They consist of a mighty branch of

pine, hollowed throughout its length with a red-hot iron. By means of this channel, one's breath is applied, from a convenient distance, to the spot which is to be revived. With a couple of stones for supports, the master's bundle of sticks and our own logs blaze and flicker, each of us having to bring a log of wood in the morning, if he would share in the treat.

For that matter, the fire was not exactly lit for us, but, above all, to warm a row of three pots in which simmered the pigs' food, a mixture of potatoes and bran. That, despite the tribute of a log, was the real object of the brushwood-fire. The two boarders, on their stools, in the best places, and we others sitting on our heels formed a semicircle around those big cauldrons, full to the brim and giving off little jets of steam, with puff-puff-puffing sounds. The bolder among us, when the master's eyes were engaged elsewhere, would dig a knife into a well-cooked potato and add it to their bit of bread; for I must say that, if we did little work in my school, at least we did a deal of eating. It was the regular custom to crack a few nuts and nibble at a crust while writing our page or setting out our rows of figures.

The Life of the Fly

We, the smaller ones, in addition to the comfort of studying with our mouths full, had every now and then two other delights, which were quite as good as cracking nuts. The back-door communicated with the yard where the hen, surrounded by her brood of chicks, scratched at the dung-hill, while the little porkers, of whom there were a dozen, wallowed in their stone trough. This door would open sometimes to let one of us out, a privilege which we abused, for the sly ones among us were careful not to close it on returning. Forthwith, the porkers would come running in, one after the other, attracted by the smell of the boiled potatoes. My bench, the one where the youngsters sat, stood against the wall, under the copper pail to which we used to go for water when the nuts had made us thirsty, and was right in the way of the pigs. Up they came trotting and grunting, curling their little tails; they rubbed against our legs; they poked their cold pink snouts into our hands in search of a scrap of crust; they questioned us with their sharp little eyes to learn if we happened to have a dry chestnut for them in our pockets. When they had gone the round, some this way and some that, they went back to the farm-yard, driven away by a friendly flick of the

master's handkerchief. Next came the visit
of the hen, bringing her velvet-coated chicks
to see us. All of us eagerly crumbled a little
bread for our pretty visitors. We vied with
one another in calling them to us and tickling
with our fingers their soft and downy backs.
No, there was certainly no lack of distractions.

What could we learn in such a school as
that! Let us first speak of the young ones, of
whom I was one. Each of us had, or rather
was supposed to have, in his hands a little
penny book, the alphabet, printed on grey
paper. It began, on the cover, with a pigeon,
or something like it. Next came a cross, followed by the letters in their order. When we
turned over, our eyes encountered the terrible
ba, be, bi, bo, bu, the stumbling-block of most
of us. When we had mastered that formidable page, we were considered to know how
to read and were admitted among the big ones.
But, if the little book was to be of any use, the
least that was required was that the master
should interest himself in us to some extent
and show us how to set about things. For this,
the worthy man, too much taken up with the
big ones, had not the time. The famous alphabet with the pigeon was thrust upon us
only to give us the air of scholars. We were

to contemplate it on our bench, to decipher it
with the help of our next neighbour, in case
he might know one or two of the letters. Our
contemplation came to nothing, being every
moment disturbed by a visit to the potatoes in
the stew-pots, a quarrel among playmates
about a marble, the grunting invasion of the
porkers or the arrival of the chicks. With
the aid of these distractions, we would wait
patiently until it was time for us to go home.
That was our most serious work.

The big ones used to write. They had the
benefit of the small amount of light in the
room, by the narrow window where the Wandering Jew and ruthless Golo faced each other,
and of the large and only table with its circle
of seats. The school supplied nothing, not
even a drop of ink; every one had to come with
a full set of utensils. The inkhorn of those
days, a relic of the ancient pencase of which
Rabelais speaks, was a long cardboard box
divided into two stages. The upper compart-
ment held the pens, made of goose- or turkey-
quills trimmed with a pen-knife; the lower
contained, in a tiny well, ink made of soot
mixed with vinegar.

The master's great business was to mend the
pens—a delicate work, not without danger for

inexperienced fingers—and then to trace at the head of the white page a line of strokes, single letters or words, according to the scholar's capabilities. When that is over, keep an eye on the work of art which is coming to adorn the copy! With what undulating movements of the wrist does the hand, resting on the little finger, prepare and plan its flight! All at once, the hand starts off, flies, whirls; and, lo and behold, under the line of writing is unfurled a garland of circles, spirals and flourishes, framing a bird with outspread wings, the whole, if you please, in red ink, the only kind worthy of such a pen. Large and small, we stood awestruck in the presence of these marvels. The family, in the evening, after supper, would pass from hand to hand the masterpiece brought back from school:

'What a man!' was the comment. 'What a man, to draw you a Holy Ghost with a stroke of the pen!'

What was read at my school? At most, in French, a few selections from sacred history. Latin recurred oftener, to teach us to sing vespers properly. The more advanced pupils tried to decipher manuscript, a deed of sale, the hieroglyphics of some scrivener.

And history, geography? No one ever

heard of them. What difference did it make
to us whether the earth was round or square!
In either case, it was just as hard to make it
bring forth anything.

And grammar? The master troubled his
head very little about that; and we still less.
We should have been greatly surprised by the
novelty and the forbidding look of such words
in the grammatical jargon as substantive, in-
dicative and subjunctive. Accuracy of language, whether of speech or writing, must be
learnt by practice. And none of us was
troubled by scruples in this respect. What was
the use of all these subtleties, when, on coming out of school, a lad simply went back to his
flock of sheep!

And arithmetic? Yes, we did a little of
this, but not under that learned name. We
called it sums. To put down rows of figures,
not too long, add them and subtract them one
from the other was more or less familiar work.
On Saturday evenings, to finish up the week,
there was a general orgy of sums. The top
boy stood up and, in a loud voice, recited the
multiplication-table up to twelve times. I
say twelve times, for in those days, because of
our old duodecimal measures, it was the custom to count as far as the twelve-times table,

instead of the ten times of the metric system. When this recital was over, the whole class, the little ones included, took it up in chorus, creating such an uproar that chicks and porkers took to flight if they happened to be there. And this went on to twelve times twelve, the first in the row starting the next table and the whole class repeating it as loud as it could yell. Of all that we were taught in school, the multiplication-table was what we knew best, for this noisy method ended by dinning the different numbers into our ears. This does not mean that we became skilful reckoners. The cleverest of us easily got muddled with the figures to be carried in a multiplication-sum. As for division, rare indeed were they who reached such heights. In short, the moment a problem, however insignificant, had to be solved, we had recourse to mental gymnastics much rather than to the learned aid of arithmetic.

When all is said, our master was an excellent man who could have kept school very well but for his lack of one thing; and that was time. He devoted to us all the little leisure which his numerous functions left him. And, first of all, he managed the property of an absentee landowner, who only occasionally set foot in the

village. He had under his care an old castle
with four towers, which had become so many
pigeon-houses; he directed the getting-in of
the hay, the walnuts, the apples and the oats.
We used to help him during the summer, when
the school, which was well-attended in winter,
was almost deserted. All that remained, because they were not yet big enough to work
in the fields, were a few children, including him
who was one day to set down these memorable
facts. Lessons at that time were less dull.
They were often given on the hay or on the
straw; oftener still, lesson-time was spent in
cleaning out the dove-cot or stamping on the
snails that had sallied in rainy weather from
their fortresses, the tall box borders of the
garden belonging to the castle.

Our master was a barber. With his light
hand, which was so clever at beautifying our
copies with curlicue birds, he shaved the nota-
bilities of the place: the mayor, the parish-
priest, the notary. Our master was a bell-
ringer. A wedding or a christening inter-
rupted the lessons: he had to ring a peal. A
gathering storm gave us a holiday: the great
bell must be tolled to ward off the lightning
and the hail. Our master was a choir-singer.
With his mighty voice, he filled the church

when he led the *Magnificat* at vespers. Our
master wound up and regulated the village-
clock. This was his proudest function. Giving a glance at the sun, to ascertain the time
more or less nearly, he would climb to the top
of the steeple, open a huge cage of rafters and
find himself in a maze of wheels and springs
whereof the secret was known to him alone.

With such a school and such a master and
such examples, what will become of my embryo tastes, as yet so imperceptible? In that
environment, they seem bound to perish,
stifled for ever. Yet no, the germ has life; it
works in my veins, never to leave them again.
It finds nourishment everywhere, down to the
cover of my penny alphabet, embellished with
a crude picture of a pigeon which I study and
contemplate much more zealously than the
A.B.C. Its round eye, with its circlet of dots,
seems to smile upon me. Its wing, of which I
count the feathers one by one, tells me of
flights on high, among the beautiful clouds;
it carries me to the beeches raising their smooth
trunks above a mossy carpet studded with white
mushrooms that look like eggs dropped by
some vagrant hen; it takes me to the snow-clad
peaks where the birds leave the starry print of
their red feet. He is a fine fellow, my pigeon-

friend: he consoles me for the woes hidden behind the cover of my book. Thanks to him, I sit quietly on my bench and wait more or less till school is over.

School out of doors has other charms. When the master takes us to kill the snails in the box borders, I do not always scrupulously fulfil my office as an exterminator. My heel sometimes hesitates before coming down upon the handful which I have gathered. They are so pretty! Just think, there are yellow ones and pink, white ones and brown, all with dark spiral streaks. I fill my pockets with the handsomest, so as to feast my eyes on them at my leisure.

On hay-making days in the master's field, I strike up an acquaintance with the Frog. Flayed and stuck at the end of a split stick, he serves as bait to tempt the crayfish to come out of his retreat by the brook-side. On the alder-trees I catch the Hoplia, the splendid Scarab who pales the azure of the heavens. I pick the narcissus and learn to gather, with the tip of my tongue, the tiny drop of honey that lies right at the bottom of the cleft corolla. I also learn that too-long indulgence in this feast brings a headache; but this discomfort in no way impairs my admiration for the glorious white flower, which wears a narrow red collar at the throat of its funnel.

When we go to beat the walnut-trees, the barren grass-plots provide me with Locusts spreading their wings, some into a blue fan, others into a red. And thus the rustic school, even in the heart of winter, furnished continuous food for my interest in things. There was no need for precept and example: my passion for animals and plants made progress of itself.

What did not make progress was my acquaintance with my letters, greatly neglected in favour of the pigeon. I was still at the same stage, hopelessly behindhand with the untractable alphabet, when my father, by a chance inspiration, brought me home from the town what was destined to give me a start along the road of reading. Despite the not insignificant part which it played in my intellectual awakening, the purchase was by no means a ruinous one. It was a large print, price six farthings, coloured and divided into compartments in which animals of all sorts taught the A.B.C. by means of the first letters of their names.

Where should I keep the precious picture? As it happened, in the room set apart for the children at home, there was a little window like the one in the school, opening in the same way out of a sort of recess and in the same way overlooking most of the village. One was on the right, the other on the left of the castle with the pigeon-house towers; both afforded an equally good view of the heights of the slanting valley. I was able to enjoy the school-window only at rare intervals, when the master left his little table; the other was at my disposal as often as I liked. I spent long hours there, sitting on a little fixed window-seat.

The view was magnificent. I could see the ends of the earth, that is to say, the hills that blocked the horizon, all but a misty gap through which the brook with the crayfish flowed under the alders and willows. High up on the sky-line, a few wind-battered oaks bristled on the ridges; and beyond there lay nothing but the unknown, laden with mystery.

At the back of the hollow stood the church, with its three steeples and its clock; and, a little higher, the village-square, where a spring, fashioned into a fountain, gurgled from one basin into another, under a wide arched roof. I could hear from my window the chatter of the women washing their clothes, the strokes of their beaters, the rasping of the pots scoured with sand and vinegar. Sprinkled over the

slopes are little houses with their garden-
patches in terraces banked up by tottering
walls, which bulge under the thrust of the
earth. Here and there are very steep lanes,
with the dents of the rock forming a natural
pavement. The mule, sure-footed though he
be, would hesitate to enter these dangerous
passes with his load of branches.

Further on, beyond the village, half-way up
the hills, stood the great ever-so-old lime-tree,
the *Tel,* as we used to call it, whose sides, hollowed out by the ages, were the favourite hiding-places of us children at play. On fair-
days, its immense, spreading foliage cast a
wide shadow over the herds of oxen and sheep.
Those solemn days, which only came once a
year, brought me a few ideas from without:
I learnt that the world did not end with my
amphitheatre of hills. I saw the inn-keeper's
wine arrive on mule-back and in goat-skin
bottles. I hung about the market-place and
watched the opening of jars full of stewed
pears, the setting-out of baskets of grapes, an
almost unknown fruit, the object of eager
covetousness. I stood and gazed in admiration at the roulette-board on which, for a sou,
according to the spot at which its needle
stopped on a circular row of nails, you won a

pink poodle made of barley-sugar, or a round jar of aniseed sweets, or, much oftener, nothing at all. On a piece of canvas on the ground, rolls of printed calico with red flowers, were displayed to tempt the girls. Close by rose a pile of beech-wood clogs, tops and box-wood flutes. Here the shepherds chose their instruments, trying them by blowing a note or two. How new it all was to me! What a lot of things there were to see in this world! Alas, that wonderful time was of but short duration! At night, after a little brawling at the inn, it was all over; and the village returned to silence for a year.

But I must not linger over these memories of the dawn of life. We were speaking of the memorable picture brought from town. Where shall I keep it, to make the best use of it? Why, of course, it must be pasted on the embrasure of my window. The recess, with its seat, shall be my study-cell; here I can feast my eyes by turns on the big lime-tree and the animals of my alphabet. And this was what I did.

And now, my precious picture, it is our turn, yours and mine. You began with the sacred beast, the ass, whose name, with a big initial, taught me the letter A. The *bœuf*, the ox, stood for B; the *canard,* the duck, told me about C; the *dindon,* the turkey, gave me the letter D. And so on with the rest. A few compartments, it is true, were lacking in clearness. I had no friendly feeling for the hippopotamus, the kamichi, or horned screamer, and the zebu, who aimed at making me say H, K and Z. Those outlandish beasts, which failed to give the abstract letter the support of a recognized reality, caused me to hesitate for a time over their recalcitrant consonants. No matter: father came to my aid in difficult cases; and I made such rapid progress that, in a few days, I was able to turn in good earnest the pages of my little pigeon-book, hitherto so undecipherable. I was initiated; I knew how to spell. My parents marvelled. I can explain this unexpected progress to-day. Those speaking pictures, which brought me amongst my friends the beasts, were in harmony with my instincts. If the animal has not fulfilled all that it promised in so far as I am concerned, I have at least to thank it for teaching me to read. I should have succeeded by other means, I do not doubt, but not so quickly nor so pleasantly. Animals for ever!

Luck favoured me a second time. As a reward for my prowess, I was given La Fontaine's Fables, in a popular, cheap edition, crammed with pictures, small, I admit, and very inaccurate, but still delightful. Here were the crow, the fox, the wolf, the magpie, the frog, the rabbit, the ass, the dog, the cat: all persons of my acquaintance. The glorious book was immensely to my taste, with its skimpy illustrations on which the animal walked and talked. As to understanding what it said, that was another story! Never mind, my lad! Put together syllables that say nothing to you as yet; they will speak to you later and La Fontaine will always remain your friend.

I come to the time when I was ten years old and at Rodez College. My functions as a serving-boy in the chapel entitled me to free instruction as a day-boarder. There were four of us in white surplices and red skull-caps and cassocks. I was the youngest of the party and did little more than walk on. I counted as a unit; and that was about all, for I was never certain when to ring the bell or move the missal. I was all of a tremble when we gathered two on this side and two on that, with genuflexions, in the middle of the sanctuary, to intone the *Domine, salvum fac regem* at the end of mass. Let me make a confes-

sion: tongue-tied with shyness, I used to leave
it to the others.

Nevertheless, I was well thought of, for, in
the school, I cut a good figure in composition
and translation. In that classical atmosphere,
there was talk of Procas, King of Alba, and of
his two sons, Numitor and Amulius. We
heard of Cynœgirus, the strong-jawed man,
who, having lost his two hands in battle, seized
and held a Persian galley with his teeth, and
of Cadmus the Phœnician, who sowed a dragon's teeth as though they were beans and
gathered his harvest in the shape of a host of
armed men, who killed one another as they
rose up from the ground. The only one who
survived the slaughter was one as tough as
leather, presumably the son of the big back
grinder.

Had they talked to me about the man in
the moon, I could not have been more startled.
I made up for it with my animals, which I was
far from forgetting amid this phantasmagoria
of heroes and demigods. While honouring
the exploits of Cadmus and Cynœgirus, I
hardly ever failed, on Sundays and Thursdays,[1] to go and see if the cowslip or the yellow

[1]The weekly half-holiday in French schools.—*Translator's Note.*

daffodil was making its appearance in the meadows, if the Linnet was hatching on the juniper-bushes, if the Cockchafers were plopping down from the wind-shaken poplars. Thus was the sacred spark kept aglow, ever brighter than before.

By easy stages, I came to Virgil and was very much smitten with Melibœus, Corydon, Menalcas, Damœtas and the rest of them. The scandals of the ancient shepherds fortunately passed unnoticed; and within the frame in which the characters moved were exquisite details concerning the Bee, the Cicada, the Turtle-dove, the Crow, the Nanny-goat and the golden broom. A veritable delight were these stories of the fields, sung in sonorous verse; and the Latin poet left a lasting impression on my classical recollections.

Then, suddenly, good-bye to my studies, good-bye to Tityrus and Menalcas. Ill-luck is swooping down on us, relentlessly. Hunger threatens us at home. And now, boy, put your trust in God; run about and earn your penn'orth of potatoes as best you can. Life is about to become a hideous inferno. Let us pass quickly over this phase.

Amid this lamentable chaos, my love for the insect ought to have gone under. Not at all.

My Schooling

It would have survived the raft of the *Medusa*. I still remember a certain Pine Cockchafer met for the first time. The plumes on her antennæ, her pretty pattern of white spots on a dark-brown ground were as a ray of sunshine in the gloomy wretchedness of the day.

To cut a long story short: good fortune, which never abandons the brave, brought me to the primary normal school at Vaucluse, where I was assured food: dried chestnuts and chick-peas. The principal, a man of broad views, soon came to trust his new assistant. He left me practically a free hand, so long as I satisfied the school curriculum, which was very modest in those days. Possessing a smattering of Latin and grammar, I was a little ahead of my fellow-pupils. I took advantage of this to get some order into my vague knowledge of plants and animals. While a dictation-lesson was being corrected around me, with generous assistance from the dictionary, I would examine, in the recesses of my desk, the oleander's fruit, the snap-dragon's seed-vessel, the Wasp's sting and the Ground-beetle's wing-case.

With this foretaste of natural science, picked up haphazard and by stealth, I left school more deeply in love than ever with insects and flowers. And yet I had to give it all up. That wider education, which would have to be my source of livelihood in the future, demanded this imperiously. What was I to take in hand to raise me above the primary school, whose staff could barely earn their bread in those days? Natural history could not bring me anywhere. The educational system of the time kept it at a distance, as unworthy of association with Latin and Greek. Mathematics remained, with its very simple equipment: a blackboard, a bit of chalk and a few books.

So I flung myself with might and main into conic sections and the calculus: a hard battle, if ever there was one, without guides or counsellors, face to face for days on end with the abstruse problem which my stubborn thinking at last stripped of its mysteries. Next came the physical sciences, studied in the same manner, with an impossible laboratory, the work of my own hands.

The reader can imagine the fate of my favourite branch of science in this fierce struggle. At the faintest sign of revolt, I lectured myself severely, lest I should let myself be seduced by some new grass, some unknown Beetle. I did violence to my feelings. My natural-history books were sentenced to oblivion, relegated to the bottom of a trunk.

And so, in the end, I am sent to teach physics and chemistry at Ajaccio College. This time, the temptation is too much for me. The sea, with its wonders, the beach, whereon the tide casts such beautiful shells, the *maquis* of myrtles, arbutus and mastic-trees: all this paradise of gorgeous nature has too much on its side in the struggle with the sine and the cosine. I succumb. My leisure-time is divided into two parts. One, the larger, is allotted to mathematics, the foundation of my academical future, as planned by myself; the other is spent, with much misgiving, in botanizing and looking for the treasures of the sea. What a country and what magnificent studies to be made, if, unobsessed by x and y, I had devoted myself whole-heartedly to my inclinations!

We are the wisp of straw, the plaything of the winds. We think that we are making for a goal deliberately chosen; destiny drives us towards another. Mathematics, the exaggerated preoccupation of my youth, did me hardly any service; and animals, which I avoided as much as ever I could, are the consolation of my old age. Nevertheless, I bear no grudge against the sine and the cosine, which I continue to hold in high esteem. They cost me many a pallid hour at one time, but they always afforded me some first-rate entertainment: they still do so, when my head lies tossing sleeplessly on its pillow.

Meanwhile, Ajaccio received the visit of a famous Avignon botanist, Requien[1] by name, who, with a box crammed with paper under his arm, had long been botanizing all over Corsica, pressing and drying specimens and distributing them to his friends. We soon became acquainted. I accompanied him in my free time on his explorations and never did the master have a more attentive disciple. To tell the truth, Requien was not a man of learning so much as an enthusiastic collector. Very few would have felt capable of competing with him when it came to giving the name or the geographical distribution of a plant. A blade of grass, a pad of moss, a scab of lichen, a thread of seaweed: he knew them all. The scientific name flashed across his mind at once. What an unerring memory, what a genius for classification amid the enormous

[1]Esprit Requien (1788-1851), a French naturalist and collector, director of the museum and botanical gardens at Avignon and author of several works on botany and conchology.—*Translator's Note.*

mass of things observed! I stood aghast at it. I owe much to Requien in the domain of botany. Had death spared him longer, I should doubtless have owed more to him, for his was a generous heart, ever open to the troubles of novices.

In the following year, I met Moquin-Tandon,[1] with whom, thanks to Requien, I had already exchanged a few letters on botany. The illustrious Toulouse professor came to study on the spot the flora which he proposed to describe systematically. When he arrived, all the hotel bedrooms were reserved for the members of the general council which had been summoned; and I offered him board and lodging: a shake-down in a room overlooking the sea; fare consisting of lampreys, turbot and sea-urchins: common enough dishes in that land of Cockayne, but possessing no small attraction for the naturalist, because of their novelty. My cordial proposal tempted him; he yielded to my blandishments; and there we were for a fortnight chatting at table *de omni*

[1] Horace Bénédict Alfred Moquin-Tandon (1804-1863), a distinguished naturalist, for twenty years director of the botanical gardens at Toulouse. He was commissioned by the French government in 1850 to compile a flora of Corsica and is the author of several important works on botany and zoology.—*Translator's Note.*

re scibili after the botanical excursion was over.

With Moquin-Tandon, new vistas opened before me. Here it was no longer the case of a nomenclator with an infallible memory: he was a naturalist with far-reaching ideas, a philosopher who soared above petty details to comprehensive views of life, a writer, a poet who knew how to clothe the naked truth in the magic mantle of the glowing word. Never again shall I sit at an intellectual feast like that:

'Leave your mathematics,' he said. 'No one will take the least interest in your formulæ. Get to the beast, the plant; and, if, as I believe, the fever burns in your veins, you will find men to listen to you.'

We made an expedition to the centre of the island, to Monte Renoso,[1] with which I was already familiar. I made the scientist pick the hoary everlasting (*Helichrysum frigidum*), which makes a wonderful patch of silver; the many-headed thrift, or mouflon-grass (*Armeria multiceps*), which the Corsicans call *erba muorone;* the downy marguerite (*Leucanthemum tomosum*), which, clad in wadding,

---

[1] A mountain of 7,730 feet, about twenty-five miles from Ajaccio.—*Translator's Note.*

shivers amid the snows; and many other rarities dear to the botanist. Moquin-Tandon was jubilant. I, on my side, was much more attracted and overcome by his words and his enthusiasm than by the hoary everlasting. When we came down from the cold mountaintop, my mind was made up: mathematics would be abandoned.

On the day before his departure, he said to me:

'You interest yourself in shells. That is something, but it is not enough. You must look into the animal itself. I will show you how it's done.'

And, taking a sharp pair of scissors from the family work-basket and a couple of needles stuck into a bit of vine-shoot which served as a makeshift handle, he showed me the anatomy of a snail in a soup-plate filled with water. Gradually he explained and sketched the organs which he spread before my eyes. This was the only, never-to-be-forgotten lesson in natural history that I ever received in my life.

It is time to conclude. I was cross-examining myself, being unable to cross-examine the silent Beetle. As far as it is possible to read within myself, I answer as follows:

'From early childhood, from the moment

of my first mental awakening, I have felt drawn towards the things of nature, or, to return to our catchword, I have the gift, the bump of observation.'

After the details which I have already given about my ancestors, it would be ridiculous to look to heredity for an explanation of the fact. Nor would any one venture to suggest the words or example of my masters. Of scientific education, the fruit of college-training, I had none whatever. I never set foot in a lecture-hall except to undergo the ordeal of examinations. Without masters, without guides, often without books, in spite of poverty, that terrible extinguisher, I went ahead, persisted, facing my difficulties, until the indomitable bump ended by shedding its scanty contents. Yes, they were very scanty, yet possibly of some value, if circumstances had come to their assistance. I was a born animalist. Why and how? No reply.

We thus have, all of us, in different directions and in a greater or lesser degree, characteristics that brand us with a special mark, characteristics of an unfathomable origin. They exist because they exist; and that is all that any one can say. The gift is not handed down: the man of talent has a fool for a son.

Nor is it acquired; but it is improved by practice. He who has not the germ of it in his veins will never possess it, in spite of all the pains of a hot-house education.

That to which we give the name of instinct when speaking of animals is something similar to genius. It is, in both cases, a peak that rises above the ordinary level. But instinct is handed down, unchanged and undiminished, throughout the sequence of a species; it is permanent and general and in this it differs greatly from genius, which is not transmissible and changes in different cases. Instinct is the inviolable heritage of the family and falls to one and all, without distinction. Here the difference ends. Independent of similarity of structure, it breaks out like genius, here or elsewhere, for no perceptible reason. Nothing causes it to be foreseen, nothing in the organization explains it. If cross-examined on this point, the Dung-beetles and the rest, each with his own peculiar talent, would answer, were we able to understand them:

'Instinct is the animal's genius.'

CHAPTER VII

THE POND

THE pond, the delight of my early childhood, is still a sight whereof my old eyes never tire. What animation in that verdant world! On the warm mud of the edges, the Frog's little Tadpole basks and frisks in its black legions; down in the water, the orange-bellied Newt steers his way slowly with the broad rudder of his flat tail; among the reeds are stationed the flotillas of the Caddis-worms, half-protruding from their tubes, which are now a tiny bit of stick and again a turret of little shells.

In the deep places, the Water-beetle dives, carrying with him his reserves of breath: an air-bubble at the tip of the wing-cases and, under the chest, a film of gas that gleams like a silver breastplate; on the surface, the ballet of those shimmering pearls, the Whirligigs, turns and twists about; hard by there skims the insubmersible troop of the Pond-skaters, who glide along with side-strokes similar to those which the cobbler makes when sewing.

The Pond

Here are the Water-boatmen, who swim on their backs with two oars spread cross-wise, and the flat Water-scorpions; here, squalidly clad in mud, is the grub of the largest of our Dragon-flies, so curious because of its manner of progression: it fills its hinder-parts, a yawning funnel, with water, spirts it out again and advances just so far as the recoil of its hydraulic cannon.

The Molluscs abound, a peaceful tribe. At the bottom, the plump River-snails discreetly raise their lid, opening ever so little the shutters of their dwelling; on the level of the water, in the glades of the aquatic garden, the Pond-snails—Physa, Limnæa and Planorbis—take the air. Dark Leeches writhe upon their prey, a chunk of Earth-worm; thousands of tiny, reddish grubs, future Mosquitoes, go spinning around and twist and curve like so many graceful Dolphins.

Yes, a stagnant pool, though but a few feet wide, hatched by the sun, is an immense world, an inexhaustible mine of observation to the studious man and a marvel to the child who, tired of his paper boat, diverts his eyes and thoughts a little with what is happening in the water. Let me tell what I remember of my first pond, at a time when ideas began to dawn in my seven-year-old brain.

How shall a man earn his living in my poor native village, with its inclement weather and its niggardly soil? The owner of a few acres of grazing-land rears sheep. In the best parts, he scrapes the soil with the swing-plough; he flattens it into terraces banked by walls of broken stones. Pannierfuls of dung are carried up on donkey-back from the cowshed. Then, in due season, comes the excellent potato, which, boiled and served hot in a basket of plaited straw, is the chief stand-by in winter.

Should the crop exceed the needs of the household, the surplus goes to feed a pig, that precious beast, a treasure of bacon and ham. The ewes supply butter and curds; the garden boasts cabbages, turnips and even a few hives in a sheltered corner. With wealth like that one can look fate in the face.

But we, we have nothing, nothing but the little house inherited by my mother and its adjoining patch of garden. The meagre resources of the family are coming to an end. It is time to see to it and that quickly. What is to be done? That is the stern question which father and mother sat debating one evening.

The Pond

Hop-o'-my-Thumb, hiding under the wood-cutter's stool, listened to his parents overcome by want. I also, pretending to sleep, with my elbows on the table, listen not to blood-curdling designs, but to grand plans that set my heart rejoicing. This is how the matter stands: at the bottom of the village, near the church, at the spot where the water of the large roofed spring escapes from its underground weir and joins the brook in the valley, an enterprising man, back from the war,[1] has set up a small tallow-factory. He sells the scrapings of his pans, the burnt fat, reeking of candle-grease, at a low price. He proclaims these wares to be excellent for fattening ducks.

"Suppose we bred some ducks," says mother. "They sell very well in town. Henri would mind them and take them down to the brook."

"Very well," says father, "let's breed some ducks. There may be difficulties in the way; but we'll have a try."

That night, I had dreams of paradise: I was with my ducklings, clad in their yellow suits; I took them to the pond, I watched them

[1]The war of 1830 with Algiers.—*Translator's Note.*

have their bath, I brought them back again, carrying the more tired ones in a basket.

A month or two after, the little birds of my dreams were a reality. There were twenty-four of them. They had been hatched by two hens, of whom one, the big, black one, was an inmate of the house, while the other was borrowed from a neighbour.

To bring them up, the former is sufficient, so careful is she of her adopted family. At first, everything goes perfectly: a tub with two fingers' depth of water serves as a pond. On sunny days, the ducklings bathe in it under the anxious eye of the hen.

A fortnight later, the tub is no longer enough. It contains neither cresses crammed with tiny Shellfish nor Worms and Tadpoles, dainty morsels both. The time has come for dives and hunts amid the tangle of the water-weeds; and for us the day of trouble has also come. True, the miller, down by the brook, has fine ducks, easy and cheap to bring up; the tallow-smelter, who has extolled his burnt fat so loudly, has some as well, for he has the advantage of the waste water from the spring at the bottom of the village; but how are we, right up there, at the top, to procure aquatic

The Pond

sports for our broods? In summer, we have hardly water to drink!

Near the house, in a freestone recess, a scanty source trickles into a basin made in the rock. Four or five families have, like ourselves, to draw their water there with copper pails. By the time that the schoolmaster's donkey has slaked her thirst and the neighbours have taken their provision for the day, the basin is dry. We have to wait for four-and-twenty hours for it to fill. No, this is not the hole in which the ducks would delight nor indeed in which they would be tolerated.

There remains the brook. To go down to it with the troop of ducklings is fraught with danger. On the way through the village, we might meet cats, bold ravishers of small poultry; some surly mongrel might frighten and scatter the little band; and it would be a hard puzzle to collect it in its entirety. We must avoid the traffic and take refuge in peaceful and sequestered spots.

On the hills, the path that climbs behind the château[1] soon takes a sudden turn and widens into a small plain beside the meadows. It skirts a rocky slope whence trickles, level with

[1] The Château de Saint-Léons, standing just outside and above the village of Saint-Léons, where the author was born in 1823.—*Translator's Note.*

the ground, a streamlet, forming a pond of some size. Here profound solitude reigns all day long. The ducklings will be well off; and the journey can be made in peace by a deserted foot-path.

You, little man, shall take them to that delectable spot. What a day it was that marked my first appearance as a herdsman of ducks! Why must there be a jar to the even tenor of such joys? The too-frequent encounter of my tender skin with the hard ground had given me a large and painful blister on the heel. Had I wanted to put on the shoes stowed away in the cupboard for Sundays and holidays, I could not. There was nothing for it but to go barefoot over the broken stones, dragging my leg and carrying high the injured heel.

Let us make a start, hobbling along, switch in hand, behind the ducks. They too, poor little things, have sensitive soles to their feet; they limp, they quack with fatigue. They would refuse to go any farther if I did not, from time to time, call a halt under the shelter of an ash.

We are there at last. The place could not be better for my birdlets; shallow, tepid water, interspersed with muddy knolls and

The Pond

green eyots. The diversions of the bath begin forthwith. The ducklings clap their beaks and rummage here, there and everywhere; they sift each mouthful, rejecting the clear water and retaining the good bits. In the deeper parts, they point their sterns into the air and stick their heads under water. They are happy; and it is a blessed thing to see them at work. We will let them be. It is my turn to enjoy the pond.

What is this? On the mud lie some loose, knotted, soot-coloured cords. One could take them for threads of wool like those which you pull out of an old ravelly stocking. Can some shepherdess, knitting a black sock and finding her work turn out badly, have begun all over again and, in her impatience, have thrown down the wool with all the dropped stitches? It really looks like it.

I take up one of those cords in my hand. It is sticky and extremely slack; the thing slips through the fingers before they can catch hold of it. A few of the knots burst and shed their contents. What comes out is a black globule, the size of a pin's head, followed by a flat tail. I recognize, on a very small scale, a familiar object: the Tadpole, the Frog's baby. I have seen enough. Let us leave the knotted cords alone.

The next creatures please me better. They spin round on the surface of the water and their black backs gleam in the sun. If I lift a hand to seize them, that moment they disappear, I know not where. It's a pity: I should have much liked to see them closer and to make them wriggle in a little bowl which I should have put ready for them.

Let us look at the bottom of the water, pulling aside those bunches of green string whence beads of air are rising and gathering into foam. There is something of everything underneath. I see pretty shells with compact whorls, flat as beans; I notice little worms carrying tufts and feathers; I make out some with flabby fins constantly flapping on their backs. What are they all doing there? What are their names? I do not know. And I stare at them for ever so long, held by the incomprehensible mystery of the waters.

At the place where the pond dribbles into the adjoining field are some alder-trees; and here I make a glorious find. It is a Scarab—not a very large one, oh no! He is smaller than a cherry-stone, but of an unutterable blue. The angels in paradise must wear dresses of that colour. I put the glorious one inside an empty snail-shell, which I plug up with a leaf. I shall admire that living jewel at my leisure, when I get back. Other distractions summon me away.

The spring that feeds the pond trickles from the rock, cold and clear. The water first collects into a cup, the size of the hollow of one's two hands, and then runs over in a stream. These falls call for a mill: that goes without saying. Two bits of straw, artistically crossed upon an axis, provide the machinery; some flat stones set on edge afford supports. It is a great success: the mill turns admirably. My triumph would be complete, could I but share it. For want of other playmates, I invite the ducks.

Everything palls in this poor world of ours, even a mill made of two straws. Let us think of something else: let us contrive a dam to hold back the waters and form a pool. There is no lack of stones for the brickwork. I pick the most suitable; I break the larger ones. And, while collecting these blocks, suddenly I forget all about the dam which I meant to build.

On one of the broken stones, in a cavity large enough for me to put my fist in, some-

thing gleams like glass. The hollow is lined
with facets gathered in sixes which flash and
glitter in the sun. I have seen something like
this in church, on the great saints'-days, when
the light of the candles in the big chandelier
kindles the stars in its hanging crystal.

We children, lying, in summer, on the straw
of the threshing-floor, have told one another stories of the treasures which a dragon guards underground. Those treasures now return to my
mind: the names of precious stones ring out
uncertainly but gloriously in my memory. I
think of the king's crown, of the princesses'
necklaces. In breaking stones, can I have
found, but on a much richer scale, the thing
that shines quite small in my mother's ring?
I want more such.

The dragon of the subterranean treasures
treats me generously. He gives me his diamonds in such quantities that soon I possess a
heap of broken stones sparkling with magnificent clusters. He does more: he gives me his
gold. The trickle of water from the rock falls
on a bed of fine sand which it swirls into bubbles. If I bent over towards the light, I see
something like gold-filings whirling where the
fall touches the bottom. Is it really the
famous metal of which twenty-franc pieces, so

The Pond

rare with us at home, are made? One would think so, from the glitter.

I take a pinch of sand and place it in my palm. The brilliant particles are numerous, but so small that I have to pick them up with a straw moistened in my mouth. Let us drop this: they are too tiny and too bothersome to collect. The big, valuable lumps must be farther on, in the thickness of the rock. We'll come back later; we'll blast the mountain.

I break more stones. Oh, what a queer thing has just come loose, all in one piece! It is turned spiral-wise, like certain flat Snails that come out of the cracks of old walls in rainy weather. With its gnarled sides, it looks like a little ram's-horn. Shell or horn, it is very curious. How do things like that find their way into the stone?

Treasures and curiosities make my pockets bulge with pebbles. It is late and the little ducklings have had all they want to eat. Come along, youngsters, let's go home. My blistered heel is forgotten in my excitement.

The walk back is a delight. A voice sings in my ear, an untranslatable voice, softer than any language and bewildering as a dream. It speaks to me for the first time of the mysteries of the pond; it glorifies the heavenly insect which I hear moving in the empty snail-shell, its temporary cage; it whispers the secrets of the rock, the gold-filings, the faceted jewels, the ram's-horn turned to stone.

Poor simpleton, smother your joy! I arrive. My parents catch sight of my bulging pockets, with their disgraceful load of stones. The cloth has given way under the rough and heavy burden.

"You rascal!" says father, at sight of the damage. "I send you to mind the ducks and you amuse yourself picking up stones, as though there weren't enough of them all round the house! Make haste and throw them away!"

Broken-hearted, I obey. Diamonds, gold-dust, petrified ram's-horn, heavenly Beetle are all flung on a rubbish-heap outside the door.

Mother bewails her lot:

"A nice thing, bringing up children to see them turn out so badly! You'll bring me to my grave. Green stuff I don't mind: it does for the rabbits. But stones, which ruin your pockets; poisonous animals, which'll sting your hand: what good are they to you, silly? There's no doubt about it: some one has thrown a spell over you!"

Yes, my poor mother, you were right, in

174

The Pond

your simplicity: a spell had been cast upon me; I admit it to-day. When it is hard enough to earn one's bit of bread, does not improving one's mind but render one more meet for suffering? Of what avail is the torment of learning to the derelicts of life?

A deal better off am I, at this late hour, dogged by poverty and knowing that the diamonds of the duck-pool were rock-crystal, the gold-dust mica, the stone horn an Ammonite and the sky-blue Beetle a Hoplia! We poor men would do better to mistrust the joys of knowledge: let us dig our furrow in the fields of the commonplace, flee the temptations of the pond, mind our ducks and leave to others, more favoured by fortune, the job of explaining the world's mechanism, if the spirit moves them.

And yet no! Alone among living creatures, man has the thirst for knowledge; he alone pries into the mysteries of things. The least among us will utter his whys and his wherefores, a fine pain unknown to the brute beast. If these questionings come from us with greater persistence, with a more imperious authority, if they divert us from the quest of lucre, life's only object in the eyes of most men, does it become us to complain? Let us be careful not to do so, for that would be denying the best of all our gifts.

Let us strive, on the contrary, within the measure of our capacity, to force a gleam of light from the vast unknown; let us examine and question and, here and there, wrest a few shreds of truth. We shall sink under the task; in the present ill-ordered state of society, we shall end, perhaps, in the workhouse. Let us go ahead for all that: our consolation shall be that we have increased by one atom the general mass of knowledge, the incomparable treasure of mankind.

As this modest lot has fallen to me, I will return to the pond, notwithstanding the wise admonitions and the bitter tears which I once owed to it. I will return to the pond, but not to that of the small ducks, the pond aflower with illusions: those ponds do not occur twice in a lifetime. For luck like that, you must be in all the new glory of your first breeches and your first ideas.

Many another have I come upon since that distant time, ponds very much richer and, moreover, explored with the ripened eye of experience. Enthusiastically I searched them with the net, stirred up their mud, ransacked their trailing weeds. None in my memories comes

up to the first, magnified in its delights and
mortifications by the marvellous perspective of
the years.

Nor would any of them suit my plans of to-
day. Their world is too vast. I should lose
myself in their immensities, where life swarms
freely in the sun. Like the ocean, they are
infinite in their fruitfulness. And then any assiduous watching, undisturbed by passers-by, is
an impossibility on the public way. What I
want is a pond on an extremely reduced scale,
sparingly stocked in my own fashion, an artificial pond standing permanently on my study-
table.

A louis has been overlooked in a corner
of the drawer. I can spend it without seriously jeopardizing the domestic balance. Let
me make this gift to Science, who, I fear,
will be none too much obliged to me. A gorgeous equipment may be all very well for laboratories wherein the cells and fibres of the dead
are consulted at great expense; but such mag-
nificence is of doubtful utility when we have to
study the actions of the living. It is the humble makeshift, of no value, that stumbles on
the secrets of life.

What did the best results of my studies of
instinct cost me? Nothing but time and,

above all, patience. My extravagant expenditure of twenty francs, therefore, will be a risky speculation if devoted to the purchase of an apparatus of study. It will bring me in nothing in the way of fresh views, of that I am convinced. However, let us try.

The blacksmith makes me the framework of a cage out of a few iron rods. The joiner, who is also a glazier on occasion—for, in my village, you have to be a Jack-of-all-trades if you would make both ends meet—sets the framework on a wooden base and supplies it with a movable board as a lid; he fixes thick panes of glass in the four sides. Behold the apparatus, complete, with a bottom of tarred sheet-iron and a trap to let the water out.

The makers express themselves satisfied with their work, a singular novelty in their respective shops, where many an inquisitive caller has wondered what use I intend to make of my little glass trough. The thing creates a certain stir. Some insist that it is meant to hold my supplies of oil and to take the place of the receptacle in general use in our parts, the urn dug out of a block of stone. What would those utilitarians have thought of my crazy mind, had they known that my costly gear would merely serve to let me watch some

The Pond

wretched animals kicking about in the water!
Smith and glazier are content with their
work. I myself am pleased. For all its rustic
air, the apparatus does not lack elegance. It
looks very well, standing on a little table in
front of a window visited by the sun for the
greater part of the day. Its holding capacity
is some ten or eleven gallons. What shall we
call it? An aquarium? No, that would be
too pretentious and would, very unjustly, suggest the aquatic toy filled with rock-work,
water-falls and gold-fish beloved of the dwell-
ers in Suburbia. Let us preserve the gravity
of serious things and not treat my learned
trough as though it were a drawing-room fu-
tility. We will call it the glass pond.

I furnish it with a heap of those limy in-
crustations wherewith certain springs in the
neighbourhood cover the dead clump of rushes.
It is light, full of holes and gives a faint sug-
gestion of a coral-reef. Moreover, it is covered with a short, green, velvety moss, a
downy sward of infinitesimal pond-weed. I
count on this modest vegetation to keep the
water in a reasonably wholesome state, without driving me to frequent renewals which
would disturb the work of my colonies. Sanitation and quiet are the first conditions of suc-

cess. Now the stocked pond will not be long
in filling itself with gases unfit to breathe, with
putrid effluvia and other animal refuse; it will
become a sink in which life will have killed
life. Those dregs must disappear as soon as
they are formed, must be burnt and purified;
and from their oxidized ruins there must even
rise a perfect life-giving gas, so that the water
may retain an unchangeable store of the
breathable element. The plant effects this puri-
fication in its sewage-farm of green cells.

When the sun beats upon the glass pond, the
work of the water-weeds is a sight to behold.
The green-carpeted reef is lit up with an infinity of scintillating points and assumes the appearance of a fairy-lawn of velvet, studded
with thousands of diamond pin's-heads. From
this exquisite jewellery pearls break loose continuously and are at once replaced by others in
the generating casket; slowly they rise, like
tiny globes of light. They spread on every
side. It is a constant display of fireworks in
the depths of the water.

Chemistry tells us that, thanks to its green
matter and the stimulus of the sun's rays, the
weeds decompose the carbonic acid gas wherewith the water is impregnated by the breathing
of its inhabitants and the corruption of the or-

The Pond

ganic refuse; it retains the carbon, which is
wrought into fresh tissues; it exhales the oxygen in tiny bubbles. These partly dissolve in
the water and partly reach the surface, where
their froth supplies the atmosphere with an excess of breathable gas. The dissolved portion
keeps the colonists of the pond alive and
causes the unhealthy products to be oxidized
and disappear.

Old hand though I be, I take an interest in
this trite marvel of a bundle of weeds per-
petuating hygienic principles in a stagnant
pool; I look with a delighted eye upon the inexhaustible spray of spreading bubbles; I see
in imagination the prehistoric times when seaweed, the first-born of plants, produced the
first atmosphere for living things to breathe
at the time when the silt of the continents was
beginning to emerge. What I see before my
eyes, between the glass panes of my trough,
tells me the story of the planet surrounding itself with pure air.

CHAPTER VIII

THE CADDIS-WORM

WHOM shall I lodge in my glass trough, kept permanently wholesome by the action of the water-weeds? I shall keep Caddis-worms, those expert dressers. Few of the self-clothing insects surpass them in ingenious attire. The ponds in my neighbourhood supply me with five or six species, each possessing an art of its own. To-day, but one of these shall receive historical honours.

I obtain it from the muddy-bottomed, stagnant pools crammed with small reeds. As far as one can judge from the habitation merely, it should be, according to the specialists, *Limnophilus flavicornis,* whose work has earned for the whole corporation the pretty name of Phryganea, a Greek term meaning a bit of wood, a stick. In a no less expressive fashion, the Provençal peasant calls it *lou porto-fais, lou porto-canèu.* This is the little grub that carries through the still waters a faggot of tiny fragments fallen from the reeds.

The Caddis-Worm

Its sheath, a travelling house, is a composite and barbaric piece of work, a megalithic pile wherein art retires in favour of amorphous strength. The materials are many and sundry, so much so that we might imagine that we had the work of dissimilar builders before our eyes, if frequent transitions did not tell us the contrary.

With the young ones, the novices, it starts with a sort of deep basket in rustic wickerwork. The twigs employed present nearly always the same characteristics and are none other than bits of small, stiff roots, long steeped and peeled under water. The grub that has made a find of these fibres saws them with its mandibles aud cuts them into little straight sticks, which it fixes one by one to the edge of its basket, always crosswise, perpendicular to the axis of the work.

Picture a circle surrounded by a bristling mass of tangents, or rather a polygon with its sides extended in all directions. On this assemblage of straight lines we place repeated layers of others, without troubling about similarity of position, thus obtaining a sort of ragged fascine, whose sticks project on every side. Such is the bastion of the child-grub, an excellent system of defence, with its continuous pile of spikes, but difficult to steer through the tangle of aquatic plants.

Sooner or later, the worm forsakes this kind of caltrop which catches on to everything. It was a basket-maker, it now turns carpenter; it builds with little beams and joists—that is to say, with round bits of wood, browned by the water, often as wide as a thick straw and a finger's-breadth long, more or less—taking them as chance supplies them.

For the rest, there is something of everything in this rag-bag: bits of stubble, fagends of rushes, scraps of plants, fragments of some tiny twig or other, chips of wood, shreds of bark, largish grains, especially the seeds of the yellow iris, which were red when they fell from their capsules and are now black as jet.

The heterogeneous collection is piled up anyhow. Some pieces are fixed lengthwise, others across, others aslant. There are angles in this direction and angles in the other, resulting in sharp little turns and twists; the big is mixed with the little, the correct rubs shoulders with the shapeless. It is not an edifice, it is a frenzied conglomeration. Sometimes, a fine disorder is an effect of art. This is not so here: the work of the Caddis-worm is not a masterpiece worth signing.

184

The Caddis-Worm

And this mad heaping-up follows straight upon the regular basket-work of the start. The young grub's fascine did not lack a certain elegance, with its dainty laths, all stacked crosswise, methodically; and, lo and behold, the builder, grown larger, more experienced and, one would think, more skilful, abandons the orderly plan to adopt another which is wild and incoherent! There is no transition-stage between the two systems. The extravagant pile rises abruptly from the original basket. But that we often find the two kinds of work placed one above the other, we would not dare ascribe to them a common origin. The fact of their being joined together is the only thing that makes them one, in spite of the incongruity.

But the two storeys do not last indefinitely. When the worm has grown slightly and is housed to its satisfaction in a heap of joists, it abandons the basket of its childhood, which has become too narrow and is now a troublesome burden. It cuts through its sheath, lops off and lets go the stern, the original work. When moving to a higher and roomier flat, it understands how to lighten its portable house by breaking off a part of it. All that remains is the upper floor, which is enlarged at the aperture, as and when required, by the same architecture of disordered beams.

Side by side with these cases, which are mere ugly faggots, we find others just as often of exquisite beauty and composed entirely of tiny shells. Do they come from the same workship? It takes very convincing proofs to make us believe this. Here is order with its charm, there disorder with its hideousness; on the one hand a dainty mosaic of shells, on the other a clumsy heap of sticks. And yet it is all produced by the same labourer.

Proofs abound. On some case which offends the eye with the want of arrangement in its bits of wood, patches are apt to appear which are quite regular and made of shells; in the same way, it is not unusual to see a horrid tangle of joists braced to a masterpiece of shell-work. One feels a certain annoyance at seeing the pretty sheath so barbarously spoilt.

This mixed construction tells us that the rustic stacker of wooden beams excels, when occasion offers, in making elegant shell-pavements and that it practises rough carpentry and delicate mosaic-work indifferently. In the latter instance, the scabbard is made, above all, of Planorbes, selected among the smaller of these Pond-snails and laid flat. Without being scrupulously regular, the work, at its best, does not lack merit. The pretty, close-whorled spirals, placed one against the other on the same level, have a very pleasing general effect. No pilgrim returning from Santiago de Compostella ever slung handsomer tippet from his shoulders.

But only too often the Caddis-worm dashes ahead, regardless of proportion. The big is joined to the small, the exaggerated suddenly stands out, to the great detriment of order. Side by side with tiny Planorbes, each at most the size of a lentil, others are fixed as large as one's finger-nail; and these cannot possibly be fitted in correctly. They overlap the regular parts and spoil their finish.

To crown the disorder, the Caddis-worm adds to the flat spirals any dead shell that comes handy, without distinction of species, provided it be not excessively large. I notice, in its collection of bric-à-brac, the Physa, the Paludina, the Limnæa, the Amber-snail and even the Pisidium,[1] that little twin-valved casket.

[1] The above are all Pond-snails, except the Pisidium, which is a Bivalve.—*Translator's Note.*

The Life of the Fly

Land-shells, swept into the ditches by the rains after the inmate's death, are accepted quite as readily. In the work made of the Mollusc's cast-off clothing, I find encrusted the spindle-shell of the Clausilium, the key-shell of the Pupa, the spiral of the smaller Helix, the yawning volute of the Vitrina, or Glass-snail, the turret-shell of the Buli-mus,[1] denizens all of the fields. In short, the Caddis-worm builds with more or less everything that comes from the plant or the dead Mollusc. Among the diversified refuse of the pond, the only materials rejected are those of a gravelly nature. Stone and pebble are excluded from the building with a care that is very rarely absent. This is a question of hydrostatics to which we will return presently. For the moment, let us try to follow the construction of the scabbard.

In a tumbler small enough to allow of easy and precise observation, I instal three or four Caddis-worms, extracted this moment from their sheaths with every possible precaution. After a number of attempts which have at last shown me the right road, I place at their disposal two kinds of materials, possessing opposite qualities; the supple and the firm,

[1]The above are all Land-snails.—*Translator's Note.*

the soft and the hard. On the one hand, we have a live aquatic plant, such as watercress, for instance, or *ombrelle d'eau*, having at its base a tufty bunch of fine white roots about as thick as a horsehair. In these soft tresses, the Caddis-worm, which observes a vegetarian diet, will find at one and the same time the wherewithal to build and eat. On the other hand, we have a little faggot of bits of wood, very dry, equal in length and each possessing the thickness of a good-sized pin. The two sorts of building-material lie side by side, mingling their threads and sticks. The animal can make its choice from the lump.

A few hours later, having recovered from the shock of losing its sheath, the Caddis-worm sets to work to manufacture a new one. It settles across a bunch of tangled rootlets, which are brought together by the builder's legs and more or less arranged by the undulating movement of the hinder-part. This gives a kind of incoherent and ill-defined suspensory belt, a narrow hammock with a number of loose catches; for the various bits of which it is made up are respected by the teeth and extended from place to place beyond the main cords of the roots. Here, without much trouble, is the support, suitably fixed by natural moorings. A few threads of silk, casually distributed, make the frail combination a trifle more secure.

And now to the work of building. Supported by the suspensory belt, the Caddis-worm stretches itself and thrusts out its middle legs, which, being longer than the others, are the grapnels intended to seize things at a distance. It meets a bit of root, fastens on to it, climbs above the point gripped, as though it were measuring the piece to a requisite length, and then, with the fine scissors of its mandibles, cuts the string.

There is at once a brief recoil, which brings the animal back to the level of the hammock. The bit detached lies across the worm's chest, held in its fore-legs, which turn it, twist it, wave it about, lay it down, lift it up, as though trying for the best position. Those fore-legs make admirably dexterous arms. Being less long than the other two pairs, they are brought into immediate contact with those primordial implements, the mandibles and the spinneret. Their delicate terminal jointing, with a movable and crooked finger, is the Caddis-worm's equivalent of our hand. They are the working-legs. The second pair, which are exceptionally long, serve to spear distant materials and to give the worker a firm footing when measuring a piece and cutting it with the pliers. Lastly, the hind-legs, of medium length, afford a support when the others are busy.

The Caddis-worm, I was saying, with the piece which it has removed held crosswise to its chest, retreats a little way along its suspensory hammock until the spinneret is level with the support furnished by the close tangle of rootlets. With a quick movement, it shifts its burden, gets it as nearly by the middle as it can, so that the two ends stick out equally on either side, and chooses the spot to place it, whereupon the spinneret sets to work at once, while the little fore-legs hold the scrap of root motionless in its transversal position. The soldering is effected with a touch of silk in the middle of the bit and along a certain distance to the right and left, as far as the bending of the head permits.

Without delay, other sticks are speared in like manner at a distance, cut off and placed in position. As the immediate neighbourhood is stripped, the material is gathered at a yet greater distance and the Caddis-worm bends even farther from its support, which now holds only its last few segments. It is a curious gymnastic display, that of this soft, hanging spine turning and swaying, while the grapnels feel in every direction for a thread.

All this labour results in a sort of casing of little white cords. The work lacks firmness and regularity. Nevertheless, judging by the builder's methods, I can see that the building would not be devoid of merit if the materials gave it a better chance. The Caddis-worm estimates the size of its pieces very fairly; it cuts them all to nearly the same length; it always arranges them crosswise on the margin of the case; it fixes them by the middle.

Nor is this all: the manner of working helps the general arrangement considerably. When the bricklayer is building the narrow shaft of a factory-chimney, he stands in the centre of his turret and turns round and round while gradually laying new rows. The Caddis-worm acts in the same way. It twists round in its sheath; it adopts without inconvenience whatever position it pleases, so as to bring its spinneret full-face with the point to be gummed. There is no straining of the neck to left or right, no throwing back of the head to reach points behind. The animal has constantly before it, within the exact range of its implements, the place at which the bit is to be fixed. When the piece is soldered, the worm turns a little aside, to a length equal to that of the last soldering, and here, along an extent which hardly ever varies, an extent determined by the swing which its head is able to give, it fixes the next piece.

These several conditions ought to result in a geometrically ordered dwelling, having a regular polygon as an opening. Then how comes it that the cylinder of bits of root is so confused, so clumsily fashioned? The reason is this: the worker possesses talent, but the materials do not lend themselves to accurate work. The rootlets supply stumps of very uneven shape and thickness. They include big and small ones, straight and bent, simple and ramified. To combine all these dissimilar pieces into an orderly whole is hardly possible, all the more so as the Caddis-worm does not appear to attach very much importance to its cylinder, which is a temporary work, hurriedly constructed to afford a speedy shelter. Matters are urgent; and very soft fibres, clipped with a bite of the mandibles, are more quickly gathered and more easily put together than joists, which require the patient work of the saw. The inaccurate cylinder, in short, held in position by numerous guy-ropes, is a base upon which a solid and definite structure will rise before long. Soon, the original work will crumble to ruins and disappear, whereas the new one, a permanent structure, will even outlast the owner.

The insects reared in a tumbler show yet another method of building the first dwelling. This time, the Caddis-worm is given a few very leafy stalks of pond-weed (*Potamogeton densum*) and a bundle of small dry twigs. It perches on a leaf, which the nippers of the mandibles cut half across. The portion left untouched will act as a lanyard and give the necessary steadiness to the early operations.

From an adjoining leaf a section is cut out entirely, an angular and good-sized piece. There is plenty of material and no need for economy. The piece is soldered with silk to the strip which was not wholly cut off. The result of three or four similar operations is to surround the Caddis-worm with a conical bag, whose wide mouth is scalloped with pointed and very irregular notches. The work of the nippers continues; fresh pieces are fixed, from one to another, inside the funnel, not far from the edge, so that the bag

The Caddis-Worm

lengthens, tapers and ends by wrapping the animal in a light and floating drapery.

Thus clad for the time being, either in the fine silk of the pond-weed or in the linsey-woolsey supplied by the roots of the water-cress, the Caddis-worm begins to think of building a more solid sheath. The present casing will serve as a foundation for the stronger building. But the necessary materials are seldom near at hand: you have to go and fetch them, you have to move your position, an effort which has been avoided until now. With this object, the Caddis-worm cuts its moorings, that is to say, the rootlets which keep the cylinder fixed, or else the half-severed leaf of pond-weed on which the cone-shaped bag has come into being.

The worm is now free. The smallness of the artificial pond, the tumbler, soon brings it into touch with what it is seeking. This is a little faggot of dry twigs, which I have selected of equal length and of slight thickness. Displaying greater care than it did when treating the slender roots, the carpenter measures out the requisite length on the joist. The distance to which it has to extend its body in order to reach the point where the break

The Life of the Fly

will be made tells it pretty accurately what
length of stick it wants.

The piece is patiently sawn off with the
mandibles; it is next taken in the fore-legs
and held crosswise below the neck. The
backward movement which brings the Caddis-worm home also brings the bit of twig to
the edge of the tube. Thereupon, the
methods employed in working with the scraps
of root are renewed in precisely the same manner. The sticks are scaffolded to the regulation height, all alike in length, amply soldered in the middle and free at either end.

With the picked materials provided, the
carpenter has turned out a work of some elegance. The joists are all arranged crosswise,
because this way is the handiest for carrying
the sticks and putting them in position; they
are fixed by the middle, because the two arms
that hold the stick while the spinneret does
its work require an equal grasp on either side;
each soldering covers a length which is seen
to be practically invariable, because it is equal
to the width described by the head in bending
first to this side and then to that when the
silk is emitted; the whole assumes a polygonal
shape, not far removed from a rectilinear
pentagon, because, between laying one piece

196

The Caddis-Worm

and the next, the Caddis-worm turns by the width of an arc corresponding with the length of a soldering. The regularity of the method produces the regularity of the work; but it is essential, of course, that the materials should lend themselves to precise coordination.

In its natural pond, the Caddis-worm does not often have at its disposal the picked joists which I give it in the tumbler. It comes across something of everything; and that something of everything it employs as it finds it. Bits of wood, large seeds, empty shells, stubble-stalks, shapeless fragments are used in the building for better or for worse, just as they occur, without being trimmed by the saw; and this jumble, the result of chance, results in a shockingly faulty structure.

The Caddis-worm does not forget its talents; but it lacks choice pieces. Give it a proper timber-yard and it at once reverts to correct architecture, of which it carries the plans within itself. With small, dead pond-snails, all of the same size, it fashions a splendid patchwork scabbard; with a cluster of slender roots, reduced by rotting to their stiff, straight, woody axis, it manufactures pretty specimens of wicker-work which could serve as models to our basket-makers.

The Life of the Fly

Let us watch it at work when it is unable to use its favourite joist. There is no point in giving it clumsy building-stones; that would only bring us back to the uncouth sheaths. Its propensity to make use of soaked seeds, those of the iris, for instance, suggests that I might try grains. I select rice, which, because of its hardness, will be tantamount to wood and, because of its clean whiteness and its oval shape, will lend itself to artistic masonry.

Obviously, my denuded Caddis-worms cannot start their work with bricks of this kind. Where would they fix their first layer? They must have a foundation, quick and easy to build. This is once more supplied by a temporary cylinder of watercress-roots. On this support follow the grains of rice, which, grouped one atop the other, straight or slanting, end by giving a magnificent turret of ivory. Next to the sheaths made of tiny snail-shells, this is the prettiest thing with which the Caddis-worm's industry has furnished me. A fine sense of order has returned, because the materials, regular and of identical character, have cooperated with the correct method of the worker.

The two demonstrations are enough.

The Caddis-Worm

Sticks and grains of rice make it plain that the Caddis-worm is not the bungler that one would expect from the monstrous buildings in the pond. Those Cyclopean piles, those mad conglomerations are the inevitable results of chance finds, which are used for the best because there is no choice. The water-carpenter has an art of its own, has method and rules of symmetry. When well-served by fortune, it is quite able to turn out good work; when ill-served, it acts like others: the work which it turns out is bad. Poverty makes for ugliness.

There is another matter wherein the Caddis-worm deserves our attention. With a perseverance which repeated trials do not tire, it makes itself a new tube when I strip it. This is opposed to the habits of the generality of insects, which do not recommence the thing once done, but simply continue it according to the usual rules, taking no account of the ruined or vanished portions. The Caddis-worm is a striking exception: it starts again. Whence does it derive this capacity?

I begin by learning that, given a sudden alarm, it readily leaves its scabbard. When I go fishing for Caddis-worms, I put them in tin boxes, containing no other moisture

than that wherewith my catches are soaked.
I heap them up loosely, to avoid any grievous
tumult and to fill the space at my disposal as
best I may. I take no further precaution. This
is enough to keep the Caddis-worms in good
condition during the two or three hours which
I devote to fishing and to walking home.

On my return, I find that a number of them
have left their houses. They are swarming
naked among the empty scabbards and those
still occupied by their inhabitants. It is a pitiful sight to see these evicted ones dragging
their bare abdomens and their frail respiratory
threads over the bristling sticks. There is no
great harm done, however; and I empty the
whole lot into the glass pond.

Not one resumes possession of an unoccupied sheath. Perhaps it would take them too
long to find one of the exact size. They
think it better to abandon the old clouts and
to manufacture cases new from top to bottom. The process is a rapid one. By the
next day, with the materials wherein the glass
trough abounds—bundles of twigs and tufts
of watercress—all the denuded worms have
made themselves at least a temporary home
in the form of a tube of rootlets.

The lack of water, combined with the ex-

The Caddis-Worm

citement of the crowding in the boxes, has upset my captives greatly; and, scenting a grave
peril, they have made off hurriedly, doffing
the cumbersome jacket, which is difficult to
carry. They have stripped themselves so as to
flee with greater ease. The alarm cannot have
been due to me: there are not many simple-
tons like myself who are interested in the af-
fairs of the pond; and the Caddis-worm has
not been cautioned against their tricks. The
sudden desertion of the crib has certainly some
other reason than man's molestations.

I catch a glimpse of this reason, the real
one. The glass pond was originally occupied by a dozen Dytisci, or Water-beetles,
whose diving-performances are so curious to
watch. One day, meaning no harm and for
want of a better receptacle, I fling among
them a couple of handfuls of Caddis-worms.
Blunderer that I am, what have I done! The
corsairs, hiding in the rugged corners of the
rock-work, at once perceive the windfall.
They rise to the surface with great strokes
of their oars; they hasten and fling themselves
upon the crowd of carpenters. Each pirate
grabs a sheath by the middle and strives to
rip it open by tearing off shells and sticks.
While this ferocious enucleation continues

with the object of reaching the dainty morsel contained within, the Caddis-worm, close-pressed, appears at the mouth of the sheath, slips out and quickly decamps under the eyes of the Dytiscus, who appears to notice nothing.

I have said before[1] that the trade of killing can dispense with intelligence. The brutal ripper of sheaths does not see the little white sausage that slips between his legs, passes under his fangs and madly flees. He continues to tear away the outer case and to tug at the silken lining. When the breach is made, he is quite crestfallen at not finding what he expected.

Poor fool! Your victim went out under your nose and you never saw it. The worm has sunk to the bottom and taken refuge in the mysteries of the rock-work. If things were happening in the large expanse of a pond, it is clear that, with their system of expeditious removals, most of the lodgers would escape scot-free. Fleeing to a distance and recovering from the sharp alarm, they would build themselves a new scabbard and all would

---

[1]In the essay on the Giant Scarites, not yet translated into English, of which the first line reads: 'The trade of war does not induce talent.'—*Translator's Note.*

be over until the next attack, which would be baffled afresh by the selfsame trick.

In my narrow trough, things take a more tragic turn. When the sheaths are done for, when the Caddis-worms that are too slow in making off have been eaten up, the Water-beetles return to the rockery at the bottom. Here, sooner or later, there are lamentable happenings. The naked fugitives are dis-covered and, succulent morsels that they are, are forthwith torn to pieces and devoured. Within twenty-four hours, not one of my band of Caddis-worms is left alive. In order to continue my studies, I had to lodge the Water-beetles elsewhere.

Under natural conditions, the Caddis-worm has its persecutors, the most formidable of whom appears to be the Water-beetle. When we consider that, to thwart the brigand's at-tacks, it has invented the idea of quitting its scabbard with all speed, its tactics are cert-ainly most appropriate; but, in that case, an exceptional condition becomes obligatory, namely, the capacity for recommencing the work. This most unusual gift of recommen-cing it possesses in a high measure. I am ready to see its origin in the persecutions of

the Dytiscus and other pirates. Necessity is
the mother of industry.

Certain Caddis-worms, of the *Sericostoma*
and *Leptocerus* species, clothe themselves in
grains of sand and do not leave the bed of the
stream. On a clear bottom, swept by the current, they walk about from one bank of ver-
dure to the other and do not think of coming
to the surface to float and sail in the sunlight. The collectors of sticks and shells are
more highly privileged. They can remain on
the level of the water indefinitely, with no
other support than their skiff, can rest in insubmersible flotillas and can even shift their
place by working the rudder.

To what do they owe this privilege? Are
we to look upon the bundle of sticks as a sort
of raft whose density is less than that of the
water? Can the shells, which are always
empty and able to contain a few bubbles of air
in their spiral, be floats? Can the big joists,
which break in so ugly a fashion the none too
great regularity of the work, serve to buoy up
the over-heavy raft? In short, is the Caddis-
worm versed in the laws of equilibrium and
does it choose its pieces, now lighter and now
heavier as the case may be, so as to constitute
a whole that is capable of floating? The fol-

lowing facts are a refutation of any such hy-
drostatic calculations in the animal.

I remove a number of Caddis-worms from
their sheaths and submit these, as they are, to
the test of water. Whether formed wholly of
fibrous remnants or of mixed materials, not
one of them floats. The scabbards made of
shells go to the bottom with the swiftness of a
bit of gravel; the others sink gently. I experiment with the separate materials one by one.
No shell remains on the surface, not even
among the Planorbes, which a many-whorled
spiral ought, one would think, to keep afloat.
The fibrous remnants must be divided into two
categories. The first, darkened by time and
soaked with moisture, sink to the bottom.
These are the most plentiful. The second,
considerably fewer in number, of more recent
date and less saturated with water, float very
well. The general result is immersion, as in
the case of the intact scabbards. I may add
that the animal, when removed from its tube,
is also unable to float.

Then how does the Caddis-worm manage
to remain on the surface without the support
of the grasses, considering that itself and its
sheath are both heavier than water? Its secret
is soon revealed. I place a few high and dry

on a sheet of blotting-paper, which will absorb the excess of liquid unfavourable to successful observation. Outside its natural environment, the animal moves about violently and restlessly. With its body half out of the scabbard, this time composed entirely of fibrous matter, it clutches with its feet at the supporting plane. Then, contracting itself, it draws the scabbard towards it, half-raising it and sometimes even making it assume a vertical position. Even so do the Bulimi move along, lifting their shell as they complete each crawling step.

After a couple of minutes in the free air, I replace the Caddis-worm in the water. This time, it floats, but like a cylinder with too much weight below. The sheath remains vertical, with its hinder orifice level with the water. Soon, an air-bubble escapes from the orifice. Deprived of this buoy, the skiff at once goes down.

The result is the same with the Caddis-worms in shell casings. At first, they float, straight up on end, and then dip under and sink, faster than the others, after sending out an air-bubble or two through the back-window.

That is enough: the secret is out. When cased in wood or in shells, the Caddis-worms,

which are always heavier than water, are able
to keep on the surface by means of a temporary air-balloon which decreases the density of
the whole structure.

This apparatus works in the simplest manner. Consider the rear of the sheath. It is
truncated, wide-open and supplied with a
membranous partition, the work of the spinneret. A round hole occupies the centre of
this screen. Beyond it lies the interior of the
scabbard, which is smoothly lined and wadded
with satin, however rough the exterior may be.
Armed at the stern with two hooks which bite
into the silky lining, the animal is able to move
backwards and forwards at will inside the
cylinder, to fix its grapnels at whatever point
it pleases and thus to keep a hold on the cylinder while the six legs and the fore-part are
outside.

When at rest, the body remains indoors entirely and the grub occupies the whole of the
tube. But let it contract ever so little towards
the front, or, better still, let it stick out a part
of its body: a vacuum is formed behind this
sort of piston, which may be compared with
that of a pump. Thanks to the rear-window,
a valve without a plug, this vacuum at once
fills, thus renewing the aerated water around

the gills, a soft fleece of hairs distributed over the back and belly.

The piston-stroke affects only the work of breathing; it does not alter the density, makes hardly any change in that which is heavier than water. To lighten the weight, the Caddis-worm must first rise to the surface. With this object, it scales the grasses of one support after the other; it clambers up, sticking to its purpose in spite of the drawback of its faggot dragging through the tangle. When it has reached the goal, it lifts the rear-end a little above the water and gives a stroke of the piston. The vacuum thus obtained fills with air. That is enough: skiff and boatman are in a position to float. The now useless support of the grasses is abandoned. The time has come for evolutions on the surface, in the glad sunlight.

The Caddis-worm possesses no great talent as a navigator. To turn round, to tack about, to shift its place slightly by a backward movement is all that it can do; and even that it does very clumsily. The front part of the body, sticking out of the case, acts as a rudder. Three or four times over, it rises abruptly, bends, comes down again and strikes the water. These paddle-strokes, repeated at intervals, carry the

The Caddis-Worm

unskilled oarsman to fresh latitudes. It becomes a voyage on the right seas when the
crossing measures a hand's-breadth.

However, tacking on the surface of the
water affords the Caddis-worm no pleasure.
It prefers to twitter in one spot, to remain
stationary in flotillas. When the time comes
to return to the quiet of the mud-bed at the
bottom, the animal, having had enough of
the sun, draws itself wholly into its sheath
again and, with a piston-stroke, expels the air
from the back-room. The normal density is
restored and it sinks slowly to the bottom.

We see, therefore, that the Caddis-worm
has not to trouble about hydrostatics when
building its scabbard. In spite of the incongruity of its work, in which the bulky and less
dense portions seem to balance the more solid,
concentrated part, it is not called upon to contrive an equipoise between the light and the
heavy. It has other artifices whereby to rise
to the surface, to float and to dive down again.
The ascent is made by the ladder of the water-
weeds. The average density of the sheath is
of no importance, so long as the burden to be
dragged is not beyond the animal's strength.
Besides, the weight of the load is greatly reduced when moved in the water.

The Life of the Fly

The admission of a bubble of air into the back-chamber, which the animal ceases to occupy, allow it, without further to-do, to remain for an indefinite period on the surface. To dive down again, the Caddis-work has only to retreat entirely into its sheath. The air is driven out; and the canoe, resuming its mean density, a greater specific density than that of water, goes under at once and descends of its own accord.

There is, therefore, no choice of materials on the builder's part, no nice calculation of equilibrium, save for one condition, that no stony matter be admitted. That apart, everything serves, large and small, joist and shell, seed and billet. Built up at haphazard, all these things make an impregnable wall. One point alone is essential: the weight of the whole must slightly exceed that of the water displaced; if not, there could be no steadiness at the bottom of the pond, without a perpetual anchorage struggling against the pull of the water. In the same manner, quick submersion would be impossible at times when the surface became dangerous and the frightened creature wanted to leave it.

Nor does this important heavier-than-water question call for lucid discernment, seeing that

The Caddis-Worm

almost the whole of the sheath is constructed at the bottom of the pond, whither all the materials picked up at random, having descended once before, are likely to descend again. In the sheaths, the parts capable of floating are very rare. Without taking their specific levity into account, simply so as not to remain idle, the Caddis-worm fixed them to its bundle when sporting on the surface of the water.

We have our submarines, in which hydraulic ingenuity displays its highest resources. The Caddis-worms have theirs, which emerge, float on the surface, dip down and even stop at mid-depth by releasing gradually their surplus air. And this apparatus, so perfectly balanced, so skilful, requires no knowledge on the part of its constructor. It comes into being of itself, in accordance with the plans of the universal harmony of things.

CHAPTER IX

THE GREENBOTTLES

I HAVE wished for a few things in my
life, none of them capable of interfering
with the common weal. I have longed to
possess a pond, screened from the indiscre-
tion of the passers-by, close to my house, with
clumps of rushes and patches of duckweed.
Here, in my leisure hours, in the shade of a
willow, I should have meditated upon aquatic
life, a primitive life, easier than our own,
simpler in its affections and its brutalities. I
should have watched the unalloyed happiness
of the Mollusc, the frolics of the Whirligig,
the figure-skating of the Hydrometra,[1] the
dives of the Dytiscus Beetle, the veering and
tacking of the Notonecta,[2] who, lying on her
back, rows with two long oars, while her
short fore-legs, folded against her chest, wait
to grab the coming prey. I should have studied the eggs of the Planorbis, a glairy nebula

[1] A Water-bug, known as the Pond-skater, who runs
about actively, on her middle- and hind-legs, on the
surface of fresh water.—*Translator's Note.*
[2] A Bug known also as the Water-boatman.—*Translator's Note.*

212

The Greenbottles

wherein focuses of life are condensed even as
suns are condensed in the nebulæ of the
heavens. I should have admired the nascent
creature that turns, slowly turns in the orb of
its egg and describes a volute, the draft, perhaps, of the future shell. No planet circles
round its centre of attraction with greater
geometrical accuracy.

I should have brought back a few ideas
from my frequent visits to the pond. Fate
decided otherwise: I was not to have my sheet
of water. I have tried the artificial pond, between four panes of glass. A poor shift!
Our laboratory aquariums are not even equal
to the print left in the mud by a mule's hoof,
when once a shower has filled the humble
basin and life has stocked it with its marvels.

In spring, with the hawthorn in flower and
the Crickets at their concerts, a second wish
often came to me. Along the road, I light
upon a dead Mole, a Snake killed with a
stone, victims both of human folly. The
Mole was draining the soil and purging it of
its vermin. Finding him under his spade, the
labourer broke his back for him and flung
him over the hedge. The Snake, roused from
her slumber by the soft warmth of April, was

coming into the sun to shed her skin and take on a new one. Man catches sight of her:

'Ah, would you?' says he. 'See me do something for which the world will thank me!'

And the harmless beast, our auxiliary[1] in the terrible battle which husbandry wages against the insect, has its head smashed in and dies.

The two corpses, already decomposing, have begun to smell. Whoso approaches with eyes that do not see turns away his head and passes on. The observer stops and lifts the remains with his foot; he looks. A world is swarming underneath; life is eagerly consuming the dead. Let us replace matters as they were and leave death's artisans to their task. They are engaged in a most deserving work.

To know the habits of those creatures charged with the disappearance of corpses, to see them busy at their work of distintegration, to follow in detail the process of transmutation that makes the ruins of what has lived return apace into life's treasure-house: these are things that long haunted my mind. I regret-

[1] The author employs the term 'auxiliaries' to denote the animals that help to protect the farmer's crops.— *Translator's Note.*

The Greenbottles

fully left the Mole lying in the dust of the road. I had to go, after a glance at the corpse and its harvesters. It was not the place for philosophizing over a stench. What would people say who passed and saw me!

And what will the reader himself say, if I invite him to that sight? Surely, to busy one's self with those squalid sextons means soiling one's eyes and mind? Not so, if you please! Within the domain of our restless curiosity, two questions stand out above all others: the question of the beginning and the question of the end. How does matter unite in order to assume life? How does it separate when returning to inertia? The pond, with its Planorbis-eggs turning round and round, would have given us a few data for the first problem; the Mole, going bad under conditions not too-repulsive, will tell us something about the second: he will show us the working of the crucible wherein all things are melted to begin anew. A truce to nice delicacy! *Odi profanum vulgus et arceo;* hence, ye profane: you would not understand the mighty lesson of the rag-tank.

I am now in a position to realize my second wish. I have space, air and quiet in the solitude of the harmas. None will come here to trouble me, to smile or to be shocked at my investigations. So far, so good; but observe the irony of things: now that I am rid of passers-by, I have to fear my cats, those assiduous prowlers, who, finding my preparations, will not fail to spoil and scatter them. In anticipation of their misdeeds, I establish workshops in mid-air, whither none but genuine corruption-agents can come, flying on their wings. At different points in the enclosure, I plant reeds, three by three, which, tied at their free ends, form a stable tripod. From each of these supports, I hang, at a man's height, an earthenware pan filled with fine sand and pierced at the bottom with a hole to allow the water to escape, if it should rain. I garnish my apparatus with dead bodies. The Snake, the Lizard, the Toad receive the preference, because of their bare skins, which enable me better to follow the first attack and the work of the invaders. I ring the changes with furred and feathered beasts. A few children of the neighbourhood, allured by pennies, are my regular purveyors. Throughout the good season, they come running triumphantly to my door, with a Snake at the end of a stick, or a Lizard in a cabbage-leaf. They bring me the Rat caught

The Greenbottles

in a trap, the Chicken dead of the pip, the
Mole slain by the gardener, the Kitten killed
by accident, the Rabbit poisoned by some
weed. The business proceeds to the mutual
satisfaction of sellers and buyer. No such
trade had ever been known before in the village nor ever will be again.

April ends; and the pans rapidly fill. An
Ant, ever so small, is the first arrival. I
thought I should keep this intruder off by
hanging my apparatus high above the ground:
she laughs at my precautions. A few hours
after the deposit of the morsel, fresh still and
possessing no appreciable smell, up comes the
eager picker-up of trifles, scales the stems of
the tripod in processions and starts the work
of dissection. If the joint suits her, she even
goes to live in the sand of the pan and digs
herself temporary platforms in order to work
the rich find more at her ease.

All through the season, from start to finish,
she will always be the promptest, always the
first to discover the dead animal, always the
last to beat a retreat when nothing more
remains than a heap of little bones bleached
by the sun. How does the vagabond, passing at a distance, know that, up there, invisible, high on the gibbet, there is some-

thing worth going for? The others, the real
knackers, wait for the meat to go bad; they
are informed by the strength of the effluvia.
The Ant, gifted with greater powers of
scent, hurries up before there is any stench
at all. But, when the meat, now two days
old and ripened by the sun, exhales its flavour, soon the master-ghouls appear upon
the scene: Dermestes and Saprini, Silphæ
and Necrophori, Flies and Staphylini,[1] who
attack the corpse, consume it and reduce it
almost to nothing. With the Ant alone, who
each time carries off a mere atom, the sanitary operation would take too long; with
them, it is a quick business, especially as certain of them understand the process of chemical solvents.

These last, who are high-class scavengers,
are entitled to first mention. They are Flies,
of many various species. If time permitted,
each of those strenuous ones would deserve a
special examination; but that would weary
the patience of both the reader and the observer. The habits of one will give us a sum-

[1]The Dermestes, or Bacon-beetle, is a small, the Sa-
prinus an exceedingly small, flesh-eating Beetle. The
Silpha is the Carrion-beetle proper; the Necrophorus, the
Burying-beetle proper. The Staphylinus, or Rove-beetle,
also lives partly upon decaying substances.—*Translator's
Note*.

mary notion of the habits of the rest. We will therefore confine ourselves to the two principal subjects, namely, the Luciliæ, or Greenbottles, and the Sarcophagæ, or Grey Flesh-flies.

The Luciliæ—Flies that glitter—are magnificent Flies known to all of us. Their metallic lustre, generally a golden green, rivals that of our finest Beetles, the Rose-chafers, Buprestes and Leaf-beetles. It gives one a shock of surprise to see so rich a garb adorn those workers in putrefaction. Three species frequent my pans: *Lucilia Cæsar*, LIN., *L. cadaverina*, LIN., and *L. cuprea*, ROB. The first two, both of whom are gold-green, are plentiful; the third, who sports a coppery lustre, is rare. All three have red eyes, set in a silver border.

Lucilia Cæsar is larger than *L. cadaverina* and also more forward in her business. I catch her in labour on the 23rd of April. She has settled in the spinal canal of a neck of mutton and is laying her eggs on the marrow. For more than an hour, motionless in the gloomy cavity, she goes on packing her eggs. I can just see her red eyes and her silvery face. At last, she comes out. I gather the fruit of her labour, an easy matter, for it all lies on the marrow, which I extract without touching the eggs.

A census would seem important. To take it at once is impracticable: the germs form a compact mass, which would be difficult to count. The best thing is to rear the family in a jar and to reckon by the pupæ buried in the sand. I find a hundred and fifty-seven. This is evidently but a minimum; for *Lucilia Cæsar* and the others, as the observations that follow will tell me, lay in packets at repeated intervals. It is a magnificent family, promising a fabulous legion to come.

The Greenbottles, I was saying, break up their laying into sections. The following scene affords a proof of this. A Mole, shrunk by a few days' evaporation, lies spread upon the sand of the pan. At one point, the edge of the belly is raised and forms a deep arch. Remark that the Greenbottles, like the rest of the flesh-eating Flies, do not trust their eggs to uncovered surfaces, where the heat of the sun's rays might endanger the existence of the delicate germs. They want dark hiding-places. The favourite spot is the lower side of the dead animal, when this is accessible.

In the present case, the only place of access is the fold formed by the edge of the belly. It is here and here alone that this day's mothers are laying. There are eight of them. After exploring the piece and recognizing its good quality, they disappear under the arch, first this one, then that, or else several at a time. They remain under the Mole for a considerable while. Those outside wait, but go repeatedly to the threshold of the cavern to take a look at what is happening within and see whether the earlier ones have finished. These come out at last, perch on the animal and wait in their turn. Others at once take their place in the recesses of the cave. They remain there for some time and then, having done their business, make room for more mothers and come forth into the sunlight. This going in and out continues throughout the morning.

We thus learn that the laying is effected by periodical emissions, broken with intervals of rest. As long as she does not feel ripe eggs coming to her oviduct, the Greenbottle remains in the sun, hovering to and fro and sipping modest mouthfuls from the carcass. But, as soon as a fresh stream descends from her ovaries, quick as lightning she makes for a propitious site whereon to deposit her burden. It appears to be the work of several days thus to divide the total laying and to distribute it at different points.

I carefully raise the animal under which these things are happening. The egg-laying mothers do not disturb themselves; they are far too busy. Their ovipositor extended telescope-fashion, they heap egg upon egg. With the point of their hesitating, groping instrument, they try to lodge each germ, as it comes, farther into the mass. Around the serious, red-eyed matrons, the Ants circle, intent on pillage. Many of them make off with a Greenbottle-egg between their teeth. I see some who, greatly daring, effect their theft under the ovipositor itself. The layers do not put themselves out, let the Ants have their way, remain impassive. They know their womb to be rich enough to make good any such larceny.

Indeed, what escapes the depredations of the Ants promises a plenteous brood. Let us come back a few days later and lift the Mole again. Underneath, in a pool of sanies, is a surging mass of swarming sterns and pointed heads, which emerge, wriggle and dive in again. It suggests a seething billow. It turns one's stomach. It is horrible, most horrible.

The Greenbottles

Let us steel ourselves against the sight: it
will be worse elsewhere.

Here is a fat Snake. Rolled into a compact whorl, she fills the whole pan. The
Greenbottles are plentiful. New ones arrive
at every moment and, without quarrel or
strife, take their place among the others,
busily laying. The spiral furrow left by the
reptile's curves is the favourite spot. Here
alone, in the narrow space between the folds,
are shelters against the heat of the sun. The
glistening Flies take their places, side by side,
in rows; they strive to push their abdomen
and their ovipositor as far forward as possible, at the risk of rumpling their wings and
cocking them towards their heads. The care
of the person is neglected amid this serious
business. Placidly, with their red eyes turned
outwards, they form a continuous cordon.
Here and there, at intervals, the rank is
broken; layers leave their posts, come and
walk about upon the Snake, what time their
ovaries ripen for another emission, and then
hurry back, slip into the rank and resume the
flow of germs. Despite these interruptions,
the work of breeding goes fast. In the course
of one morning, the depths of the spiral furrow are hung with a continuous white bark,

the heaped-up eggs. They come off in great slabs, free of any stain; they can be shovelled up, as it were, with a paper scoop. It is a propitious moment if we wish to follow the evolution at close quarters. I therefore gather a profusion of this white manna and lodge it in glass tubes, test-tubes and jars, with the necessary provisions.

The eggs, about a millimetre[1] long, are smooth cylinders, rounded at both ends. They hatch within twenty-four hours. The first question that presents itself is this: how do the Greenbottle-grubs feed? I know quite well what to give them, but I do not in the least see how they manage to consume it. Do they eat, in the strict sense of the word? I have reasons to doubt it.

Let us consider the grub grown to a sufficient size. It is the usual Fly-larva, the common maggot, shaped like an elongated cone, pointed in front, truncated behind, where two little red spots show, level with the skin: these are the breathing-holes. The front, which is called the head by stretching a word—for it is little more than the entrance to an intestine—the front is armed with two little black hooks, which slide in a translucent sheath, pro-

[1] About .039 inch.—*Translator's Note.*

ject a little way outside and go in turn by
turn. Are we to look upon these as mandibles? Not at all, for, instead of having their
points facing each other, as would be required
in a real mandibular apparatus, the two hooks
work in parallel directions and never meet.
What they are is ambulatory organs, grapnels
assisting locomotion, which give a purchase
on the plane and enable the animal to advance
by means of repeated contractions. The maggot walks with the aid of what a superficial
examination would pronounce to be a machine
for eating. It carries in its gullet the equivalent of the climber's alpenstock.

Let us hold it, on a piece of flesh, under
the lens. We shall see it walking about, raising and lowering its head and, each time,
stabbing the meat with its pair of hooks.
When stationary, with its crupper at rest, it
explores space with a continual bending of its
fore-part; its pointed head pokes about, jabs
forward, goes back again, producing and
withdrawing its black mechanism. There is
a perpetual piston-play. Well, look as carefully and conscientiously as I please, I do not
once see the weapons of the mouth tackle a
particle of flesh that is torn away and swallowed. The hooks come down upon the meat

at every moment, but never take a visible mouthful from it. Nevertheless, the grub waxes big and fat. How does this singular consumer, who feeds without eating, set about it? If he does not eat, he must drink; his diet is soup. As meat is a compact substance, which does not liquefy of its own accord, there must, in that case, be a certain recipe to dissolve it into a fluid broth. Let us try to surprise the maggot's secret.

In a glass tube, sealed at one end, I insert a piece of lean flesh, the size of a walnut, which I have drained of its juices by squeezing it in blotting-paper. On the top of this, I place a few slabs of Greenbottle-eggs collected a moment ago from the Snake in my earthen pan. The number of germs is, roughly, two hundred. I close the tube with a cotton plug, stand it upright, in a shady corner of my study, and leave things to take their course. A control-tube, prepared like the first, but not stocked with maggots, is placed beside it.

As early as two or three days after the hatching, I obtain a striking result. The meat, which was thoroughly drained by the blotting-paper, has become so moist that the young vermin leave a wet mark behind them as they crawl over the glass. The swarming brood creates a sort of mist with the crossing and criss-crossing of its trails. The control-tube, on the contrary, keeps dry, proving that the moisture in which the worms move is not due to a mere exudation from the meat.

Besides, the work of the maggot becomes more and more evident. Gradually, the flesh flows in every direction like an icicle placed before the fire. Soon, the liquefaction is complete. What we see is no longer meat, but fluid Liebig's extract. If I overturned the tube, not a drop of it would remain.

Let us clear our minds of any idea of solution by putrefaction, for in the second tube a piece of meat of the same kind and size has remained, save for colour and smell, what it was at the start. It was a lump and it is a lump, whereas the piece treated by the worms runs like melted butter. Here we have maggot chemistry able to rouse the envy of physiologists when studying the action of the gastric juice.

I obtain better results still with hard-boiled white of egg. When cut into pieces the size of a hazel-nut and handed over to the Greenbottle's grubs, the coagulated albumen dissolves into a colourless liquid which the eye might mistake for water. The fluidity becomes so great that, for lack of a support, the worms perish by drowning in the broth; they are suffocated by the immersion of their hind-part, with its open breathing-holes. On a denser liquid, they would have kept at the surface; on this, they cannot.

A control-tube, filled in the same way, but not colonized, stands beside that in which the strange liquefaction takes place. The hard-boiled white of egg retains its original appearance and consistency. In course of time, it dries up, if it does not turn mouldy; and that is all.

The other quaternary compounds performing the same functions as albumen—the gluten of cereals, the fibrin of blood, the casein of cheese and the legumin of chick-peas—undergo a similar modification, in varying degrees. Fed, from the moment of leaving the egg, on any one of these sub-stances, the worms thrive very well, provided that they escape drowning when the gruel becomes too clear; they would not fare better on a corpse. And, as a general rule, there is not much danger of going under: the matter only half-liquefies; it becomes a running pea-soup, rather than an actual fluid.

Even in this imperfect case, it is obvious

that the Greenbottle-grubs begin by liquefying their food. Incapable of taking solid
nourishment, they first transform the spoil
into running matter; then, dipping their heads
into the product, they drink, they slake their
thirst, with long sups. Their dissolvent, comparable in its effects with the gastric juice of
the higher animals, is, beyond a doubt,
emitted through the mouth. The piston of
the hooks, continually in movement, never
ceases spitting it out in infinitesimal doses.
Each spot touched receives a grain of some
subtle pepsin, which soon suffices to make that
spot run in every direction. As digesting,
when all is said, merely means liquefying, it
is no paradox to assert that the maggot digests its food before swallowing it.

These experiments with my filthy, evil-
smelling tubes have given me some delightful
moments. The worthy Abbé Spallanzani[1]
must have known some such when he saw
pieces of raw meat begin to run under the
action of the gastric juice which he took, with
pellets of sponge, from the stomachs of
Crows. He discovered the secrets of digest-

[1]Lazaro Spallanzani (1729-1799), the Italian naturalist, author of important works on the circulation of the
blood, digestion, generation and microscopic animals.—
Translator's Note.

ion; he realized in a glass tube the hitherto unknown labours of gastric chemistry. I, his distant disciple, behold once more, under a most unexpected aspect, what struck the Italian scientist so forcibly. Worms take the place of the Crows. They slaver upon meat, gluten, albumen; and those substances turn to fluid. What our stomach does within its mysterious recesses the maggot achieves outside, in the open air. It first digests and then imbibes.

When we see it plunging into the carrion broth, we even wonder if it cannot feed itself, at least to some extent, in a more direct fashion. Why should not its skin, which is one of the most delicate, be capable of absorbing? I have seen the egg of the Sacred Beetle and other Dung-beetles[1] growing considerably larger—I should like to say, feeding—in the thick atmosphere of the hatching-chamber. Nothing tells us that the grub of the Greenbottle does not adopt this method of growing. I picture it capable of feeding all over the surface of its body. To the gruel absorbed by the mouth it adds the balance of what is gath-

[1] Cf. *Insect Life:* chaps. i and ii; and *The Life and Love of the Insect:* chaps. i to iv and vii.—*Translator's Note.*

ered and strained through the skin. This would explain the need for provisions lique-fied beforehand.

Let us give one last proof of this prelimi-nary liquefaction. If the carcass—Mole, Snake or another—left in the open air have a wire-gauze cover placed over it, to keep out the Flies, the game dries under a hot sun and shrivels up without appreciably wetting the sand on which it lies. Fluids come from it, certainly, for every organized body is a sponge swollen with water; but the liquid discharge is so slow and restricted in quantity that the heat and the dryness of the air di-sperse it as it appears, while the underlying sand remains dry, or very nearly so. The car-cass becomes a sapless mummy, a mere bit of leather. On the other hand, do not use the wire-gauze cover, let the Flies do their work unimpeded; and things forthwith assume an-other aspect. In three or four days, an ooz-ing sanies appears under the animal and soaks the sand to some distance.

I shall never forget the striking spectacle with which I conclude this chapter. This time, the dish is a magnificent Æsculapius' Snake, a yard and a half long and as thick as a wide bottle-neck. Because of its size, which exceeds the dimensions of my pan, I roll the reptile in a double spiral, or in two storeys. When the copious joint is in full process of dissolution, the pan becomes a puddle wherein wallow, in countless numbers, the grubs of the Greenbottle and those of *Sarcophaga carnaria,* the Grey or Chequered Flesh-fly, which are even mightier liquefiers. All the sand in the apparatus is saturated, has turned into mud, as though there had been a shower of rain. Through the hole at the bottom, which is protected by a flat pebble, the gruel trickles drop by drop. It is a still at work, a mortuary still, in which the Snake is being drawn off. Wait a week or two; and the whole will have disappeared, drunk up by the sun: naught but the scales and bones will remain on a sheet of mud.

To conclude: the maggot is a power in this world. To give back to life, with all speed, the remains of that which has lived, it macerates and condenses corpses, distilling them into an essence wherewith the earth, the plant's foster-mother, may be nourished and enriched.

CHAPTER X

THE GREY FLESH-FLIES

HERE the costume changes, not the manner of life. We find the same frequenting of dead bodies, the same capacity for the speedy liquefaction of the fleshy matter. I am speaking of an ash-grey Fly, the Greenbottle's superior in size, with brown streaks on her back and silver gleams on her abdomen. Note also the blood-red eyes, with the hard look of the knacker in them. The language of science knows her as *Sarcophaga*, the flesh-eater; in the vulgar tongue she is the Grey Flesh-fly, or simply the Flesh-fly.

Let not these expressions, however accurate, mislead us into believing for a moment that the Sarcophagæ are the bold company of master-tainters who haunt our dwellings, more particularly in autumn, and plant their vermin in our ill-guarded viands. The author of those offences is *Calliphora vomitoria*, the Bluebottle, who is of a stouter build and arrayed in darkest blue. It is she who buzzes against our window-panes, who craftily besieges the

233

meat-safe and who lies in wait in the darkness
for an opportunity to outwit our vigilance.
The other, the Grey Fly, works jointly with
the Greenbottles, who do not venture inside
our houses and who work in the sunlight.
Less timid, however, than they, should the
outdoor yield be small, she will sometimes
come indoors to perpetrate her villainies.
When her business is done, she makes off as
fast as she can, for she does not feel at home
with us.

At this moment, my study, a very modest
extension of my open-air establishments, has
become something of a charnel-house. The
Grey Fly pays me a visit. If I lay a piece of
butcher's meat on the window-sill, she hastens
up, works her will on it and retires. No
hiding-place escapes her notice among the jars,
cups, glasses and receptacles of every kind
with which my shelves are crowded.

With a view to certain experiments, I collected a heap of wasp-grubs, asphyxiated in
their underground nests. Stealthily she arrives,
discovers the fat pile and, hailing as treasure-
trove this provender whereof her race perhaps
has never made use before, entrusts to it an
instalment of her family. I have left at the
bottom of a glass the best part of a hard-

The Grey Flesh-Flies

boiled egg from which I have taken a few bits of white intended for the Greenbottle-maggots. The Grey Fly takes possession of the remains, recks not of their novelty and colonizes them. Everything suits her that falls within the category of albuminous matters: everything, down to dead Silk-worms; everything, down to a mess of kidney-beans and chick-peas.

Nevertheless, her preference is for the corpse: furred beast and feathered beast, reptile and fish, indifferently. Together with the Greenbottles, she is sedulous in her attendance on my pans. Daily she visits my Snakes, takes note of the condition of each of them, savours them with her proboscis, goes away, comes back, takes her time and at last proceeds to business. Still, it is not here, amid the tumult of callers, that I propose to follow her operations. A lump of butcher's meat laid on the window-sill, in front of my writing-table, will be less offensive to the eye and will facilitate my observations.

Two Flies of the genus Sarcophaga frequent my slaughter-yard: *Sarcophaga carnaria* and *Sarcophaga hæmorrhoidalis,* whose abdomen ends in a red speck. The first species, which is a little larger than the second, is more numerous and does the best part of the work in the

open-air shambles of the pans. It is this Fly
also who, at intervals and nearly always alone,
hastens to the bait exposed on the window-
sill.

She comes up suddenly, timidly. Soon she
calms herself and no longer thinks of fleeing
when I draw near, for the dish suits her. She
is surprisingly quick about her work. Twice
over—buzz! Buzz!—the tip of her abdomen
touches the meat; and the thing is done: a
group of vermin wriggles out, releases itself
and disperses so nimbly that I have no time
to take my lens and count then accurately. As
seen by the naked eye, there were a dozen of
them. What has become of them? One
would think that they had gone into the flesh,
at the very spot where they were laid, so
quickly have they disappeared. But that dive
into a substance of some consistency is impossible to these new-born weaklings. Where
are they? I find them more or less everywhere in the creases of the meat; singly and already groping with their mouths. To collect
them in order to number them is not practicable, for I do not want to damage them.
Let us be satisfied with the estimate made at a
rapid glance: there are a dozen or so, brought

The Grey Flesh-Flies

into the world in one discharge of almost in-
appreciable length.

Those live grubs, taking the place of the
usual eggs, have long been known. Everybody is aware that the Flesh-flies bring forth
living maggots, instead of laying eggs. They
have so much to do and their work is so
urgent! To them, the instruments of the
transformation of dead matter, a day means a
day, a long space of time which it is all-im-
portant to utilize. The Greenbottle's eggs,
though these are of very rapid development,
take twenty-four hours to yield their grubs.
The Flesh-flies save all this time. From their
matrix, labourers flow straightway and set to
work the moment they are born. With these
ardent pioneers of sanitation, there is no rest
attendant upon the hatching, there is not a
minute lost.

The gang, it is true, is not a numerous one;
but how often can it not be renewed! Read
Réaumur's[1] description of the wonderful pro-
creating-machinery boasted by the Flesh-flies.
It is a spiral ribbon, a velvety scroll whose
nap is a sort of fleece of maggots set closely

[1] Réne Antoine Ferchault de Réaumur (1683-1757), the
inventor of the Réaumur thermometer and author of
Mémoires pour servir à l'histoire naturelle des insectes
(1734-42).—*Translator's Note.*

237

together and each cased in a sheath. The patient biographer counted the host: it numbers, he tells us, nearly twenty thousand. You are seized with stupefaction at this anatomical fact.

How does the Grey Fly find the time to settle a family of such dimensions, especially in small packets, as she has just done on my window-sill? What a number of dead Dogs, Moles and Snakes must she not visit before exhausting her womb! Will she find them? Corpses of much size do not abound to that extent in the country. As everything suits her, she will alight on other remains of minor importance. Should the prize be a rich one, she will return to it to-morrow, the day after and later still, over and over again. In the course of the season, by dint of packets of grubs deposited here, there and everywhere, she will perhaps end by housing her entire brood. But then, if all things prosper, what a glut, for there are several families born during the year! We feel it instinctively: there must be a check to these generative enormities.

Let us first consider the grub. It is a sturdy maggot, easy to distinguish from the Greenbottle's by its larger girth and especially by the way in which its body terminates behind.

The Grey Flesh-Flies

There is here a sudden breaking-off, hollowed into a deep cup. At the bottom of this crater are two breathing-holes, two stigmata with amber-red tips. The edge of the cavity is fringed with half a score of pointed, fleshy festoons, which diverge like the spikes of a coronet. The creature can close or open this diadem at will by bringing the denticulations together or by spreading them out wide. This protects the air-holes which might otherwise be choked up when the maggot disappears in the sea of broth. Asphyxia would supervene, if the two breathing-holes at the back became obstructed. During the immersion, the festooned coronet shuts like a flower closing its petals and the liquid is not admitted to the cavity.

Next follows the emergence. The hind-part reappears in the air, but appears alone, just at the level of the fluid. Then the coronet spreads out afresh, the cup gapes and assumes the aspect of a tiny flower, with the white denticulations for petals and the two bright red dots, the stigmata at the bottom, for stamens. When the grubs, pressed one against the other, with their heads downwards in the fetid soup, make an unbroken shoal, the sight of those breathing-cups incessantly opening and closing, with a little clack like a valve, almost makes one forget the horrors of the charnel-yard. It suggests a carpet of tiny Sea-anemones. The maggot has its beauties after all.

It is obvious, if there be any logic in things, that a grub so well-protected against asphyxiation by drowning must frequent liquid surroundings. One does not encircle one's hindquarters with a coronet for the sole satisfaction of displaying it. With its apparatus of spokes, the Grey Fly's grub informs us of the dangerous nature of its functions: when working upon a corpse, it runs the risk of drowning. How is that? Remember the grubs of the Greenbottle, fed on hard-boiled white of egg. The dish suits them; only, by the action of their pepsin, it becomes so fluid that they die submerged. Because of their hinder stigmata, which are actually on the skin and devoid of any defensive machinery, they perish when they find no support apart from the liquid.

The Flesh-fly's maggots, though incomparable liquefiers, know nothing of this peril, even in a puddle of carrion broth. Their bulky hind-part serves as a float and keeps the air-holes above the surface. When, for further investigation, they must needs go

The Grey Flesh-Flies

under completely, the anemone at the back shuts and protects the stigmata. The grubs of the Grey Fly are endowed with a life-buoy because they are first-class liquefiers, ready to incur the danger of a ducking at any moment.

When high and dry on the sheet of cardboard where I place them to observe them at my ease, they move about actively, with their breathing-rose wide-spread and their stigmata rising and falling as a support. The cardboard is on my table, at three steps from an open window, and lit at this time of day only by the soft light of the sky. Well, the maggots, one and all of them, turn in the opposite direction to the window; they hastily, madly take to flight.

I turn the cardboard round, without touching the runaways. This action makes the creatures face the light again. Forthwith, the troop stops, hesitates, takes a half turn and once more retreats towards the darkness. Before the end of the race-course is reached, I again turn the cardboard. For the second time, the maggots veer round and retrace their steps. Repeat the experiment as often as I will, each time the squad wheels about in the opposite direction to the window and persists

The Life of the Fly

in avoiding the trap of the revolving cardboard.

The track is only a short one: the cardboard measures three hand's-breadths in length. Let us give more space. I settle the grubs on the floor of the room; with a hairpencil, I turn them with their heads pointing towards the lighted aperture. The moment they are free, they turn and run from the light. With all the speed whereof their cripple's shuffle allows, they cover the tiled floor of the study and go and knock their heads against the wall, twelve feet off, skirting it afterwards, some to the right and some to the left. They never feel far enough away from that hateful illuminated opening.

What they are escaping from is evidently the light, for, if I make it dark with a screen, the troop does not change its direction when I turn the cardboard. It then progresses quite readily towards the window; but, when I remove the screen, it turns tail at once.

That a grub destined to live in the darkness, under the shelter of a corpse, should avoid the light is only natural; the strange part is its very perception. The maggot is blind. Its pointed fore-part, which we hesitate to call a head, bears absolutely no trace of any optical

apparatus; and the same with every other part
of the body. There is nothing but one bare,
smooth, white skin. And this sightless creature, deprived of any special nervous points
served by ocular power, is extremely sensitive
to the light. Its whole skin is a sort of retina,
incapable of seeing, of course, but able, at any
rate, to distinguish between light and darkness.
Under the direct rays of a searching sun, the
grub's distress could be easily explained. We
ourselves, with our coarse skin, in comparison
with that of the maggot, can distinguish between sunshine and shadow without the help
of the eyes. But, in the present case, the problem becomes singularly complicated. The subjects of my experiment receive only the diffused light of the sky, entering my study
through an open window; yet this tempered
light frightens them out of their senses. They
flee the painful apparition; they are bent upon
escaping at all costs.

Now what do the fugitives feel? Are they
physically hurt by the chemical radiations?
Are they exasperated by other radiations,
known or unknown? Light still keeps many
a secret hidden from us and perhaps our optical science, by studying the maggot, might
become the richer by some valuable informa-

tion. I would gladly have gone farther into the question, had I possessed the necessary apparatus. But I have not, I never have had and of course I never shall have the resources which are so useful to the seeker. These are reserved for the clever people who care more for lucrative posts than for fair truths. Let us continue, however, within the measure which the poverty of my means permits.

When duly fattened, the grubs of the Flesh-flies go underground to transform themselves into pupæ. The burial is intended, obviously, to give the worm the tranquillity necessary for the metamorphosis. Let us add that another object of the descent is to avoid the importunities of the light. The maggot isolates itself to the best of its power and withdraws from the garish day before contracting into a little keg. In ordinary conditions, with a loose soil, it goes hardly lower than a hand's-breadth down, for provision has to be made for the difficulties of the return to the surface when the insect, now full-grown, is impeded by its delicate Fly-wings. The grub, therefore, deems itself suitably isolated at a moderate depth. Sideways, the layer that shields it from the light is of indefinite thickness; upwards, it measures about four inches. Behind through something tantamount to tufa, that is to say, through earth which a shower has rendered compact. For the descent, the grub has its fangs; for the assent, the Fly has nothing. Only that moment come into existence, she is a weakling, with tissues still devoid of any firmness. How does she manage to get out? We shall know by watching a few pupæ placed at the bottom of a test-tube filled with earth. The method of the Flesh-flies will teach us that of the Greenbottles and the other Flies, all of whom make use of the same means.

Enclosed in her pupa, the nascent Fly begins by bursting the lid of her casket with a hernia which comes between her two eyes and doubles or trebles the size of her head. This cephalic blister throbs: it swells and subsides by turns, owing to the alternate flux and reflux of the blood. It is like the piston of an hydraulic press opening and forcing back the front part of the keg.

The head makes its appearance. The hydrocephalous monster continues the play of her forehead, while herself remaining stationary. Inside the pupa, a delicate work is being performed: the casting of the white nymphal tunic. All through this operation, the hernia

this screen reigns utter darkness, the buried one's delight. This is capital.

What would happen if, by an artifice, the sideward layer were nowhere thick enough to satisfy the grub? Now, this time, I have the wherewithal to solve the problem, in the shape of a big glass tube, open at both ends, about three feet long and less than an inch wide. I use it to blow the flame of hydrogen in the little chemistry-lessons which I give my children.

I close one end with a cork and fill the tube with fine, dry, sifted sand. On the surface of this long column, suspended perpendicularly in a corner of my study, I install some twenty Sarcophaga-grubs, feeding them with meat. A similar preparation is repeated in a wider jar, with a mouth as broad as one's hand. When they are big enough, the grubs in either apparatus will go down to the depth that suits them. There is no more to be done but to leave them to their own devices.

The worms at last bury themselves and harden into pupæ. This is the moment to consult the two apparatus. The jar gives me the answer which I should have obtained in the open fields. Four inches down, or thereabouts, the worms have found a quiet lodging, radiations capable of acting upon this lover of darkness? They are certainly not the simple luminous rays, for a screen of fine, heaped-up earth, nearly half an inch in thickness, is perfectly opaque. Then, to alarm the grub, to warn it of the over-proximity of the exterior and send it to mad depths in search of isolation, other radiations, known or unknown, must be required, radiations capable of penetrating a screen against which ordinary radiations are powerless. Who knows what vistas the natural philosophy of the maggot might open out to us? For lack of apparatus, I confine myself to suspicions.

To go underground to a yard's depth—and farther if my tube had allowed it—is on the part of the Flesh-fly's grub a vagary provoked by unkind experiment: never would it bury itself so low down, if left to its own wisdom. A hand's-breadth thickness is quite enough, is even a great deal when, after completing the transformation, it has to climb back to the surface, a laborious operation absolutely resembling the task of an entombed well-sinker. It will have to fight against the sand that slips and gradually fills up the small amount of empty space obtained; it will perhaps, without crowbar or pickaxe, have to cut itself a gallery is still projecting. The head is not the head of a Fly, but a queer, enormous mitre, spreading at the base into two red skull-caps, which are the eyes. To split her cranium in the middle, shunt the two halves to the right and left and send surging through the gap a tumour which staves the barrel with its pressure: this constitutes the Fly's eccentric method.

For what reason does the hernia, once the keg is staved, continue swollen and projecting? I take it to be a waste-pocket into which the insect momentarily forces back its reserves of blood in order to diminish the bulk of the body to that extent and to extract it more easily from the nymphal slough and afterwards from the narrow channel of the shell. As long as the operation of the release lasts, it pushes outside all that it is able to inject of its accumulated humours; it makes itself small inside the pupa and swells into a bloated deformity without. Two hours and more are spent in this laborious stripping.

At last, the Fly comes into view. The wings, mere scanty stumps, hardly reach the middle of the abdomen. On the outer edge, they have a deep notch similar to the waist of a violin. This diminishes by just so much the surface and the length, an excellent device for decreasing the friction along the earthy column which has next to be scaled. The hydro-cephalous one resumes her performance more vigorously than ever; she inflates and deflates her frontal knob. The pounded sand rustles down the insect's sides. The legs play but a secondary part. Stretched behind, motion-less, when the piston-stroke is delivered, they furnish a support. As the sand descends, they pile it and nimbly push it back, after which they drag along lifelessly until the next ava-lanche. The head advances each time by a length equal to that of the sand displaced. Each stroke of the frontal swelling means a step forward. In a dry, loose soil, things go pretty fast. A column six inches high is trav-ersed in less than a quarter of an hour.

As soon as it reaches the surface, the insect, covered with dust, proceeds to make its toilet. It thrusts out the blister of its forehead for the last time and brushes it carefully with its front tarsi. It is important that the little pounding-engine should be carefully dusted before it is taken inside to form a forehead that will open no more: this lest any grit should lodge in the head. The wings are carefully brushed and polished; they lose their curved notches; they

Problem No. 2. A map has a graphical scale on which 1.5 inches reads 500 strides. 1. What is the R. F. of the map? 2. How many miles are represented by 1 inch?

Problem No. 3. The Leavenworth map in back of this book has a graphical scale and a measured distance of 1.25 inches reads 1,100 yards. Required: 1. The R. F. of the map; 2. Number of miles shown by 1 inch on the map.

Problem No. 4. 1. Construct a scale to read yards for a map of R. F. = $\frac{1}{21120}$. 2. How many inches represent 1 mile?

547. Scaling Distances From a Map. There are four methods of scaling distances from maps:

1st. Apply a piece of straight edged paper to the distance between any two points, A and B, for instance, and mark the distance on the paper. Now, apply the paper to the graphical scale,(Figure 2, Par. 540), and read the number of yards on the main scale and add the number indicated on the extension. For example: $600 + 75 = 675$ yards.

2nd. By taking the distance off with a pair of dividers and applying the dividers thus set to the graphical scale, the distance is read.

3rd. By use of an instrument called a map measurer, Figure 11, set the hand on the face to read zero, roll the small wheel over the distance; now roll the wheel in an opposite direction along the graphical scale, noting the number of yards passed over. Or, having rolled over the distance, note the number of inches on the dial and multiply this by the number of miles or other units per inch. A map measurer is valuable for use in solving map problems in patrolling, advance guard, outpost, etc.

Figure 11.

4th. Apply a scale of inches to the line to be measured, and multiply this distance by the number of miles per inch shown by the map.

Having learned how to take off distances on the map, the next step in map reading is to determine differences of elevation.

548. Method of Representing Differences of Elevation. Since maps are representations on paper of ground which has size not only in a horizontal (level) but in a vertical (up and down) direction, it is necessary to have some means of rapidly determining elevations. This is accomplished in one of three ways:

1st. By means of contours. A contour line is the line in which a horizontal (level) plane cuts the surface of the ground. It may also be said

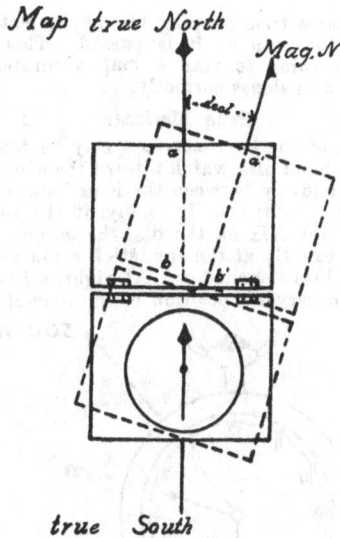

Map true North

Mag. N

true South
Figure 17.

meridian of the map and move the map around horizontally until the north end of the needle points toward the north of its circle, whereupon the map is oriented. If there is a true meridian on the map, but not a magnetic meridian, one may be constructed as follows, if the magnetic declination is known:

(Figure 17): Place the true meridian of the map directly under the magnetic needle of the compass and then move the compass box until the needle reads an angle equal to the magnetic declination. A line in extension of the sighting line a'-b' will be the magnetic meridian. If the magnetic declination of the observer's position is not more than 4° or 5°, the orientation will be given closely enough for ordinary purposes by taking the true and magnetic meridians to be identical.

2d. If neither the magnetic nor the true meridian is on the map, but the observer's position on the ground is known: Move the map horizontally until the direction of some definite point on the ground is the same as its direction on the map; the map is then oriented. For example, suppose you are standing on the ground at 8, q k' (Fort Leavenworth Map), and can see the U. S. penitentiary off to the south. Hold the map in front of you and face toward the U. S. penitentiary, moving the map until the line joining 8 and the U. S. penitentiary (on the map) lies in the same direction as the line joining those two points on the ground. The map is now oriented.

Having learned to orient a map and to locate his position on the map, the noncommissioned officer should then practice moving over the ground and at the same time keeping his map oriented and noting each ground feature on the map as it is passed. This practice is of the greatest value in learning to read a map accurately and to estimate distances, directions and slopes correctly.

True Meridian

553. The position of the true meridian may be found as follows (Fig. 18): Point the hour hand of a watch toward the sun; the line joining the pivot and the point midway between the hour hand and XII on the dial, will point toward the south; that is to say, if the observer stands so as to face the sun and the XII on the dial, he will be looking south. To point the hour hand exactly at the sun, stick a pin as at (a) Fig. 18 and bring the hour hand into the shadow. At night, a line drawn toward the north star from the observer's position is approximately a true meridian.

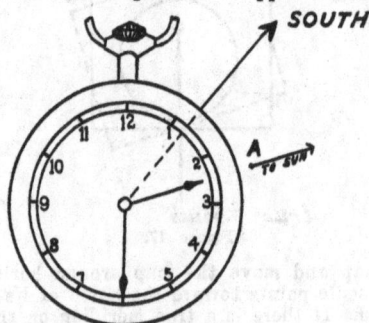

SOUTH

A
TO SUN

Figure 18

The line joining the "pointers" of the Great Bear or Dipper, prolonged about five times its length passes nearly through the North Star, which can be recognized by its brilliancy.

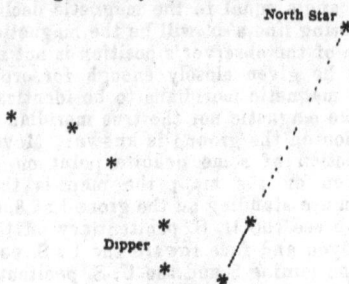

North Star

Dipper

Figure 19.

Conventional Signs

554. Rivers, lakes, mountains, forests, roads, houses, telegraph lines, etc., are represented on maps by symbols called Conventional Signs, in which an effort is made to imitate the general appearance of the objects as seen from a high point directly overhead. On account of this similarity of the object to its sign or symbol on the map, the noncommissioned officer will usually have no trouble in deciding at once the meaning of a new symbol. Fig. 21 gives Conventional Signs used on military maps, and they should be thoroughly learned so that their meaning will be known at a glance.

There is a constant tendency to simplify the Conventional Signs, and very often simply the outline of an object, such as a forest, cultivated ground, etc., is indicated with the name of the object printed within the outline. Thus:

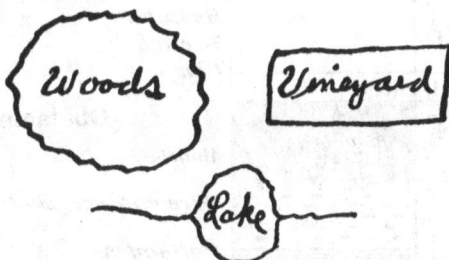

Figure 20.

Such means are used very frequently in rapid sketching, on account of the time that they save.

By reference to the map of Fort Leavenworth, the meaning of all its symbols is at once evident from the names printed thereon; for example, that of a city, woods, roads, streams, railroad, etc.; where no Conventional Sign is used on any area, it is to be understood that any growths thereon are not high enough to furnish any cover. As an exercise, pick out from the map the following conventional signs: Unimproved road, cemetery, railroad track, hedge, wire fence, orchard, streams, lake. The numbers on the various road crossings have no equivalent on the ground, but are placed on the maps to facilitate description of routes, etc. Often the numbers at road crossings on other maps denote the elevation of these points.

Visibility

555. The problem of visibility is based on the relations of contours and map distances previously discussed, and includes such matters as the determination of whether a point can or can not be seen from another; whether a certain line of march is concealed from the enemy; whether a particular area is seen from a given point.

On account of the necessary inaccuracy of all maps it is impossible to determine exactly how much ground is visible from any given point—that is, if a correct reading of the map shows a certain point to be just barely visible, then it would be unsafe to say positively that on the ground this point could be seen or could not be seen. It is, however, of great importance for the noncommissioned officer to be able to determine

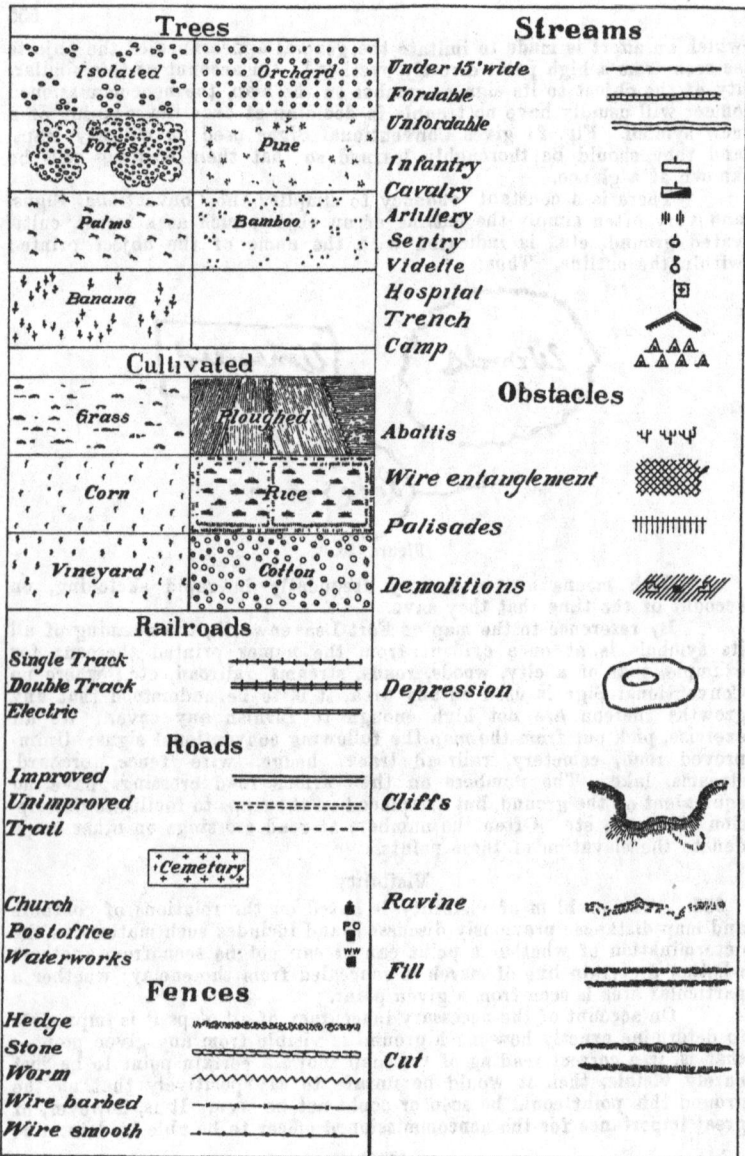

Trees

Isolated	Orchard
Forest	Pine
Palms	Bamboo
Banana	

Cultivated

Grass	Ploughed
Corn	Rice
Vineyard	Cotton

Railroads

Single Track
Double Track
Electric

Roads

Improved
Unimproved
Trail

Cemetery

Church
Postoffice
Waterworks

Fences

Hedge
Stone
Worm
Wire barbed
Wire smooth

Streams

Under 15' wide
Fordable
Unfordable

Infantry
Cavalry
Artillery
Sentry
Vidette
Hospital
Trench
Camp

Obstacles

Abattis

Wire entanglement

Palisades

Demolitions

Depression

Cliffs

Ravine

Fill

Cut

Figure 21.

at a glance, within about one contour interval, whether or not such and such a point is visible; or whether a given road is generally visible to a certain scout, etc. For this reason no effort is made to give an exact mathematical solution of problems in visibility further than would be useful in practical work with a map in the solution of map problems in patrolling.

In the solution of visibility problems, it is necessary that the

Figure 22.

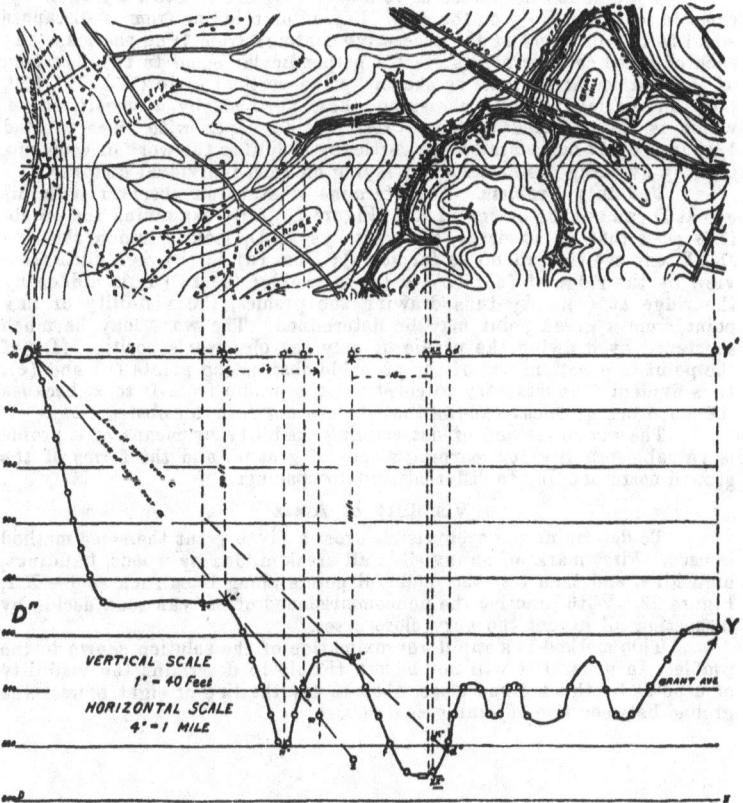

Fig. 23

noncommissioned officer should thoroughly understand the meaning of profiles and their construction. A profile is the line supposed to be cut from the surface of the earth by an imaginary vertical (up and down) plane. (See Fig. 23.) The representation of this line to scale on a sheet of paper is also called a profile. Figure 23 shows a profile on the line D-y (Figure 22) in which the horizontal scale is the same as that of the map (Figure 22) and the vertical scale is 1 inch = 40 feet. It is customary to draw a profile with a greater vertical than horizontal scale in order to make the slopes on the profile appear to the eye as they exist on the ground. Consequently, always note especially the vertical scale in examining any profile; the horizontal scale is usually that of the map from which the profile is taken.

A profile is constructed as follows: (Fig. 23): Draw a line D'—y' equal in length to D—y on the map. Lay off on this line from D' distances equal to the distances of the successive contours from D on the map. At each of these contour points erect a perpendicular equal to the elevation of this particular contour, as shown by the vertical scale (960, 940, 920, etc.) on the left. Join successively these verticals by a smooth curve, which is the required profile. Cross section paper with lines printed 1/10 inch apart horizontally and vertically simplifies the work of construction, by avoiding the necessity of laying off each individual distance.

Visibility Problem. To determine whether an observer with his eye at D can see the bridge at XX (Figure 22). By examining the profile it is seen that an observer, with his eye at D, looking along the line D-XX, can see the ground as far as (a); from (a) to (b), is hidden from view by the ridge at (a); (b) to (c) is visible; (c) to (d) is hidden by the ridge at (c). By thus drawing the profiles, the visibility of any point from a given point may be determined. The work may be much shortened by drawing the profile of only the observer's position (D) of the point in question, and of the probable obstructing points (a) and (c). It is evidently unnecessary to construct the profile from D to x, because the slope being concave shows that it does not form an obstruction.

The above method of determining visibility by means of a profile is valuable practice for learning slopes of ground, and the forms of the ground corresponding to different contour spacings.

Visibility of Areas

To determine the area visible from a given point the same method is used. First mark off as invisible all areas hidden by woods, buildings, high hills, and then test the doubtful points along lines such as D—XX, Figure 22. With practice the noncommissioned officer can soon decide by inspection all except the very close cases.

This method is a rapid approximation of the solution shown in the profile. In general it will not be practicable to determine the visibility of a point by this method closer than to say the line of sight pierces the ground between two adjoining contours.

CHAPTER IX

MILITARY SKETCHING

(While this chapter presents the principal features of military sketching in a simple, clear manner, attention is invited to the fact that the only way that any one who has never done any sketching can follow properly the statements made, is to do so with the instruments and the sketching material mentioned at hand. In fact, the only way to learn how to sketch is *to sketch*.)

556. A military sketch is a rough map showing the features of the ground that are of military value.

Military sketching is the art of making such a military sketch.

Military sketches are of three kinds:

Position sketches, Fig. 1;

Outpost sketches;

Road sketches.

All kinds of military sketches are intended to give a military commander detailed information of the ground to be operated over, when th's is not given by the existing maps, or when there are no maps of the area.

The general methods of sketching are:

(1)   The location of points by intersection.

(2)   The location of points by traversing.

(3)   The determination of the heights of hills, shapes of the ground, etc., by contours.

(1)   To locate a point by intersection proceed as follows: Set up, level and orient the sketching board (Par. 552), (or Sketching Case, Fig. 3), at A, Fig. 1.   The board is said to be oriented when the needle is parallel to the lines across the face of the compass, Fig. 3, of the cavalry case, or parallel to the sides of the compass through of the drawing board, Fig. 4.   (At every station the needle must have this position, so that every line on the sketch will be parallel to the corresponding line or direction on the ground.)   Assume a point (A) on the paper, Fig 1 Y, in such a position that the ground to be sketched will fall on the sheet.   Lay the ruler on the board and point it to the desired point (C), all the while keeping the edge of the ruler on the point (A), Fig. 1 Y.   Draw an

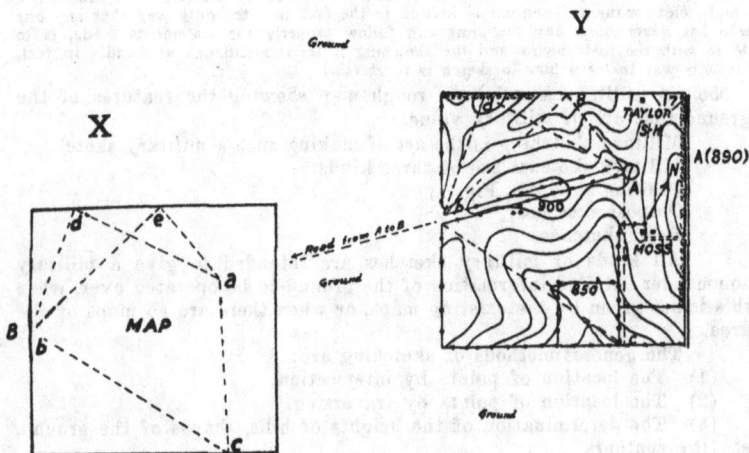

X

Y

C

(Note: This diagram represents the sketching board.)

Fig 1

indefinite line along the edge. Now move to (B), Fig. 1 X, plotted on the map in (b), Fig. 1 X, and having set up, leveled and oriented as at (A) Fig. 1 Y, sight toward (C) as before. The intersection (crossing) of the two lines locates (C) on the sketch at (c), Fig. 1 X.

(2) To locate a point by traversing is done as follows: With the board set up, leveled and oriented at A, Fig. 1 Y, as above, draw a line in the direction of the desired point B, Fig. 1 X, and then move to B, counting strides, keeping record of them with a tally register, Fig. 5, if one is available. Set up the board at B, Fig. 1 X, and orient it by laying

the ruler along the line (a)-(b), Fig. 1 X, and moving the board until the ruler is directed toward A, Fig. 1 Y, on the ground; or else orient by the needle as at A. With the scale of the sketcher's strides on the

(Sketching Case)—Figure 3.

ruler, Fig. 3, lay off the number of strides found from A, Fig. 1 Y, to B, Fig. 1 X, and mark the point (b), Fig. 1 X. Other points, such as C, D, etc., would be located in the same way.

[503]

556 (contd.)

(3) To draw in contours on a sketch, the following steps are necessary:

(a) From the known or assumed elevation of a located station as A, Fig. 1 Y, (elevation 890), the elevations of all hill tops, stream junctures, stream sources, etc., are determined.

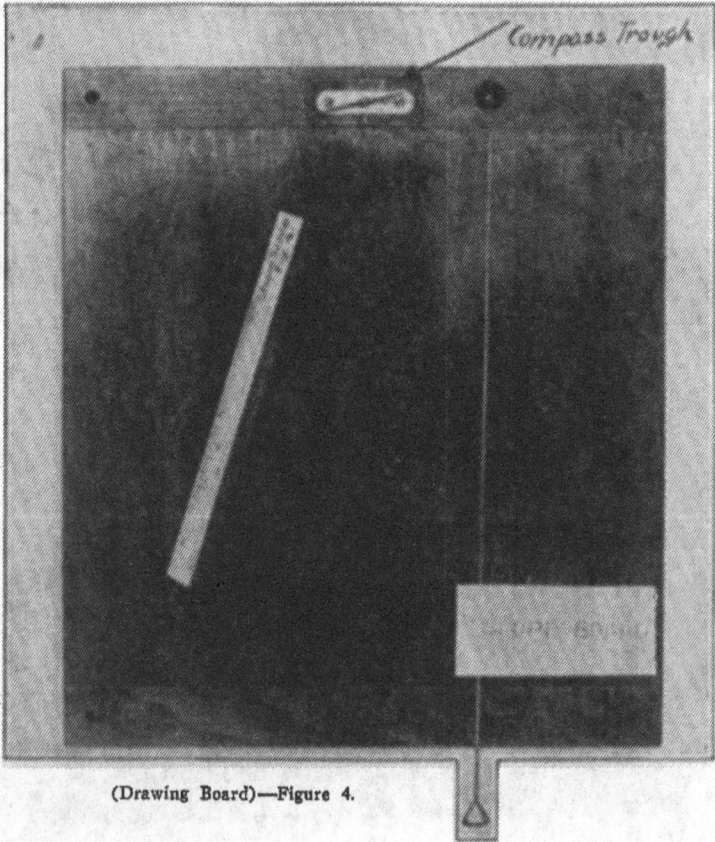

(Drawing Board)—Figure 4.

(b) Having found the elevations of these critical points the contours are put in by spacing them so as to show the slope of the ground along each line such as (a)-(b), (a)-(c), etc., Fig. 1 Y, as these slopes actually are on the ground.

To find the elevation of any point, say C (shown on sketch as c), proceed as follows:

(Tally Register)—Figure 5.   (Clinometer)—Figure 6.

Read the vertical angle with slope board. Fig. 4, or with a clinometer, Fig. 6. Suppose this is found to be 2 degrees; lay the scale of M. D.* (ruler, Fig. 4) along (a)-(c), Fig. 1 Y, and note the number of divisions of—2 degrees (minus 2°) between (a) and (c). Suppose there are found to be 5½ divisions; then, since each division is 10 feet, the total height of A above C is 55 feet (5½ × 10). C is therefore 835 ft. elev. which is written at (c), Fig. 1 Y. Now looking at the ground along A-C, suppose you find it to be a very decided concave (hollowed out) slope, nearly flat at the bottom and steep at the top. There are to be placed in this space (a)-(c), Fig. 1 Y, contours 890, 880, 870, 860 and 850, and they would be spaced close at the top and far apart near (c), Fig. 1 Y, to give a true idea of the slope.

The above is the entire principle of contouring in making sketches and if thoroughly learned by careful repetition under different conditions, will enable the student to soon be able to carry the contours with the horizontal locations.

557.   **Position Sketching**

Instruments used in Position and Outpost sketching:

1. Drawing board with attached compass, Fig. 4.
2. Loose ruler (on board, Fig. 3 and 4).
3. Rough tripod or camera tripod.
4. Scale of M. D.'s (shown on ruler, Fig. 4).
5. Scale of the sketcher's strides (at 6″ to 1 mile), shown on ruler, Fig. 3.
6. Clinometer (not absolutely necessary if board has slope board), Fig. 6.
7. Scale of hundreds of yards (shown on ruler, Fig. 3).

Methods to be used:

1. Select a Base Line, that is, a central line ¼ to ½ mile long in the area to be sketched. The base should have at its end some plainly marked objects, such as telegraph poles, trees, corners of buildings, etc., and from its ends a good view of the area should be possible.

---

*The construction of a scale of M D's is described under map reading, par. 543 Scales of M. D.'s can be obtained from the Secretary, Army Service Schools, Fort Leavenworth, Kans.

2. Set up, level and orient the drawing board at one end of the base (A), Fig. 1 Y. Draw a meridian on the sheet parallel to the position of the magnetic needle.

Assume a point (A), Fig. 1 Y, corresponding to the ground point (A), 890, on the sheet in such a position that the area to be sketched will lie on the sheet.

3. Sight at hill tops, stream junctures, stream heads, etc., to begin the location of these points by intersection.

4. Traverse to B and complete the locations by intersection as previously explained.

5. Draw the details of country between A and B and in the vicinity of this line, using the conventional signs for roads, houses, etc.

6. The lines from station (b), Fig. 1 X, to any of the other located points may now be used as a new base line to carry the work over additional area.

7. In case parts of the area are not visible from a base line, these parts are located by traversing as before explained.

8. Having learned by several repetitions the above steps, the sketcher will then combine contouring (see contouring above) with his horizontal locations.

558. Outpost Sketching

The methods of Outpost Sketching are the same as for position sketching, except that the sketcher can not advance toward the supposed position of the enemy beyond the outpost line. Therefore a base line must be selected on or in rear of the line of observation. From this base line all points visible toward the enemy are located by intersection or by traverse along the base line, details being shown by conventional signs and contours as for the position sketch.

559. Road Sketching

Instruments used:

1. Drawing board or sketching case.
2. Loose ruler.
3. Scales of strides, if made dismounted; scale of time, trotting or walking, if mounted.
4. Scale of hundreds of yards, at 3″ to 1 mile.
5. Scale of M. D.
6. Clinometer (if slope board not available).

Figure 7.

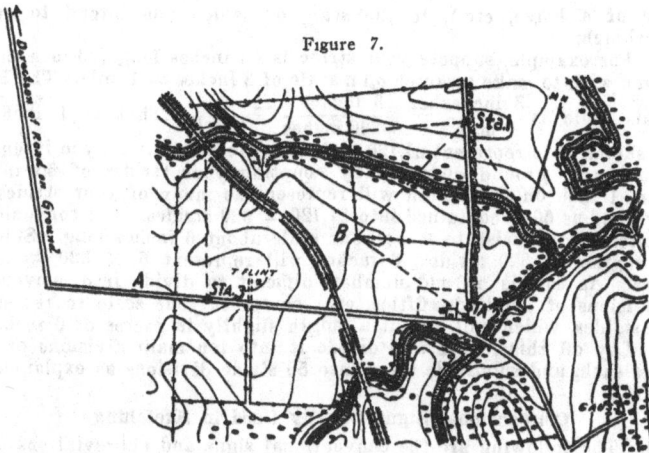

Methods:

1. At station 1, Fig. 7, orient the board as described under "To locate a point by intersection," page 501, holding the board in the hands in front of the body of the sketcher who faces toward station 2.

2. Important points in the vicinity, such as the railroad bridge, the stream juncture, hill tops, are sighted for intersections, lines drawn as shown and the sketcher traverses (as under traverse above) to station 2.

3. At station 2, he locates and draws in all details between station 1 and 2 to include about 300 yards on each side of the road.

4. The traverse is then continued forward as described for 1 and 2.

5. After some practice in horizontal sketching, as just described, the sketcher will be able to take up contouring in combination. The methods are as described in paragraph on contouring.

6. Method to follow when the traverse runs off the paper as at A, Fig. 7; re-orient the board so that the road forward will lie across the long dimensions of the paper; draw a meridian parallel to the compass needle and assume a point on the new sheet corresponding to the last point (A) plotted on the first sheet.

7. On completion of the sketch the various sections will be pasted together, so that all the meridians are parallel.

Construction of Working Scales.* The construction of scales has already been explained under Map Reading. To make a working scale (one that is used by the person making a sketch), construct a scale of convenient length, about 6 inches, as described in Par. 546a, to read in the units you intend to measure your distance with (your stride, pace,

---

*Sheets of working scales reading in paces, strides, minutes, etc., at a scale of 3 and 6 inches to the mile, can be obtained at little cost from the Secretary, Army Service School, Fort Leavenworth, Kansas.

stride of a horse, etc.), to the scale on which you intend to make your sketch.

For example, suppose your stride is 66 inches long (33-inch pace) and you wish to make a sketch on a scale of 3 inches = 1 mile. The R. F. of this scale is $\dfrac{3 \text{ inches}}{1 \text{ mile}} = \dfrac{3 \text{ inches}}{63360 \text{ inches}} = \dfrac{1}{21120}$ That is 1 inch on your sketch is to represent 21,120 inches on the ground. As you intend to measure your ground distances by counting your strides of 66 inches length, 1 inch on the sketch will represent as many of your strides on the ground as 66 is contained into 21,120 = 320 strides. For convenience in sketching you wish to make your scale about 6 inches long. Since 1 inch represents 320 strides, 6 inches will represent 6 × 320 = 1,920 strides. As this is an odd number, difficult to divide into convenient subdivisions of hundreds, fifties, etc., construct your scale to represent 2,000 strides, which will give it a length slightly in excess of 6 inches— 6.25. Lay off this length and divide it into ten main divisions of 200 strides each, and subdivide these into 50 stride divisions as explained in Par. 540.

Conventional Signs Usually Used in Sketching

560. The following are the conventional signs and abbreviations used in military sketching, that are authorized by the Field Service Regulations:

The following abbreviations are authorized for use on field maps and sketches. When these words are used they must be written in full or abbreviated as shown. The abbreviations must not be used for words other than those in the table. Words not in the table are not as a rule abbreviated

abut	abutment	P O	post Office
B S	blacksmith shop	Pt	point
bot	bottom	Q.	quarry
Cr	creek	q p.	queen post
cul	culvert	R	river
cult	cultivated	R H	round house
d	deep	R R	railroad
E	east	S	south
f	fordable	s	steel
gir	girder	S H	school house
G M.	grist mill	S M	saw mill
i	iron	Sta.	station
I	island	st	stone
Jc,	junction	str	stream
kp.	king-post	tres	trestle
L	lake	tr	truss
Mt.	mountain	W T	water tank
N	north	W W	water works
n f	not fordable	W	west
p.	pier	wd	wide
pk	plank	w	wood

Telegraph

Railroads — Single track / Double track / Trolley — Elec.

Roads — Improved / Unimproved / Trail

Fences — barbed wire / smooth wire / wood / stone / hedge

Bridge

Indicate character and span by abbreviations

Example.

$$\frac{40 \times 20}{10}$$

Meaning wooden king post bridge, 40 feet long, 20 feet wide, and 10 feet above the water

Streams

Indicate character by abbreviations.

Example:

Meaning a stream 15 feet wide, 8 feet deep, and not fordable.

House •   Church ♣   School house = S.H.

Woods { Woods }   Orchards   Cultivated Land Cult.

If boundary lines are fences they are indicated as such.

Brush, crops or grass, important as cover or forage   Brush, corn, grass, etc.

Cemetery   Trees, isolated

Cut and fill —   Cut / 10'   cut 10 feet deep

                  Fill / 10'   fill 10 feet high

For more elaborate map work the authorized conventional signs are used.

Points for Beginners to Remember

561.   1.   Always keep your pencils sharpened and have an eraser handy. No one but an expert can sketch with a dull pencil.

2.   Use hard pencils when learning to sketch—4H to 6H—and go over your work afterwards with a softer pencil—2H.

3.   Do not try to put down on your sketch a mass of small details that are too small to be shown on the scale at which you are sketching. For example, if you are making a sketch on a scale of 3 inches = 1 mile, do not try to show each house in a row of houses; simply indicate that there is a row of houses, by putting down several distinct conventional signs for houses in a row; nor should you try to show every little "cut" through which the road may run.   Only use about one sign to the inch for telegraph or telephone lines, for wire fences, etc.

4.   When first practicing sketching only plot the route over which you walk, indicating it by a single line.   When you can do this with facility, go back over one of these plotted routes and fill in the woods, houses, streams and the other large features.

5.   The beginner should sketch the same ground several times over— at least three or four times.   Practice alone will make perfect.

6.   Always try to compare your finished sketch with an accurate map of the ground, if one is obtainable.   Try to practice on ground of which you can obtain a map.

7.   Make each course (the distance you go between points where the direction of your route changes) as long as possible.

8.   Do not try to contour until you are expert at making a sketch showing all the flat details (roads, streams, woods, houses, etc.).

9.   Never try to "sketch in" the contours until you have plotted the stream lines or the direction of the valleys, ravines, etc.   The contours are fitted to or sketched around the drainage system; not the drainage system to the contours.

10.   Always "size up" ground before you sketch it; that is, take a general view of it, noticing the drainage system (the direction in which the streams flow or ravines run), the prominent hills and ridges, the direction the roads run, etc.

CHAPTER X

LOADING WAGONS

562. The property to be loaded should be carefully inspected before any is loaded, to see that everything is in good order and properly boxed, crated or tied.

Large, heavy boxes should be avoided.

The following general rules must be observed:

1. Heavy stuff must go on the bottom (and forward rather than rear) and light stuff on top—thus, heavy articles will not crush light ones and the center of gravity will be nearer the axles, making the turning over of the load more difficult.

2. Things needed first upon reaching camp must be placed on top or in rear.

The following method of loading a wagon is in accordance with the general principles cited above:

Ammunition. Ordinarily just back of the forward axle. In case of possible need, however, the ammunition should be placed where it could be gotten at immediately.

Axes, Spades, Shovels, and (Unhandled) Picks. Should be outside of wagonbed, in leather pockets or strong bags, or stood on end at rear of wagon. They should not be placed between the sides of the wagon and the load.

Blanket Rolls. If to be carried on wagon, they should be rolled tightly and left straight—not tied in a circle—and loaded on top, crosswise.

Camp Kettles and Buckets. Under the wagon, suspended from the reach pole.

Field Desk. To be placed on or near bottom and well forward, as it is seldom required early.

Field Range. On bottom, at rear of wagon. (The Infantry Equipment Board recommended that the field range be carried on tail gate of the wagon, lowered to a position of about 30 degrees from the horizontal.)

Forage. If to be carried on wagon, in front of ammunition.

Lashing. Use two pieces of ¾-inch rope about 75 feet long, passing over load first from front to rear diagonally, and finally secured by being tied to rings on the rear bolster standards—never to the end gate rods. The rope should be passed through strong hooks securely clinched to the body of the wagon, and not passed around the ends of the bows.

Officers' Bedding Rolls. To be on top of load.

Rations. Surplus rations (not required for next camp) in bottom of wagon, between ammunition and ration box.

Bacon should be on the bottom of wagon, where the grease will do no harm.

Ration Box. Next to field range, toward front of wagon. After the field range has been unloaded, the ration box is readily accessible and need not be unloaded.

[512]

At every camp the ration box should be restocked for the next camp.

Sibley Stoves. Slung on chain, just outside of feed box and below the Buzzacott oven.

Stove Pipe. Should be crated and lashed on in rear of a wagon.

Tentage. Should be rolled and not folded, except in places where absolutely necessary—and placed across wagon, on top of boxes, etc.

(Attention is invited to the fact that canvas becomes unserviceable more from handling and transportation than from wear when in actual use in sheltering troops.)

The tents, properly dried out, should be laid out smoothly on the ground; the part of the wall appearing uppermost should be folded over toward the peak of the tent; that underneath should be (by lifting the lower part of the tent) in like manner folded under and toward the peak; then by commencing at the peak, at the final folding, the wall of the tent will appear on the outside of the completed roll.

Ropes not required for securing the bundle should be folded inside.

Tent Pins. On top, in sacks.

Tent Poles. Should be tied with a rope and placed just inside the bows so as to extend above the wagon bed side; or carried in two iron hooks suspended from side of wagon bed, about four feet apart.

NOTES

1. **Pots and Kettles.** Should be in gunny sacks so as not to dirty everything.

2. The quartermaster sergeant should ride on one of the wagons.

3. A noncommissioned officer should personally superintend the loading of every wagon, the same noncommissioned officer always having charge of the same wagon.

4. The jockey box should be left entirely for use of teamster, and in which should be kept wrench, grease, spare bolts, mule shoes, etc.

5. A detail of men, the size of which depends upon the number of wagons, should accompany the train. Often the guard, or old guard performs this duty, but it is preferable to detail men who know how to meet emergencies such as a wagon tipping over on a hillside, wagons requiring repacking, mule down and hurt, etc.

CHAPTER XI
MARCHES

(Based on Infantry Drill Regulations)

563. Marching constitutes the principal occupation of troops in campaign and is one of the heaviest causes of loss. This loss may be materially reduced by proper training and by the proper conduct of the march.

The training of infantry should consist of systematic physical exercises to develop the general physique and of actual marching to accustom men to the fatigue of bearing arms and equipment.

Before mobilization troops should be kept in good physical condition and so practiced as to teach them thoroughly the principles of marching. At the first opportunity after mobilization the men should be hardened to cover long distances without loss.

With new or untrained troops, the process of hardening the men to this work must be gradual. Immediately after being mustered into the service the physical exercises and marching should be begun. Ten-minute periods of vigorous setting-up exercises should be given three times a day to loosen and develop the muscles. One march should be made each day with full equipment, beginning with a distance of 2 or 3 miles and increasing the distance daily as the troops become hardened, until a full day's march under full equipment may be made without exhaustion.

A long march should not be made with untrained troops. If a long distance must be covered in a few days, the first march should be short, the length being increased each succeeding day.

Special attention should be paid to the fitting of shoes and the care of feet. Shoes should not be too wide or too short. Sores and blisters on the feet should be promptly dressed during halts. At the end of the march feet should be bathed and dressed; the socks and, if practicable, the shoes should be changed.

The drinking of water on the march should be avoided. The thirst should be thoroughly quenched before starting on the march and after arrival in camp. On the march the use of water should, in general, be confined to gargling the mouth and throat or to an occasional small drink at most.

Except for urgent reasons, marches should not begin before an hour after daylight, but if the distance to be covered necessitates either breaking camp before daylight or making camp after dark, it is better to do the former.

Night marching should be avoided when possible.

A halt of 15 minutes should be made after the first half or three-quarters of an hour of marching, thereafter a halt of 10 minutes is made in each hour. The number and length of halts may be varied, according to the weather, the condition of the roads, and the equipment carried by the men. When the day's march is long a halt of an hour should be made at noon and the men allowed to eat.

The rate of march is regulated by the commander of the leading company of each regiment, or, if the battalions be separated by greater than normal distances, by the commander of the leading company of each battalion. He should maintain a uniform rate, uninfluenced by the movements of troops or mounted men in front of him.

The position of companies in the battalion and of battalions in the regiment is ordinarily changed daily so that each in turn leads.

The marching efficiency of an organization is judged by the amount of straggling and elongation and the condition of the men at the end of the march.

An officer of each company marches in its rear to prevent undue elongation and straggling.

When necessary for a man to fall out on account of sickness, he should be given a permit to do so. This is presented to the surgeon, who will admit him to the ambulance, have him wait for the trains, or follow and rejoin his company at the first halt.

Special attention should be paid to the rate of march. It is greater for trained than for untrained troops; for small commands than for large ones; for lightly burdened than for heavily burdened troops. It is greater during cool than during hot weather. With trained troops, in commands of a regiment or less, marching over the average roads, the rate should be from 2¾ to 3 miles per hour. With larger commands carrying full equipment, the rate will be from 2 to 2½ miles per hour.

The marching capacity of trained infantry in small commands is from 20 to 25 miles per day. This distance will decrease as the size of the command increases. For a complete division the distance can seldom exceed 12½ miles per day unless the division camps in column.

In large commands the marching capacity of troops is greatly reduced by faulty march orders and poor march discipline.

The march order should contain such instructions as will enable the troops to take their proper places in column promptly. Delay or confusion in doing so should be investigated. On the other hand, organization commanders should be required to time their movements so that the troops will not be formed sooner than necessary.

The halts and starts of the units of a column should be regulated by the watch and be simultaneous.

Closing up during a halt, or changing gait to gain or lose distance should be prohibited.

Quartermaster sergeants, cooks, artificers, and company clerks march with the field train, under command of the supply officer in charge of the train.

(All of the above is from the Infantry Drill Regulations.)

In time of war protection for troops on the march is provided by means of advance guards, flanking parties, and rear guards.

When practicable, marches should begin in the morning[1] after the men have had their breakfast, and the following general rules should be observed:

1. The canteens should be filled before the march begins.

2. The pace at the head of the column must be steady and the column must be kept closed up throughout its length.

3. After the first half or three-quarters of an hour's march, the command should be halted for about fifteen minutes to allow the men to relieve themselves and to adjust their clothing and accoutrements.

---

[1] If considerable distance is to be marched without water, the start should be made late in the afternoon and continued until night and then again early the next morning, halting before the sun gets hot.

563 (contd.)

4. After the first rest, there should be a halt of ten minutes every hour.

Immediately upon halting, the company should be cautioned, "Any man wishing to relieve himself, do so at once"—otherwise some will wait until the halt is nearly over.

5. Indiscriminate rushing for water upon halting should not be allowed—one or more men from every squad should be designated to fill the canteens of the squad.

6. No man should be allowed to leave the ranks without permission of his company commander.

Men allowed to fall out on account of sickness should be given notes to the surgeon. If a man be very sick a noncommissioned officer or reliable private should fall out with him.

7. Whenever a stream is forded or any obstacle passed, the head of the column should be halted a short distance beyond, so as to enable the rest of the column to close up.

8. In crossing shallow streams, the men should be kept closed up and not allowed to pick their way.

9. All men should be made to keep their places in column.

10. A lieutenant or the first sergeant should march in rear of the company to look after stragglers.

11. Nibbling while actually marching should be prohibited.

12. When the troops march for the greater part of the day, a halt of an hour should be made about noon, near wood and water, if practicable.

13. The halt for the night should be made in plenty of time to allow tents to be pitched, supper cooked, etc., before dark.

14. Since marching at the rear of the column is more disagreeable and fatiguing than marching at the front, organizations should take daily turns in leading.

ARTICLES OF WAR

Art. 54. Every officer commanding in quarters, garrison or on the march, shall keep good order, and, to the utmost of his power, redress all abuses or disorders which may be committed by any officer or soldier under his command; and if, upon complaint made to him of officers or soldiers beating or otherwise ill-treating any person, disturbing fairs or markets, or committing any kind of riot, to the disquieting of the citizens of the United States, he refuses or omits to see justice done to the offender, and reparation made to the party injured, so far as part of the offender's pay shall go toward such reparation, he shall be dismissed from the service, or otherwise punished as a court martial may direct.

Art. 55. All officers and soldiers are to behave themselves orderly in quarters and on the march; and whoever commits any waste or spoil, either in walks or trees, parks, warrens, fish ponds, houses, gardens, grain fields, inclosures, or meadows, or maliciously destroys any property whatsoever belonging to inhabitants of the United States (unless by order of a general officer commanding a separate army in the field) shall, besides such penalties as he may be liable to by law, be punished as a court martial may direct.

CHAPTER XII

CARE OF THE HEALTH AND FIRST AID TO THE SICK AND INJURED

CARE OF THE HEALTH *

564. 1. A soldier should endeavor to be always at his best. He should avoid all exposures, not in line of duty, which he knows would be likely to injure his health, for if he is from any cause below par he is liable to break down under influences which otherwise might have had but little effect on him.

2. Even in garrison, in time of peace, soldiers often expose themselves unnecessarily by going out without overcoats when the weather is such as to require their use, or by failing to remove damp socks or other clothing on their return to barracks.

3. At rests on the march one should sit down or lie down if the ground is suitable, for every minute so spent refreshes more than five minutes standing or loitering about.

4. At the midday rest lunch should be eaten, but it should always be a light meal.

5. On the march or during exercise in hot weather the body loses water continuously by the skin and lungs and this loss must be replaced as it occurs to keep the blood in proper condition. Only a few swallows should be taken at a time, no matter how plentiful the water supply may be. When exceedingly thirsty after a long, dry stretch, water should not be taken freely at once, but in smaller drinks at intervals, until the desire for more is removed.

6. Smoking in the heat of the day or on the march is depressing and increases thirst.

7. On hot marches water should be taken quite frequently, but as already stated, in small quantities at a time, to replace the loss by perspiration. This will often prevent attacks of heat exhaustion and sunstroke.

8. On a hurried or forced march, particularly in sultry weather, a soldier may become faint and giddy from the heat and fatigue. His face becomes pale, his lips lead-colored, his skin covered with clammy perspiration, and he trembles all over. His arms and equipments should be removed and his clothing loosened at the neck, while he is helped to the nearest shade to lie down, with his head low, until the ambulance train or wagons come up. Meanwhile, fan him, moisten his forehead and face with water and, if conscious, make him swallow a few sips from time to time.

9. If the soldier comes into camp much exhausted, a cup of hot coffee is the best restorative. When greatly fatigued it is dangerous to eat heartily.

10. When the tents have been arranged for the night and the duties of the day are practically over, the soldier should clean himself and his

---

* From THE SOLDIER'S HANDBOOK, by N Hershler, Chief Clerk, General Staff Corps, U. S. Army.

564 (contd.)

clothes as thoroughly as the means at hand will permit. No opportunity of taking a bath nor of washing socks and underclothing should be lost. In any event the feet should be bathed or mopped with a wet towel every evening to invigorate the skin.

11. In the continued absence of opportunity for bathing it is well to take an air bath and a moist or dry rub before getting into fresh underclothes and, in this case, the soiled clothes should be freely exposed to the sun and air, when the blanket roll is unpacked.

12. By attention to cleanliness of the person and of the clothing, the discomforts of prickly heat, chafing, cracking, blistering, and other irritations of the skin will be avoided. If chafings do occur apply to the surgeon for a healing remedy, for, if neglected, they may fester and cause much trouble.

13. A hearty meal should be eaten when the day's work is over, but the soldier should eat slowly, chewing every mouthful into a smooth pulp before swallowing; and it is good when one can rest a while after his meal. Hard bread and beans when not thoroughly chewed give rise to diarrhoea, one of the most dangerous of camp diseases. Fresh meat should be eaten sparingly when used for the first time after some days on salt rations.

14. The soldier would do well to restrict himself to the company dietary. Particularly should he avoid the articles of food or drink for sale by hawkers and peddlers. Green fruit and overripe fruit are dangerous, as is also fruit to which the individual is unaccustomed. Unpeeled fruit should never be eaten, for it may have been handled by persons suffering from dangerous infectious diseases.

15. It should be unnecessary to speak of the danger from the use of intoxicating liquors, for every soldier knows something of this. The mind of a man under the influence of these liquors is so befogged that he is unable to protect himself from accidents and exposures. How many men have passed from this world because of exposures during intoxication! How many have lost their health and strength and become wretched sufferers during the remainder of a shortened existence! Besides, for days after indulgence in liquor the system is broken down and the individual less able to stand the fatigues, exposures or wounds of the campaign.

16. If filtered or condensed water is not furnished to the troops, and spring water is not to be had, each soldier should fill his canteen over night with weak coffee or tea for the next day's march. This involves boiling, and the boiling destroys all dangerous substances in water. Typhoid fever, cholera, and dysentery are caused by impure water.

17. All the belongings of the soldier should be taken under shelter at night to protect them from rain or heavy dews.

18. When not prevented by the military conditions, soldiers should sleep in their shirts and drawers, removing their shoes, socks, and other clothing.

19. In the morning wash the head, face and neck with cold water. With the hair kept closely cut, this can be done even when the water supply is limited.

20. In hot climates, where marches are made or other military work performed in the early morning or late in the evening, a sleep should be taken after the midday meal to make up for the shortened rest at night. Everyone, to keep in good condition, should have a total of eight hours' sleep in the twenty-four.

21. If the march is not to be resumed, the soldier should take the first opportunity of improving his sleeping accommodations by building a bunk, raised a foot and a half, or more, from the ground. This is of the first importance when the ground is damp. The poncho, or slicker, must be relied upon as a protection in marching camps, but when the camp is to be occupied for some days, bunks should be built.

22. In hot climates this raising of the bunks from the ground lessens the danger from malarial fevers.

23. When malarial fevers are prevalent, hot coffee should be taken in the morning immediately after roll call, and men going on duty at night should have a lunch and coffee before starting.

24. The soldier should never attempt to dose himself with medicine. He should take no drugs except such as are prescribed by the surgeon.

25. No matter how short a time the camp is to be occupied its surface should not be defiled. The sinks should be used by every man, and the regulations concerning their use should be strictly complied with. Waste water and refuse of food should be deposited in pits or other receptacles designed to receive them. Attention to these points will prevent foul odors and flies.

26. When there are foul odors and flies in a camp the spread of typhoid fever, cholera, dysentery, and yellow fever is likely to occur.

27. When any of these diseases are present in a command every care should be taken to have the hands freshly washed at meal times.

28. In the camps of field service the interior of tents should be sunned and aired daily, and efforts should be made by every soldier to have his bunk, arms, equipments, and clothing in as neat and clean condition as if he were in barracks at a permanent station.

29. Harmful exposures are more frequent in hot than in cold weather. Soldiers seek protection against cold, but in seeking shade, coolness, and fresh breezes in hot weather they often expose themselves to danger from diarrhoea, dysentery, pneumonia, rheumatism, and other diseases. A chill is an exciting cause of these affections; it should be avoided as much as possible.

30. When the feet become wet the first opportunity should be taken of putting on dry socks.

31. When the clothing becomes wet in crossing streams or in rain storms there is little danger so long as active exercise is kept up, but there is great danger if one rests in the wet clothing.

32. When the underclothes are wet with perspiration the danger is from chill after the exercise which caused the perspiration is ended. If the soldier can not give himself a towel rub and a change of underclothing, he should put on his blouse and move about until his skin and clothes become dry.

33. To rest or cool off, and particularly to fall asleep, in a cool, shady place in damp clothes is to invite suffering, perhaps permanent disability or death.

34. When an infectious disease is known to be present among the civil population in the neighborhood of a military camp or station, care should be taken by every member of the command to avoid exposure to the infection. Scarlet fever, measles, and diphtheria, are met with in the United States, but in some localities our troops may have to guard against smallpox, yellow fever, cholera, and bubonic plague. The careless or reckless individual will be the first to suffer, but he may not suffer alone; many of his comrades may become affected and die through his fault.

35. Such infections prevail mostly among the lower classes of a community who have no knowledge of the difference between healthful and unhealthful conditions of life. Communication with them should therefore be avoided.

36. The soldier should remember that association with lewd women may disable him for life.

THE CARE OF THE FEET

565. The feet should be kept clean and the nails cut close and square. An excellent preventative against sore feet is to wash them every night in hot (preferably salt) water and then dry thoroughly.

Rubbing the feet with hard soap, grease or oil of any kind before starting on a march is also good.

Sore or blistered feet should be rubbed with tallow from a lighted candle and a little common spirits (whiskey or alcohol in some other form) and the socks put on at once.

Blisters should be perforated and the water let out, but the skin must not be removed.

A little alum in warm water is excellent for tender feet.

Two small squares of zinc oxide plaster, one on top of the other, will prevent the skin of an opened blister from being pulled off. Under no circumstances should a soldier ever start off on a march with a pair of new shoes.

FIRST AID TO THE SICK AND INJURED

566. In operating upon a comrade, the main things are to keep cool, act promptly, and make him feel that you have no doubt that you can pull him through all right. Place him in a comfortable position, and expose the wound. If you cannot otherwise remove the clothing quickly and without hurting him, rip it up the seam. First stop the bleeding, if there is any; then cleanse the wound; then close it, if a cut or torn wound; then apply a sterilized dressing; then bandage it in place.

As for the patient himself, let him never say die. Pluck has carried many a man triumphantly through what seemed the forlornest hope.

As most of the first-aid work in war under present conditions will be done by the individual soldier acting alone and not by a squad of

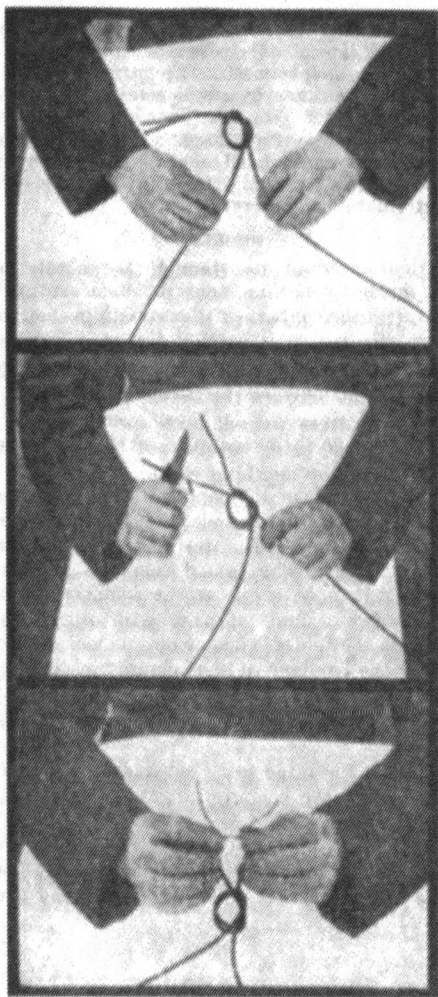

MAKING TEMPORARY SPLICE.

two or more men, it is important that his training should be largely
individual and such as will develop self-reliance and resourcefulness.

The object of any teaching upon first aid, or early assistance of the injured or sick, is not only to enable one person to help another, but in some measure to help himself. The purpose of these directions is to show how this may be done by simple means and by simple methods. It is a mistake to think that you must know many things to be helpful, but you must understand a few things clearly in order to assist the patient in the severest cases until he can be seen by the surgeon or those who are thoroughly trained. In ordinary cases what you can do may often be all that is necessary.

WOUNDS

567. When a ball enters or goes through the muscles or soft parts of the body alone, generally nothing need be done except to protect the wound or wounds with the contents of the first-aid packet, used as follows:

1. If there is one wound, carefully remove the paper from one of the two packages without unfolding compress or bandage and hold by grasping the outside folds between the thumb and fingers.

When ready to dress wound, open compress by pulling on the two side folds of bandage, being careful not to touch the inside of the compress with the fingers or anything else.

Still holding one roll of the bandage in each hand, apply the compress to the wound and wrap the ends of the bandage around the limb or part until near the ends, when the ends may be tied together or fastened with safety pins. The second compress and bandage may be applied over the first or may, if the arm is wounded, be used as a sling.

2. If there are two wounds opposite each other, use one compress opened out—but with the folded bandage on the back—for one wound, and hold it in place by the bandage of the compress used to cover the other wound.

3. If there are two wounds, not opposite each other, apply a compress to each.

4. If the wound is too large to be covered by the compress, find and break the stitch holding the compress together, unfold it, and apply as directed above.

Be careful not to touch the wound with your fingers nor handle it in any way, for the dirt of your hands is harmful and you must disturb a wound as little as possible. Never wash the wound except under the orders of a medical officer.

FIG. 1.—SLEEVE AS SLING.

BANDAGES AND SLINGS

In addition to the slings made with the bandage, two forms of slings furnished by the ordinary clothing are here shown. (Figs. 1 and 2.)

The bandaging will stop all ordinary bleeding. Generally this is all that is necessary for the first treatment, and sometimes it is all that is needed for several days. The importance of the care with which this first dressing is made can not be too seriously insisted upon. It is better to leave a wound undressed than to dress it carelessly or ignorantly, so that the dressing must soon be removed.

FIG. 2.—FLAP OF COAT AS SLING.

BLEEDING FROM WOUNDS

Now and then a wound will bleed very freely, because a large blood vessel has been wounded, and you must know how to stop the bleeding, or hemorrhage as it is called. Remember that all wounds bleed a little, but that as a rule this bleeding will stop in a few minutes if the patient is quiet, and that the firm pressure of the pads and bandage will keep it controlled. Occasionally, but not often, something else must be done.

FIG 3.

Looking upon the heart as a pump, you will understand that to stop the current of blood pumped through the arteries you must press upon the blood vessel between the wound and the heart. This pressure stops the current of blood in the same way that you would stop the flow of water in a leaky rubber hose or tube by pressing upon it between the leak and the pump, or other source of power. The points or places where you can best do this for the different parts of the body are illustrated in the woodcuts. These points are chosen for pressure because the blood vessels which you wish to control there lie over a bone against which effective pressure can be made.

POINTS FOR COMPRESSION WITH THUMB AND FINGERS

The temporal artery is reached by pressure in front of the ear just above where the lower jaw can be felt working in its socket. A branch of this artery crosses the temple on a line from the upper border of the ear to above the eyebrow.

The carotid artery may be compressed by pressing the thumb or fingers deeply into the neck in front of the strongly marked muscle which reaches from behind the ear to the upper part of the breastbone. Fig. 3 shows pressure on the carotid of the left side.

FIG 4.—SUBCLAVIAN RIGHT SIDE

In bleeding from wounds of the shoulder or armpit the subclavian artery may be reached by pressing the thumb deeply into the hollow behind the middle of the collar bone.

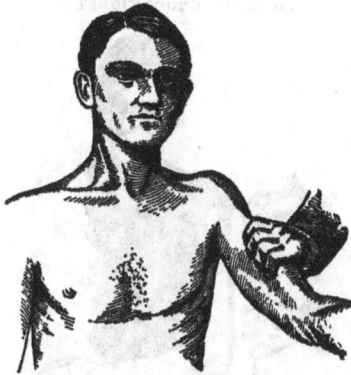

FIG. 5.—BRACHIAL, LEFT SIDE.

In bleeding from any part of the arm or hand the brachial artery should be pressed outwards against the bone just behind the inner border of the larger muscle of the arm.

In bleeding from the thigh, leg, or foot press backward with the thumbs on the femoral artery at the middle of the groin where the artery passes over the bone. The point is a little higher up than that indicated in Fig. 6.

There are two other simple means for helping to stop bleeding—such as elevating or holding an arm or leg upright when those parts are wounded, and by applying cold to the wound; but you will find the compress and

FIG. 6.

bandage, or the pressure made by your fingers, as described to be most useful in the great majority of cases.

IMPROVISED TOURNIQUET

FIG. 7.—COMPRESSION OF RIGHT BRACHIAL.

When, however, the bleeding continues after you have used these simpler means, or your fingers become tired in making the pressure,

[526]

which they may do after ten or fifteen minutes, you will have to use what is called a "tourniquet," and generally will be obliged to improvise one out of material at hand. The principle of such a tourniquet is easily understood—a pad or compress placed on the line of the artery and a strap or band to go over the pad and around the limb so that, when tightened, it will press the padd upon the artery and interrupt the flow of the blood. In the arm apply the tourniquet over the point shown for compression by the fingers; in the thigh, four or

IMPROVISED TOURNIQUET

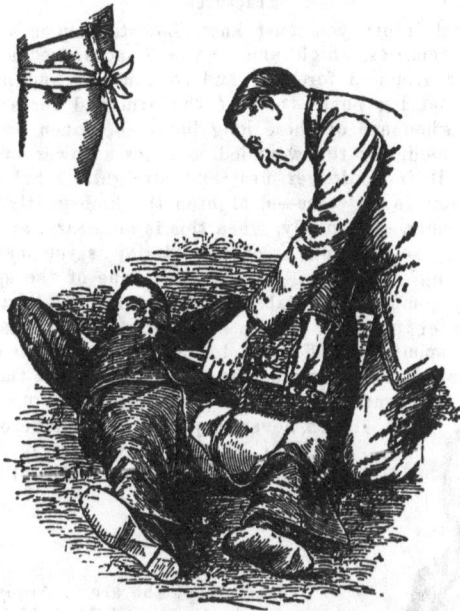

FIG 8 —COMPRESSION OF LEFT FEMORAL.

five inches below the groin, as it can not be applied higher up. (See Fig. 8.)

The pad or compress may be made of such an object as a cork, or smooth round stone wrapped in some material to make it less rough; the bandage folded, a handkerchief, or a cravat being used for the strap. After tying the band closely around the limb any degree of pressure may be made by passing under it a stick, bayonet, or something of that kind, and twisting or turning it around so that the pad is pressed firmly in place. Turn the stick slowly and stop at once when the blood ceases to flow, fixing the stick in place in another bandage. Remember that you may do harm in two ways in using this rough tourniquet. First by bruising the flesh and muscles if you use too much force, and, second, by keeping this pressure up too long and thus strangling the limb. It is a good rule to relax or ease up on this or any other tourniquet at the end of an hour, and allow it to remain loose but in place, if no bleeding appears. By watching you can tighten the tourniquet at any time if necessary.

Fractures

568. The next injury you must know how to help is a broken bone. The lower extremities, thigh and leg, are more frequently wounded than the upper arm and forearm; and so you will find more fractures of the thigh and leg bones than of the arm and forearm. You will usually know when one of these long bones is broken by the way the arm or leg is held, for the wounded man loses power of control over the limb, and it is no longer firm and straight. What you must do is much the same in all cases—straighten the limb gently, pulling upon the end of it firmly, and quietly, when this is necessary, and fix or retain it in position by such splints or other material as you may have. This is called "setting" the bone. If you have none of the splint material supplied, many common materials will do for immediate and temporary use—a shingle or piece of board, a carbine boot, a scabbard, a tin gutter or rain spout cut and fitted to the limb, a bunch of twigs, etc. Whatever material you chose must be well padded upon the side next to the limb, and afterwards secured or bound firmly in place, care being taken never to place the bandage over the fracture, but always above and below. Some of these methods are shown in the following figures:

Fracture of the arm: Apply two splints, one in front, the other behind, if the lower part of the bone is broken; or to the inner and outer sides, if the fracture is in the middle or upper part; support by sling as in Fig. 9.

FIG 9.

Fig. 10.

Fracture of the forearm: Place the forearm across the breast, thumb up, and apply a splint to the outer surface extending to the wrist, and to the inner surface extending to the tips of the fingers; support by slings as in Fig. 10.

Fig 11.

Fracture of the thigh: Apply a long splint, reaching from the armpit to beyond the foot on the outside, and a short splint on the inside (Fig. 11). The military rifle may be used as an outside splint, but its application needs care. A blanket rolled into two rolls, forming a trough for the limb, is useful.

568 (contd.)

FIG. 12.

The carbine boot may be used to advantage in splinting fractures of the thigh and leg, as illustrated in Figs. 12 and 13.

FIG. 13.

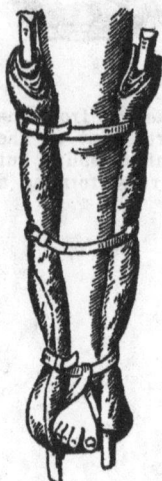

FIG. 14.

Fracture of the leg: Apply two splints, one on the outside, the other on the inside of the limb. When nothing better can be had, support may be given by a roll of clothing and two sticks, as shown in Fig. 14

FIG. 15.—FRACTURE OF LEFT LEG, SUPPORTED BY SOUND LEG

Many surgeons think that the method of fixing the wounded leg to its fellow, and of binding the arm to the body, is the best plan for the field, as the quickest and as serving the immediate purpose.

The object of all this is to prevent, as far as possible, any motion of the broken bone, and so limit the injury to the neighboring muscles, and to lessen the pain.

Be very careful always to handle a broken limb gently. Do not turn or twist it more than is necessary to get it straight, but secure it quickly and firmly in one of the ways shown, and so make the patient comfortable for carriage to the dressing station or hospital. Time. is not to be wasted in complicated dressings.

OTHER WOUNDS

There are, of course, many wounds of the head, face, and of the body, but for the most part you will have little to do with these except to protect the wound itself with the contents of the first-aid packet, or, if bleeding makes it necessary, use in addition several of the packet compresses to control it. As the surface blood vessels of the head and face lie over the bones and close to them, it will generally not be difficult to stop the bleeding by this means or by the pressure with the fingers, as already shown. Remember, as you were told, to make the pressure between the heart and the bleeding point.

With wounds about the body, the chest and abdomen, you must not meddle, except to protect them, when possible without much handling, with the materials of the packet.

CAUTIONS

You have already been warned to be gentle in the treatment of the wounded, and the necessity for not touching the wound must always be in your mind; but there are some other general directions which you will do well to remember:

1. Act quickly but quietly.
2. Make the patient sit down or lie down.
3. See an injury clearly before treating it.
4. Do not remove more clothing than is necessary to examine the injury, and keep the patient warm with covering if needed. Always

[531]

rip, or, if you can not rip, cut the clothing from the injured part, and pull nothing off.

5. Give alcoholic stimulants cautiously and slowly, and only when necessary. Hot drinks will often suffice when obtainable.

6. Keep from the patient all persons not actually needed to help him.

The Diagnosis Tag

569. The diagnosis tag is very important in preventing unnecessary handling of the wounded man and interference with his dressing on the field. When available, it is to be attached by the person who applies the first dressing and is not to be removed until the patient reaches the field hospital. When a patient has a tag on it is to be carefully read before additional treatment is given, and will usually indicate that no further treatment is needed before reaching the hospital.

Poisoned Wounds

570. When a wound is known to be poisoned, such as one infected by the venom of a snake or a rabid animal, the treatment should be directed toward preventing the passage of the poison into the circulation. In snake bites the poison acts quickly; to prevent its absorption a bandage should be carried around the limb between the wound and the heart, tight enough to compress the veins; then gèt the poison out of the wound by laying it open and sucking the poison out (if there is no crack in the mouth or lips) and destroying what is left by cauterization with fire or caustic. Stimulants may be freely given if the heart is weak.

In the bite of a rabid animal the poison is for a long time localized in the wound and there is no danger of immediate absorption. Do not use a tourniquet, but use the other local measures advised for snake bite.

571. Bite of Rabid Animal. The bite of a mad dog, wolf, skunk, or other animal subject to rabies, requires instant and hero'c treatment. Immediately twist a tourniquet very t'ght above the wound, and then cut out the whole wound with a knife, or cauterize it to the bottom with a hot iron; then drink enough whiskey to counteract the shock.

572. Bruises. Ordinary bruises are best treated with cold, wet cloths. Raw, lean meat applied to the part will prevent discoloration. Severe bruises, which are likely to form abscesses, should be covered with cloths wrung out in water as hot as can be borne, to be reheated as it cools; afterwards with hot poultices.

Burns. If clothing sticks to the burn, do not try to remove it, but cut around it and flood it with oil. Prick blisters at both ends with a perfectly clean needle, and remove the water by gentle pressure, being careful not to break the skin. A good application for a burn, including sun burn, is carron oil (equal parts linseed oil and limewater). Druggists supply an ointment known as "solidified carron oil" that is easier to carry. A three per cent solution of carbolic acid, applied with absorbent cotton or a bandage, is an excellent application. Better still is the salve known as ungentine. Lacking these, the next best thing is common baking soda. (Baking soda is the bicarbonate; washing soda, or plain soda, is the carbonate; do not confuse them.) Dissolve in as little water as is required to take it up; saturate a cloth with this and apply. Another good application for burns is the scrapings of a raw potato, renewed when it feels hot. If you have none of these, use any kind of clean oil or unsalted grease, or dust flour over the burn, or use moist earth, preferably clay; then cover with cotton cloth. Do not remove the dead skin until new skin has formed underneath.

573. Burning Clothes, particularly that of females, has been the unnecessary cause of many horrible deaths, either from ignorance of the proper means of extingushing the flames, or from lack of presence of m'nd to apply them. A person whose clothing is blazing should (1) immediately be made to lie down—be thrown if necessary. The tendency of flames is upward, and when the patient is lying down, they not only have less to feed upon, but the danger of their reaching the face, with the possibility of choking and of ultimate deformity, is greatly diminished. (2) The person should then be quickly wrapped up in a coat, shawl, rug, blanket or any similar article, preferably woolen, and never cotton, and the fire completely smothered by pressing and patting upon the burning points from the outside of the envelope.

The flames having been controlled in this way, when the wrap is removed, great care should be taken to have the slightest sign of a blaze immediately and completely stifled. This is best done by pinching it, but water may be used. Any burns and any prostration or shock should be treated in the manner prescribed for them.

It is always dangerous for a woman to attempt to smother the burning clothing of another, on account of the danger to her own clothing. If she attempts it, she should always carefully hold between them the rug in which she is about to wrap the sufferer.

Chigers. Apply sodium hyposulphate ("Hypo"). Bacon is also excellent.

Choking. Foreign Body in the Throat. The common practice of slapping the back often helps the act of coughing to dislodge choking bodies in the pharynx or windpipe.

When this does not succeed, the patient's mouth may be opened and two fingers passed back into the throat to grasp the object. If the effort to grasp the foreign body is not successful, the act will produce vomiting, which may expel it.

A wire, such as a hairpin, may be bent into a loop and passed into the pharynx to catch the foreign body and draw it out. The utmost precautions must be taken neither to harm the throat nor to lose the loop.

In children, and even in adults, the expulsion of the body may be facilitated by lifting a patient up by the heels and slapping his back in this position.

Summon a physician, taking care to send him information as to the character of the accident, so that he may bring with him the instruments needed for removing the obstruction.

574. Colds. Put on warm, dry clothing. Drink freely of hot ginger tea; cover well at night; give dose of quinine every six hours; loosen the bowels.

575. Constipation. Give doses compound cathartic pills, eat freely of preserves; drink often.

576. Convulsions. Give hot baths at once; rub well the lower parts of the body to stimulate; keep water as hot as possible without scalding, then dry and wrap up very warm.

577. Cramps and Chills. Mix pepper and ginger in very hot water and drink. Give dose of cramp tablets.

A hot stone makes a good foot warmer.

578. Diarrhoea. Apply warm bandages to stomach; fire brown a little flour to which two teaspoonfuls of vinegar and one teaspoonful of salt are added; mix and drink. This is a cure, nine cases out of ten. A tablespoonful of warm vinegar and teaspoonful of salt will cure most severe cases. Don't eat fruit. A hot drink of ginger tea is good. Repeat every few hours the above.

579. Dislocations. A dislocation of a finger can generally be reduced by pulling strongly and at the same time pushing the tip of the finger backward.

If a shoulder is thrown out of joint, have the man lie down, place a pad in his arm pit, remove your shoe, and seat yourself by his side, facing him; then put your foot in his armpit, grasp the dislocated arm in both hands, and simultaneously push with your foot, pull on his arm, and swing the arm toward his body till a snap is heard or felt.

For any other dislocation, if you can possibly get a surgeon, do not meddle with the joint, but surround it with flannel cloths, wrung out in hot water, and support with soft pads.

580. Fainting. Lay the patient on his back, with feet higher than his head. Loosen tight clothing, and let him have plenty of fresh air. Sprinkle his face with cold water and rub his arms with it. When consciousness returns, give him a stimulant. For an attack of dizziness bend the head down firmly between the knees.

581. Drowning. Being under water for four or five minutes is generally fatal, but an effort to revive the apparently drowned should always be made unless it is known that the body has been under water for a very long time. The attempt to revive the patient should not be delayed for the purpose of removing his clothes or placing him in the ambulance. Begin the procedure as soon as he is out of the water, on the shore or in the boat. The first and most important thing is to start artificial respiration without delay.

The Schaefer method is preferred because it can be carried out by one person without assistance, and because its procedure is not exhausting to the operator, thus permitting him, if required, to continue it for one or two hours. Where it is known that a person has been under water for but a few minutes, continue the artificial respiration for at least one and a half to two hours before considering the case hopeless. Once that patient has begun to breathe, watch carefully to see that he does not stop again. Should the breathing be very faint, or should he stop breathing, assist him again with artificial respiration. After he starts breathing do not lift him, nor permit him to stand until the breathing has become full and regular.

Fig 16.

As soon as the patient is removed from the water, turn him face to the ground, clasp your hands under his waist and raise the body so any water may drain out of the air passages while the head remains low. (Fig. 16.)

Schaefer Method

The patient is laid on his stomach, arms extended from his body beyond his head, face turned to one side so that the mouth and nose do not touch the ground. This position causes the tongue to fall forward of its own weight and so prevents its falling back into the air passages. Turning the head to one side prevents the face coming into

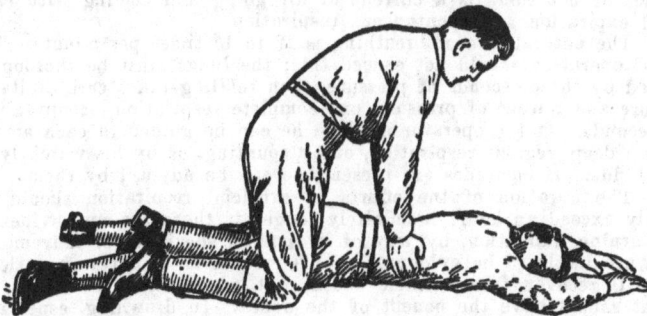

Fig. 17.

contact with mud or water during the operation. This position also facilitates the removal from the mouth of foreign bodies such as to-bacco, chewing gum, false teeth, etc., and favors the expulsion of mucus, blood, vomitus, serum, or any liquid that may be in the air passages. (Fig. 17.)

The operator kneels, straddles one or both of the patient's thighs, and faces his head. Locating the lowest rib, the operator, with his thumbs nearly parallel to his fingers, places his spread hands so that the little finger curls over the twelfth rib. If the hands are on the pelvic bones the object of the work is defeated; hence the bones of the pelvis are first located in order to avoid them. The hand must be free from the pelvis and resting on the lowest rib. By operating on the bare back it is easier to locate the lower ribs and avoid the pelvis. The nearer the ends of the ribs the hands are placed without sliding off, the better. The hands are thus removed from the spine, the fingers being nearly out of sight.

The fingers help some, but the chief pressure is exerted by the heels (thenar and hypothenar eminences) of the hands, with the weight coming straight from the shoulders. It is a waste of energy to bend the arms at the elbows and shove in from the sides, because the muscles of the back are stronger than the muscles of the arms.

The operator's arms are held straight, and his weight is brought from his shoulders by bringing his body and shoulders forward. This weight is gradually increased until at the end of the three seconds of vertical pressure upon the lower ribs of the patient the force is felt to be heavy enough to compress the parts; then the weight is suddenly removed; if there is danger of not returning the hands to the right position again they can remain lightly in place, but it is usually better to remove the hands entirely. If the operator is light, and the patient an overweight adult, he can utilize over 80 per cent of his weight by raising his knees from the ground, and supporting himself entirely on his toes and the heels of his hands, the latter properly placed on the ends of the floating ribs of the patient. In this manner he can work as effectively as a heavy man.

A light feather, or a piece of absorbent cotton drawn out thin and held near the nose by some one, will indicate by its movements whether or not there is a current of air going and coming with each forced expiration and spontaneous inspiration.

The natural rate of breathing is 12 to 15 times per minute. The rate of operation should not exceed this; the lungs must be thoroughly emptied by three seconds of pressure, then refilling takes care of itself. Pressure and release of pressure, one complete respiration, occupy about five seconds. If the operator is alone he can be guided in each act by his own deep, regular respiration, or by counting, or by his watch lying by his side. If comrades are present, he can be advised by them.

The duration of the efforts at artificial respiration should ordinarily exceed an hour; indefinitely longer if there are any evidences of returning animation, by way of breathing, speaking, or movements. There are liable to be evidences of life within 25 minutes in patients who will recover from electric shock, but where there is doubt, the patient should have the benefit of the doubt. In drowning, especially, recoveries are on record after two hours or more of unconsciousness; hence, the Schaefer method, being easy of operation, is more likely to be persisted in.

Aromatic spirits of ammonia may be poured on a handkerchief and held continuously within three inches of the face and nose; if other ammonia preparations are used, they should be diluted or held farther away. Try it on your own nose first.

When the operator is a heavy man it is necessary to caution him not to bring force too violently upon the ribs, as one of them might be broken.

Do not attempt to give liquids of any kind to the patient while unconscious. Apply warm blankets and hot water bottles as soon as they can be obtained.

The Schaefer method of artificial respiration is also applicable in cases of electric shock, asphyxiation by gas, and of failure of respiration following concussion of the brain.

584. Drunkenness. Cold water dashed in the face often proves a most satisfactory awakener.

Cause vomiting by tickling the pharynx with a feather or something of the kind; by administering a tablespoonful of salt or mustard in a cup of warm water. Aromatic spirits of ammonia is very efficient in sobering a drunken man—a teaspoonful in half a cup of water.

A cup of hot coffee after vomiting will aid to settle the stomach and clear the mind.

Lay the subject in a comfortable position, applying hot, dry fomentations, if there is marked coldness.

585. Ear, Foreign Body in. In case of living insect, (a) hold a bright light to the ear. The fascination which a light has for insects will often cause them to leave the ear to go to the light. If this fails, (b) syringe the ear with warm salt and water, or (c) pour in warm oil from a teaspoon, and the intruder will generally be driven out.

If the body be vegetable, or any substance liable to swell, do not syringe the ear, for the fluid will cause it to swell, and soften and render it much more difficult to extract. In a case of this kind, where a bean, a grain of corn, etc., has gotten into the ear, the body may be jerked out by bending the head to the affected side and jumping repeatedly.

If the body is not liable to swell, syringing with tepid water will often wash it out.

If these methods fail, consult a medical man. The presence of a foreign body in the ear will do no immediate harm, and it is quite possible to wait several days, if a surgeon cannot be gotten before.

586. Earache. A piece of cotton sprinkled with pepper and moistened with oil or fat will give almost instant relief. Wash with hot water.

587. Eyes, inflamed. Bind on hot tea leaves or raw fresh meat. Leave on over night. Wash well in morning with warm water.

588. Eye, Foreign Body in. Close the eye for a few moments and allow the tears to accumulate; upon opening it, the body may be washed out by them. Never rub the eye.

If the body lies under the lower lid, make the patient look up, and at the same time press down upon the lid; the inner surface of the lid will be exposed, and the foreign body may be brushed off with the corner of a handkerchief.

If the body lies under the upper lid, (1) grasp the lashes of the upper lid and pull it down over the lower, which should at the same time, with the other hand, be pushed up under the upper. Upon repeating this two or three times, the foreign body will often be brushed out on the lower lid. (2) If this fails, the upper lid should be turned up; make the patient shut his eye and look down; then with a pencil or some similar article press gently upon the lid at about its middle, and grasping the lashes with the other hand, turn the lid up over the pencil, when its inner surface will be seen, and the foreign body may readily be brushed off.

If the body is firmly imbedded in the surface of the eye, a careful attempt may be made to lift it out with the point of a needle. If not at once successful, this should not be persisted in, as the sight may be injured by injudicious efforts.

After the removal of a foreign body from the eye, a sensation as if of its presence often remains. People not infrequently complain of a foreign body when it has already been removed by natural means. Sometimes the body has excited a little irritation, which feels like a foreign body. If this sensation remains over night, the eye needs attention, and a surgeon should be consulted; for it should have passed away if no irritating body is present.

After the removal of an irritating foreign body from the eye, some bland fluid should be poured into it. Milk, thin mucilage of gum arabic, sweet oil, or salad oil are excellent for this purpose.

589. Famishing. Do not let a starved person eat much at a time. Prepare some broth, or a gruel of corn meal or oatmeal thoroughly cooked, and feed but a small spoonful, repeating at intervals of a few minutes. Give very little the first day, or there will be bloating and nausea.

590. Fatigue, excessive. Take a stimulant or hot drink when you get to camp (but not until then), and immediately eat something. Then rest between blankets to avoid catching cold.

591. Fevers. Give doses of quinine tablets; loosen bowels if necessary; keep dry and warm.

592. Freezing. Keep away from heat. To toast frost bitten fingers or toes before the fire would bring chilblains, and thawing out a badly frozen part would probably result in gangrene, making amputation necessary. Rub the frozen part with snow, or with ice cold water, until the natural color of the skin is restored. Then treat as a burn.

Chilblains should be rubbed with whiskey or alum water.

Freezing to Death. At all hazards keep awake. Take a stick and beat each other unmercifully; to restore circulation to frozen limbs rub with snow; when roused again don't stop or fall asleep—it is certain death. Remember this and rouse yourself.

593. Head, How to Keep Cool. By placing wet green leaves inside of hat.

594. Insect Stings. Extract the sting, if left in the wound, and apply a solution of baking soda, or a slice of raw onion, or a paste of clay, mixed with saliva, or a moist quid of tobacco. Ammonia is the common remedy, but oil of sassafras is better. A watch key or other small hollow tube pressed with force over the puncture and held there several minutes will expel a good deal of the poison.

595. Ivy Poison. Relieved with solution of baking soda and water; use freely as a cooling wash. Keep the bowels open.

596. Lightning, Struck by. Dash cold water on body continually; if severe case, add salt to water; continue for hours if necessary. If possible submerge body in running water up to neck.

597. Nose, Foreign Body in. Close the clear side of the nose by pressure with a finger, and make the patient blow the nose hard. This will usually dislodge the object.

If this fails, induce sneezing either by tickling the nose with a feather or something of the kind, or by administering snuff.

The nasal douche, where a syringe or a long rubber tube suitable for a siphon is available, may be used in case the body is not liable to swell, injecting luke warm water into the clear nostril with the expectation that it will push the body out of the other.

If these fail, and the body can be seen clearly, an effort may be made to fish it out by passing a piece of wire, bent into a little hook, back into the nostril close to the wall, and catching the body with it. A hairpin may be bent straight and the hook formed at one end. Do not continue these maneuvers very long nor let them be rough in the slightest degree.

All simple efforts having failed, send for a physician. There is no danger in leaving the foreign body in place for some days if it is impossible to consult a physician in less time.

598. Nosebleed is sometimes uncontrollable by ordinary means. Try lifting the arms above the head and snuffing up alum water or salt water. If this fails, make a plug by rolling up part of a half-inch strip of cloth, leaving one end dangling. Push this plug as far up the nose as it will go, pack the rest of the strip tightly into the nostril, and let the end protrude. If there is leakage backward into the mouth, pack the lower part of plug more tightly. Leave the plug in place several hours; then loosen with warm water or oil, and remove very gently.

599. Ointment for Bruises, Etc. Wash with hot water; then anoint with tallow or candle grease.

600 Piles. Men with piles should take special pains to keep their bowels open and to bathe the parts with cold water.

601. Poisons. In all cases of poisoning there should be no avoidable delay in summoning a physician. The most important thing is that the stomach should be emptied at once. If the patient is able to swallow this may be accomplished by emetics, such as mustard and water, a teaspoonful of mustard to a glass of water, salt and water, powdered ipecac and copious draughts of luke warm water. Vomiting may also be induced by tickling the back of the throat with a feather. When the patient begins to vomit, care should be taken to support the head in order that the vomited matter may be ejected at once, and not swallowed again or drawn into the wind pipe.

602. Poultices. Poultices may be needed not only for bruises but for felons, boils, carbuncles, etc. They are easily made from corn meal or oat meal. Mix by adding a little at a time to boiling water and stirring to a thick paste; then spread on cloth. Renew from time to time as it cools.

To prevent a poultice from sticking, cover the under surface with clean mosquito netting, or smear the bruise with oil. It is a good idea to dust some charcoal over a sore before putting the poultice on. The woods themselves afford plenty of materials for good poultices. Chief of these is slippery elm, the mucilaginous innerbark of which, boiled in water and kneaded into a poultice, is soothing to inflammation and softens the tissues. Good poultices can also be made from the soft rind of tamarack, the rootbark of basswood or cottonwood, and many other trees or plants. Our frontiersmen, like the Indians, often treated wounds by merely applying the chewed fresh leaves of alder, striped maple (moosewood), or sassafras.

603. Salves. Balsam obtained by pricking the little blisters on the bark of balsam firs is a good application for a wound; so is the honey like gum of the liquidambar or sweet gum tree, raw turpentine from any pine tree, and the resin procured by "boxing" (gashing) a cypress or hemlock tree, or by boiling a knot of the wood and skimming off the surface. All of these resins are antiseptics and soothing to a wound.

604. Scalds. Relieve instantly with common baking soda and soaking wet rags—dredge the soda on thick and wrap wet clothes thereon. To dredge with flour is good also.

605. Shock. In case of collapse following an accident, operation, fright, treat first as for fainting. Then rub the limbs with flannel stroking the extremities toward the heart. Apply hot plates, stones, or bottles of hot water, wrapped in towels, to the extremities and over the stomach. Then give hot tea or coffee, or if there is no bleeding, a tablespoonful of whiskey and hot water, repeating three or four times an hour.

606. Skin, protection of, in cold weather. Smear the face, ears and hands with oil or grease. The eyes may be protected from the reflection of the sun on snow by blackening the nose and cheeks.

607. Snake Bite. When a man is bitten he should instantly twist a tourniquet very tightly between the wound and the heart, to keep the poison, as far as possible, from entering the system. Then cut the wound wide open, so it may bleed freely, and suck the wound, if practicable (the poison is harmless if swallowed, but not if it gets into the circulation through an abrasion in the mouth or through a hollow tooth). Loosen the ligature before long to admit fresh blood to the injured part, but tighten it again very soon, and repeat this alternate tightening and loosening for a considerable time. The object is to admit only a little of the poison at a time into the general circulation. Meantime drink whiskey in moderate doses, but at frequent intervals. If a great quantity is guzzled all at once it will do more harm than good. Whiskey is not an antidote; it has no effect at all on the venom; its service is simply as a stimulant for the heart and lungs, thus helping the system to throw off the poison, and as a bracer to the victim's nerves, helping him over the crisis.

608. Snow or Sun Blindness. Smear the nose and face about the eyes with charcoal.

609. Sore Throat. Fat bacon or pork tied on with a dry stocking; keep on until soreness is gone then remove fat and keep covering on a day longer. Tincture of Iron diluted; swab the throat. Gargling with salt and hot water is effective. Listerine, used as a gargle, is also good.

610. Sprains. The regular medical treatment is to plunge a sprained ankle, wrist or finger, into water as hot as can be borne at the start, and to raise the heat gradually thereafter to the limit of endurance. Continue for half an hour, then put the joint in a hot, wet bandage, reheat from time to time, and support the limb in an elevated position, the leg being stretched as high as the hip, or the arm carried in a sling. In a day or two begin gently moving and kneading the joint, and rub with liniment, oil, or vaselin.

Sprains may also be treated by the application of cold water and cloths.

As a soothing application for sprains, bruises, etc., the virtues of witch hazel are well known. A decoction (strong tea) of the bark is easily made, or a poultice can be made from it. The inner bark of kinnikinick, otherwise known as red willow or silky cornel, makes an excellent astringent poultice for sprains. The pain and inflammation of a sprained ankle are much relieved by dipping tobacco leaves in water and binding them around the injured part.

611. Stunning. Concussion of the brain: lay the man on his back, with head somewhat raised. Apply heat as for shock, but keep the head cool with wet cloths. Do not give any stimulant—that would drive blood to the brain, where it is not wanted.

612. Sunstroke. Lay the patient in a cool place, position same as for stunning. If the skin is hot, remove clothing, or at least loosen it. Hold a vessel or hat full of cold water four or five feet above him and pour a stream first on his head, then on his body, and last on his extremities. Continue until consciousness returns. Renew if symptoms recur.

If the skin is cool (a bad sign) apply warmth, and give stimulating drinks.

613. Thirst. Allow the sufferer only a spoonful of water at a time, but at frequent intervals. Bathe him if possible.

To quench thirst. Don't drink too often, better rinse out the mouth often, taking a swallow or two only. A pebble or button kept in the mouth will help quench that dry and parched tongue.

614. Toothache. Warm vinegar and salt. Hold in mouth around tooth until pain ceases, or plug cavity with cotton mixed with pepper and ginger.

NOTE

The only way to learn how to use bandages, slings and splints; how to make tourniquets, and how to handle fractures, is to have someone who thoroughly understands these things show you in person how to do them and then for you to do them yourself. It is, therefore, suggested that, if practicable, such instruction be received from some noncommissioned officer of the Hospital Corps.

CHAPTER XIII

MILITARY COURTESY

615. Its Importance. Some soldiers do not see the necessity for saluting, standing at attention, and other forms of courtesy, because they do not understand their significance—their object. It is a well-known fact that military courtesy is a very important part of the education of the soldier, and there are good reasons for it.

General Orders No. 183, Division of the Philippines, 1901, says: "In all armies the manner in which military courtesies are observed and rendered by officers and soldiers, is the index to the manner in which other duties are performed."

The Army Regulations tells us, "Courtesy among military men is indispensable to discipline; respect to superiors will not be confined to obedience on duty, but will be extended on all occasions."

THE NATURE OF SALUTES AND THEIR ORIGIN

The Civilian Salute

616. When a gentleman raises his hat to a lady he is but continuing a custom that had its beginning in the days of knighthood, when every knight wore his helmet as a protection against foes. However, when coming among friends, especially ladies, the knight would remove his helmet as a mark of confidence and trust in his friends. In those days failure to remove the helmet in the presence of ladies signified distrust and want of confidence—today it signifies impoliteness and a want of good breeding.

The Military Salute

617. From time immemorial subordinates have always uncovered before superiors, and equals have always acknowledged each other's presence by some courtesy—this seems to be one of the natural, nobler instincts of man. It was not so many years ago when a sentinel saluted not only with his gun but by taking off his hat also. However, when complicated headgear like the bearskin and the helmet came into use, they could not be readily removed and the act of removing the hat was finally conventionalized into the present salute—into the movement of the hand to the visor as if the hat were going to be removed.

Every once in a while a man is found who has the mistaken idea that he smothers the American spirit of freedom, that he sacrifices his independence, by saluting his officers. Of course, no one but an anarchist or a man with a small, shrivelled-up mind can have such ideas.

Manly deference to superiors, which in military life is merely recognition of constituted authority, does not imply admission of inferiority any more than respect for law implies cowardice.

The recruit should at once rid himself of the idea that saluting and other forms of military courtesy are un-American. The salute is the soldier's claim from the very highest in the land to instant recognition as a soldier. The raw recruit by his simple act of saluting, commands like honor from the ranking general of the Army—aye, from even the President of the United States.

While the personal element naturally enters into the salute to a certain extent, when a soldier salutes an officer he is really saluting the office rather than the officer personally—the salute is rendered as a mark of respect to the rank, the position that the officer holds, to the authority with which he is vested. A man with the true soldierly instinct never misses an opportunity to salute his officers.

As a matter of fact, military courtesy is just simply an application of common, every-day courtesy and common sense. In common, every-day courtesy no man with the instincts of a gentleman ever thinks about taking advantage of this thing and that thing in order to avoid paying to his fellow-man the ordinary, conventional courtesies of life, and if there is ever any doubt about the matter, he takes no chances but extends the courtesy. And this is just exactly what the man who has the instincts of a real soldier does in the case of military courtesy. The thought of "Should I salute or should I not salute" never enters the mind of a soldier just because he happens to be in a wagon, in a post office, etc.

In all armies of the world, all officers and soldiers are required to salute each other whenever they meet or pass, the subordinate saluting first. The salute on the part of the subordinate is not intended in any way as an act of degradation or a mark of inferiority, but is simply a military courtesy that is as binding on the officer as it is on the private, and just as the enlisted man is required to salute the officer first, so is the officer required to salute his superiors first. It is a bond uniting all in a common profession, marking the fact that above them there is an authority that both recognize and obey—the Country! Indeed, by custom and regulations, it is as obligatory for the ranking general of the Army to return the salute of the recruit, as it is for the latter to give it.

Let it be remembered that the military salute is a form of greeting that belongs exclusively to the Government—to the soldier, the sailor, the marine—it is the mark and prerogative of the military man and he should be proud of having the privilege of using that form of salutation—a form of salutation that marks him as a member of the Profession of Arms—the profession of Napoleon, Wellington, Grant, Lee, Sherman Jackson and scores of others of the greatest and most famous men the world has ever known. The military salute is ours, it is ours only. Moreover, it belongs only to the soldier who is in good standing, the prisoner under guard, for instance, not being allowed to salute. Ours is a grand fraternity of men-at-arms, banded together for national defense, for the maintenance of law and order—we are bound together by the love and respect we bear the flag—we are pledged to loyalty, to one God, one country—our lives are dedicated to the defense of our country's flag—the officer and the private belong to a brotherhood whose regalia is the uniform of the American soldier, and they are known to one another and to all men, by an honored sign and symbol of knighthood that has come down to us from the ages—THE MILITARY SALUTE!

Whom To Salute

618. When covered, all enlisted men within saluting distance and not in ranks, salute all officers (if uncovered, a soldier stands at attention, without saluting).

Soldiers at all times and in all situations pay the same compliments to officers of the Army, Navy, Marine Corps, and volunteers, and to officers of the Organized Militia in uniform as to officers of their own regiment, corps, or arm of service.

The commander of a body of troops salutes all general officers and his regimental, post, battalion, or company commander, by bringing his command to attention and saluting in person. The troops are brought to attention in time to permit the salute to be rendered at the prescribed distance; they are held at attention until after the salute has been acknowledged.

When an officer entitled to the salute passes in rear of a body of troops, it is brought to attention while he is opposite the post of the commander.

The commander of a body of troops salutes in person all other officers senior to him in rank; the troops are not brought to attention except that a noncommissioned officer commanding a detachment less than a company will bring it to attention before saluting an officer.

The commander of a body of troops exchanges salutes with the commanders of other bodies of troops; the troops are brought to attention during the exchange.

An officer commanding a body of troops is saluted by all officers junior to him in rank and by all enlisted men. He acknowledges the salutes in person; the command is not brought to attention.

The commanding officer is saluted by all commissioned officers in command of troops or detachments. Troops under arms will salute as prescribed in drill regulations.

The Manual of Interior Guard Duty requires sentinels to salute foreign naval and military officers, but there are no instructions about other enlisted men saluting them. However, as an act of courtesy they should be saluted the same as our own officers.

Respect to Be Paid the National Air, the Flag and Colors and Standards

619. The National Air. Whenever "The Star Spangled Banner" is played at a military station, or at any place where persons belonging to the military service are present in their official capacity or present unofficially but in uniform, all officers and enlisted men present will stand at attention, facing toward the music, except at retreat, when they will face toward the flag, retaining that position until the last note of

the air, and then salute. With no arms in hand the salute will be the hand salute. The same respect will be observed toward the national air of any other country, when it is played as a compliment to official representatives of such country.

620. **The Flag.** The flag is lowered at the sounding of the last note of the retreat, and while it is being lowered the band plays "The Star Spangled Banner," or, if there is no band present, the field music sounds "to the color."

When "to the color" is sounded by the field music while the flag is being lowered the same respect will be observed as when "The Star Spangled Banner" is played by the band, and in either case officers and enlisted men out of ranks will face toward the flag, stand at attention, and render the prescribed salute at the last note of the music. Flags on flag staffs and other permanent poles are not saluted.

621. **Colors and Standards.** The prescribed salute must always be rendered when passing the national or regimental color or standard *uncased.* Colors and standards that are cased, that is to say, that are in their waterproof case, are not saluted. (Note: By "the prescribed salute" is meant, if unarmed, the "hand salute;" if armed with the rifle, the "rifle salute" (for sentries on post the "present arms"); if armed with a drawn saber, the "present saber;" if wearing a sheathed saber or other side arm, the "hand salute."

By "Colors" and "Standards" is meant the national flags and the regimental flags that are carried by regiments and also by engineer battalions. They may be of either silk or bunting. In the Army Regulations the word "color" is used in referring to regiments of infantry, battalions of engineers and Philippine scouts, and the coast artillery, while "Standard" is used in reference to regiments of cavalry and field artillery.

By "Flag" is meant the national emblem that waves from flag staffs and other stationary poles. They are always of bunting.

When to Salute

622. Soldiers salute officers day and night, and whether either or both are in uniform or civilian dress.

Salutes are not rendered when marching in double time or at the trot or gallop. The soldier must first come to quick time or walk before saluting.

When making or receiving official reports all officers will salute, if covered; if uncovered, they stand at attention. When under arms, the salute is made with the sword or saber, if drawn; otherwise with the hand.

On meeting, all officers salute when covered; when uncovered, they exchange the courtesies observed between gentlemen. Military courtesy requires the junior to salute first, but when the salute is introductory to a a report made at a military ceremony or formation to the representative of a common superior—as, for example, to the adjutant, officer of the day, etc.—the officer making the report, whatever his rank, will salute first; the officer to whom the report is made will acknowledge, by saluting, if covered, or verbally, if uncovered, that he has received and understood the report.

How to Salute

623. For the manner of making the hand salute, see Par. 82; the rifle salute, Par. 124; and the saber salute, Par. 322a.

In saluting, the hand or weapon is held in the position of salute until the salute has been acknowledged or until the officer has passed or has been passed.

On all occasions outdoors, and also in public places, such as stores, theaters, railway and steamboat stations, and the like, the salute to any person whatever by officers and enlisted men in uniform, with no arms in hand, whether on or off duty, shall be the hand salute, the right hand being used, the headdress not to be removed.

When an enlisted man with no arms in hand passes an officer he salutes with the right hand. Officers are saluted whether in uniform or not.

An enlisted man, armed with the saber and out of ranks, salutes all officers with the saber if drawn; otherwise he salutes with the hand. If on foot and armed with a rifle, he makes the rifle salute.

A noncommissioned officer or private in command of a detachment without arms salutes all officers with the hand, but if the detachment be on foot and armed with the rifle, he makes the rifle salute, and if armed with a saber he salutes with it.

Enlisted men out of doors and armed with the rifle, salute with the piece on the right shoulder; if indoors, the rifle salute is rendered at the order or trail.

In approaching or passing each other within saluting distance (about 30 paces), individuals or bodies of troops exchange salutes when at a distance of about 6 paces. If they do not approach each other that closely the salute is exchanged at the point of nearest approach.

If the officer and soldier are approaching each other on the same walk, for instance, the hand is brought up to the headdress when six paces from the officer. If they are on opposite sides of the street, the hand is brought up when about ten paces in advance of the officer. If the officer and soldier are not going in opposite directions and the officer does not approach within six paces, the salute is rendered when the officer reaches the nearest point to the soldier. If a soldier passes an officer from the rear, the hand is raised as he reaches the officer; if an officer passes a soldier from the rear, the soldier salutes just as the officer is about to pass him.

A soldier salutes with the "present arms" only when on post as a sentinel. At all other times when armed with the rifle he salutes with the prescribed rifle salute. If in the open, he renders the "rifle salute;" if indoors, he salutes from the order.

Prisoners do not salute officers. They merely stand at attention. It is customary for paroled prisoners and others who are not under the immediate charge of sentinels, to fold their arms when passing or addressing officers.

Miscellaneous

624. Saluting distance is that within which recognition is easy. In general it does not exceed 30 paces.

"Eyes right" and "present arms" are not executed by troops except in the ceremonies and in saluting the color.

It is very unmilitary to salute with the hand in the pocket, or a cigarette, cigar or pipe in the mouth.

Officers should at all times acknowledge the courtesies of enlisted men by returning, in the manner prescribed, the salutes given, and salutes should be returned smartly and promptly. When several officers in company are saluted, all return it.

Soldiers actually at work do not cease work to salute an officer unless addressed by him.

Before addressing an officer, an enlisted man makes the prescribed salute with the weapon with which he is armed, or, if unarmed and covered, with the right hand. He also makes the same salute after receiving a reply. If uncovered, he stands at attention without saluting.

Indoors, an unarmed enlisted man uncovers and stands at attention upon the approach of an officer. If armed, he salutes as heretofore prescribed.

(According to custom, the term "indoors" is interpreted as meaning military offices, barracks, quarters and similar places—it does not mean such places as stores, storehouses, riding halls, stables, post exchange buildings, hotels, places of amusement, depots and exhibition halls, etc. In such places an unarmed soldier remains either covered, or uncovered according to the custom of the place, and whether or not he salutes depends upon circumstances, the occasion for saluting being determined by common sense and military spirit.

For instance, an enlisted man riding in a street car, or in the act of purchasing goods in a store, or eating in a hotel, would not salute unless addressed by the officer. However, in case of a soldier occupying a seat in a crowded street car, if he recognized a person standing to be an officer, it would be but an act of military courtesy for him to rise, salute and offer the officer his seat.)

When an officer enters a room where there are soldiers, the word **Attention** is given by some one who perceives him, when all rise and remain standing in the position of a soldier until the officer leaves the room. Soldiers at meals do not rise.

An enlisted man, if seated, rises on the approach of an officer, faces toward him, and, if covered, salutes; if uncovered, he stands at attention. Standing, he faces an officer for the same purpose. If the parties remain in the same place or on the same ground, such compliments need not be repeated.

Uncovering is not a form of the prescribed salute, and the hand salute is executed only when covered.

A soldier, if covered, should always salute before addressing an officer. Likewise he should always salute if addressed by an officer. A soldier, if uncovered, always salutes before leaving an officer.

A soldier salutes an officer passing in double time or at a trot or gallop—the question of gait applies to the one who salutes and not the one who is saluted.

When an officer approaches a number of enlisted men out of doors, the word **ATTENTION** should be given by someone who perceives him, when all stand at attention and all salute. It is customary for all to salute at or about the same instant, taking the time from the soldier nearest the officer, and who salutes when the officer is six paces from him.

A soldier riding in a wagon should salute officers that he passes. He would salute without rising. Likewise, a soldier driving a wagon should salute, unless both hands are occupied.

A junior who is mounted dismounts before addressing a senior who is dismounted.

A junior walks or rides on the left of a senior and keeps step with him. A soldier accompanying an officer walks on the officer's left and about one pace to his rear.

The following are the mistakes usually made by soldiers in rendering salutes:—

1. They do not begin the salute soon enough; often they do not raise the hand to the headdress until they are only a pace or two from the officer—the salute should always begin when at least six paces from the officer.

2. They do not turn the head and eyes toward the officer who is saluted—the head and eyes should always be turned toward the officer saluted and kept turned as long as the hand is raised.

3. The hand is not kept to the headdress until the salute is acknowledged by the officer—the hand should always be kept raised until the salute has been acknowledged, or it is evident the officer has not seen the saluter.

5. The salute is often rendered in an indifferent, lax manner—the salute should always be rendered with life, snap and vim; the soldier should always render a salute as if he MEANT IT.

For salutes by the guard and detachments of the guard, see Par. 368a.

No honors are paid by troops when on the march or in trenches, except that they may be called to attention, and no salute is rendered when marching in double time or at the trot or gallop.

CHAPTER XIV

MILITARY DEPORTMENT AND APPEARANCE—PERSONAL
CLEANLINESS—CARE OF CLOTHING AND OTHER EQUIP-
MENT—CARE AND PRESERVATION OF SHOES—FORMS OF
SPEECH—DELIVERY OF MESSAGES, ETC.

625. Military Deportment and Appearance. The enlisted man is no
longer a civilian but a soldier. He is, however, still a citizen of the
United States and by becoming a soldier also he is in no way relieved
of the responsibilities of a citizen; he has merely assumed in addition
thereto the responsibilities of a soldier. For instance, if he should visit
an adjoining town and become drunk and disorderly while in uniform,
not only could he be arrested and tried by the civil authorities, but he
could also be tried by the summary court at his post for conduct to the
prejudice of good order and military discipline. Indeed, his uniform is
in no way whatsover a license for him to do anything contrary to law
and be protected by the government.

Being a soldier, he must conduct himself as such at all times,
that he may be looked upon not only by his superior officers as a soldier,
but also by the public as a man in every way worthy of the uniform of
the American soldier.

Whether on or off duty, he should always look neat and clean,
ever remembering that in bearing and in conversation he should be every
inch a soldier—shoes must be clean and polished at all times; no chewing,
spitting, gazing about, or raising of hands in ranks—he should know his
drill, his orders and his duties—he should always be ready and willing
to learn all he can about his profession—he should never debase himself
with drink.

It should be remembered that the soldiers of a command can
make the uniform carry distinction and respect, or they can make it a
thing to be derided.

The soldier should take pride in his uniform.

A soldier should be soldierly in dress, soldierly in carriage,
soldierly in courtesies.

A civilian owes it to himself to be neat in dress. A soldier owes
it to more than himself—he owes it to his comrades, to his company—
he owes it to his country, for just so far as a soldier is slack so far
does his company suffer; his shabbiness reflects first upon himself then
upon his company and finally upon the entire Army.

It is a fact known to students of human nature that just in
proportion as a man is neatly and trimly dressed is he apt to conduct
himself with like decency. The worst vagabonds in our communities
are the tramps, with their dirty bodies and dirty clothes; the most
brutal deeds in all history were those of the ragged, motley mobs of
Paris in the days of the French revolution; the first act of the mutineer
has ever been to debase and deride his uniform.

It is also a well known fact that laxity in dress and negligence
in military courtesy run hand in hand with laxity and negligence in

almost everything else, and that is why we can always look for certain infallible sysmptoms in the individual dress, carriage and courtesies of soldiers.

Should a soldier give care and attention to his dress?

Yes, sir, not only should a soldier be always neatly dressed, but he should also be properly dressed—that is, he should be dressed as required by regulations. A soldier should always be neat and trim, precise in dress and carriage and punctilious in salute. Under no circumstances should the blouse or overcoat be worn unbuttoned, or the cap back or on the side of the head. His hair should be kept properly trimmed, his face clean shaved or beard trimmed and his shoes polished, his trousers pressed, the belt accurately fitted to the waist so that it does not sag, his leggins cleaned, his brass letters, numbers and crossed rifles polished, and his white gloves immaculate.

Should a man ever be allowed to leave the post on pass if not properly dressed?

No, sir; never. The Army Regulations require that chiefs of squads shall see that such members of their squads as have passes leave the post in proper dress.

Should a soldier ever stand or walk with his hands in his pockets?

No, sir; never. There is nothing more unmilitary than to see a soldier standing or walking with his hands in his pockets.

The real soldier always stands erect. He never slouches.

Is it permissible, while in uniform, to wear picture buttons, chains, watch charms, etc., exposed to view?

No, sir; it is not.

May the campaign hat or any other parts of the uniform be worn with civilian dress?

No, sir; this is prohibited by the uniform order, which especially states that when the civilian dress is worn it will not be accompanied by any mark or part of the uniform.

May a mixed uniform be worn—for example, a khaki coat and olive drab trousers?

No, sir; under no circumstances.

626. Personal Cleanlinesss. Is personal cleanliness a matter of importance?

Yes, and the army regulations require soldiers to bathe frequently. In this company soldiers are required to bathe at least once a week. They are also required to brush their teeth and comb their hair daily. The Army Regulations require that the hair be kept short and the beard neatly trimmed, and that all soiled clothing be kept in the barrack bag. It is also required that in garrison, and whenever practicable in the field, soldiers wash their hands thoroughly after going to the latrines and before each meal, in order to prevent the transmission of typhoid fever and other diseases by germs taken into the mouth with food from unclean hands.

What may be done to a soldier who persists in being filthy?

He may be scrubbed by order of the captain.

Who is immediately responsible for the cleanliness of the soldiers?

According to the Army Regulations, each chief of squad is held responsible for the cleanliness of his men.

627. **Care of Clothing and Other Equipment.** These articles are given the soldier by the government for certain purposes, and he has, therefore, no right to be in any way careless or neglectful of them.

Clothing, Ornaments and Buttons.

Every article of clothing in the hands of an enlisted man should receive as much care as he gives to his person.

Spots should always be removed as soon as possible. Preparations for this purpose can always be obtained from any drug store at small cost.

Turpentine will take out paint. Grease spots can be removed by placing a piece of brown paper, newspaper, or other absorbent paper over the stain, and the pressing with a hot iron.

Chevrons and stripes can be cleaned by moistening a clean woolen rag with gasoline and rubbing the parts and then pressing with a hot iron.

Blue clothing should be thoroughly brushed and pressed once a week—two pressing irons and boards in a company, troop or battery will provide for this.

All gilt ornaments and buttons should be polished once a week—one button stick and brush per squad should be provided for this purpose. "Polishine" is recommended as a suitable polish—although there are many others just as good, but none better. Olive drab clothing should also be pressed weekly. This will stimulate a desire in the men to take better care of their clothing, as a wrinkled or soiled article is thrown around carelessly—while a pressed article is laid away to prevent its wrinkling, thus lasting longer.

Soiled khaki clothing and leggings should be washed by the men—they can generally do it better than the laundry. Khaki and leggings require little or no pressing, if not wrung out before being placed out to dry. Khaki so washed wears about twice as long as when washed by a steam laundry.

The service hat and blue and olive drab cap require nothing but brushing. The cover for the khaki cap should be washed as often as necessary—not oftener, perhaps, than every two weeks and always in cold water and dried on the cap itself.

No article should be worn without first being brushed.

Shirts, underwear, socks, etc., should be laid away neatly. Articles of clothing soiled from wear or from long standing in the locker should be sent to the laundry immediately.

A special suit of clothing should be set aside for inspections, parades, and other ceremonies and the uniform worn at these formations should not be worn in barracks—each man invariably has sufficient old garments for barrack use. A change of clothing after formation will be found to be a great help in preserving clothing The special suit mentioned should be kept well brushed, pressed and neatly folded.

Russet Leather Equipment

628. To preserve the life of russet leather equipments they should be cleaned whenever dirt, grit or dust has collected on them or when they

have become saturated with the sweat of a horse. In cleaning them the parts should first be separated and each part sponged, using a lather of castile soap and warm water. When nearly dry a lather of Crown soap and warm water should be used. If the equipment is cared for frequently this method is sufficient; but if the leather has become hard and dry a little neatsfoot oil should be applied after washing with castile soap. When the oil is dry the equipments should be sponged lightly with Crown soap and water, which will remove the surplus remaining on the surface. If a polish is desired a thin coat of russet leather polish issued by the Ordnance Department should be applied and rubbed briskly with a dry cloth.

Particular care should be taken not to use too much Crown soap or water, as the result will be detrimental to the life of the leather. In no case should leather be dipped in water or be placed in the sun to dry.

Special care should be taken to use as little water as possible and in applying the lather of soap and warm water to have the sponge moistened only.

Camp Equipment

629. The shelter tent half should never be scrubbed with soap and brush—the lye in the soap eats the fiber, thus causing the tent to leak. Rinsing in cold water will accomplish all that is necessary and never render the tent unserviceable.

All articles of equipment, viz: the shelter tent half, haversack, canteen, field belt and suspenders should be neatly marked, with the letter of the company, number of the regiment and company number of man in whose possession the articles are placed and when turned in and reissued this number should become the number of the man to whom they are issued. The soldier is thus inspired to neatness by the fact that his eye falls upon a neatly marked set of equipments and he will give accordingly more care to his equipment.

The pins and pole should be washed in hot water—never scraped—immediately upon return from a march where they have been used.

The mess pan, tin cup, knife, fork and spoon should be sterilized in hot water after each meal in camp and weekly in garrison.

The Care and Preservation of Shoes

630. Shoes should at all times be kept polished, by being so kept they are made more pliable and wear longer.

Shoes must withstand harder service than any other article worn, and more shoes are ruined through neglect than by wear in actual service.

Proper care should be taken in selecting shoes to secure a proper fit, and by giving shoes occasional attention much discomfort and complaint will be avoided.

Selection. A shoe should always have ample length, as the foot will always work forward fully a half size in the shoe when walking, and sufficient allowance for this should be made. More feet are crippled and distorted by shoes that are too short than for any other reason. A shoe should fit snug yet be comfortable over ball and instep, and when first worn should not lace close together over instep Leather always stretches and loosens at instep and can be taken up by lacing.

The foot should always be held firmly but not too tightly in proper position. If shoes are too loose, they allow the foot to slip around, causing the foot to chafe; corns, bunions, and enlarged joints are the result.

Repairs. At the first sign of a break shoes should be repaired, if possible. Always keep the heels in good condition. If the heel is allowed to run down at side, it is bad for the shoe and worse for the foot; it also weakens the ankle and subjects the shoe to an uneven strain, which makes it more liable to give out. Shoes if kept in repair will give double the service and comfort.

Shoe Dressing. The leather must not be permitted to become hard and stiff. If it is impossible to procure a good shoe dressing, neatsfoot oil or tallow are the best substitutes; either will soften the leather and preserve its pliability. Leather requires oil to preserve its pliability, and if not supplied will become brittle, crack, and break easily under strain. Inferior dressings are always harmful, and no dressing should be used which contains acid or varnish. Acid burns leather as it would the skin, and polish containing varnish forms a false skin which soon peels off, spoiling the appearance of the shoe and causing the leather to crack. Paste polish containing turpentine should also be avoided.

Perspiration. Shoes becoming damp from perspiration should be dried naturally by evaporation. It is dangerous to dry leather by artificial heat. Perspiration contains acid which is harmful to leather, and shoes should be dried out as frequently as possible.

Wet Shoes. Wet or damp shoes should be dried with great care. When leather is subjected to heat, a chemical change takes place, although no change in appearance may be noted at the time. Leather when burnt becomes dry and parched and will soon crack through like pasteboard when strained. This applies to leather both in soles and uppers. When dried, the leather should always be treated with dressing to restore its pliability. Many shoes are burned while on the feet without knowledge of the wearer by being placed while wet on the rail of a stove or near a steam pipe. Care should be taken while shoes are being worn never to place the foot where there is danger of their being burned.

Keep Shoes Clean. An occasional application of soap and water will remove the accumulations of old dressing and allow fresh dressing to accomplish its purpose.

Directions for Polishing. Russet calf leather should be treated with great care. Neither acid, lemon juice, nor banana peel should be used for cleaning purposes. Only the best liquid dressing should be used and shoes should not be rubbed while wet.

Black calf shoes should be cleaned frequently and no accumulation of old blacking allowed to remain. An occasional application of neatsfoot oil is beneficial to this leather, and the best calf blacking only should be used to obtain polish.

Liquid Dressing. Care should be taken in using liquid dressing. Apply only a light even coat and allow this to dry into the leather before rubbing with a cloth When sufficiently dry to rub, a fine powdery substance remains on the surface. This, when rubbed with a soft cloth, produces a high polish that lasts a long time and which is quickly renewed by an occasional rubbing. Too much dressing is useless and injurious. (Quartermaster General's Office, June 16, 1889.)

631. Forms of Speech. In speaking to an officer it is not proper for a soldier to say, "You, etc.," but the third person should always be used, as, for example, "Does the captain want his horse this morning?"— do not say, "Do you want your horse this morning?"

In beginning a conversation with an officer, a soldier should use the third person in referring to himself instead of the pronouns "I" and "me." However, when the conversation has commenced, it is perfectly proper, and usual, for the soldier to use the pronouns "I" and "me," but an officer is *always* addressed in the third person and never as "you."

In speaking to an officer, an enlisted man should refer to another enlisted man by proper title, as, "Sergeant Richards," "Corporal Smith," "Private Wilson."

Privates and others should always address noncommissioned officers by their titles. For example, "Sergeant Smith," "Corporal Jones," etc., and not "Smith," "Jones," etc.

When asked his name, a soldier should answer, for instance, "Private Jones, sir"

When given an order or instructions of any kind by an officer, a soldier should always say, "Yes, sir," thus letting the officer know that the soldier understands the order or instructions.

Short, direct answers should be made in the form of, "No, sir," "Yes, sir," "I don't know, sir," "I will try, sir," etc.

After a soldier has finished a thing that he was ordered to do, he should always report to the officer who gave him the order. For example, "The captain's message to Lieutenant Smith has been delivered."

If ordered to report to an officer for any purpose, do not go away without first ascertaining if the officer is through with you, as it often happens he has something else he would like to have you do. After having finished the work given in the beginning, report, for instance, "Sir, is the captain through with me?"

When an officer calls a soldier who is some distance away, the soldier should immediately salute, and say, "Yes, sir," and, if necessary, approach the officer with a quickened step. If the officer is waiting on the soldier, the latter should take up the double time.

MISCELLANEOUS

632. How to Enter an Office. In entering an office a soldier should give two or three knocks at the door (whether it be open or closed); when told to come in, enter, taking off the hat (if unarmed), close the door and remain just inside the door until asked what is wanted; then go within a short distance of the officer, stand at attention, salute and make known your request in as few words as possible. On completion, salute, face toward the door, and go out, being careful to close the door if it was closed when you entered. If it was not closed, leave it open.

633. Complaints to the Captain. Complaints must never be made directly to the captain unless the soldier has the captain's permission to do so, or the first sergeant refuses to have the matter reported. If dissatisfied with his food, clothing, duties, or treatment, the facts should be reported to the first sergeant, with the request, if necessary, to see the captain.

It is also customary for soldiers who wish to speak to the captain about anything to see the first sergeant first, and when speaking to the captain to inform him that he has the first sergeant's permission to do so. Thus: "Private Smith has the first sergeant's permission to speak to the captain," etc.

634. How the Soldier is Paid. As soon as the company is formed in column of files, take off your right hand glove, and fold it around your belt in front of the right hip. When your name is called, answer "Here," step forward and halt directly in front of the paymaster, who will be directly behind the table; salute him. When he spreads out your pay on the table in front of you, count it quickly, take it up with your un-gloved hand, execute a left or right face and leave the room and building, unless you wish to deposit, in which case, you will remain in the hall outside the pay-room, until the company has been paid, when you enter the pay-room. Men wishing to deposit money with the paymaster, will always notify the first sergeant before the company is marched to the pay table.

635. Delivery of Messages. When an enlisted man receives a message, verbal or written, from an officer for delivery, he will in case he does not understand his instructions, ask the officer to repeat them, saying, for instance, "Sir, Private Smith does not understand; will the captain please repeat?" When he has received his instructions, and understands them, he will salute, and say: "Yes, sir," execute an about face, and proceed immediately to the officer for whom the message is intended. He will halt three or four paces directly in front of the officer and if the officer be junior to the officer sending the message, he will say, "Sir, Captain Smith presents his compliments," etc., and then deliver the message, or, "The commanding officer presents his compliments to Lieutenant Smith and would like to see him at headquarters." He will salute immediately before he begins to address the officer and will hold his hand at the position of salute while he says, "Sir, Captain Smith presents his compliments," or "The commanding officer presents his compliments to Lieutenant Smith." If the officer sending the message be much junior to the one receiving it, the soldier will not present his compliments, but will say, for instance, "Sir, Lieutenant Smith directed me to hand this letter to the captain," or "Sir, Lieutenant Smith directed me to say to the captain," etc. As soon as the message has been delivered, the soldier will salute, execute an about-face, and proceed at once to the officer who sent the message, and will similarly report to him, "Sir, the lieutenant's message to Captain Smith has been delivered," and leave.

Before leaving always ascertain whether there is an answer.

636. Appearance as Witness. The uniform is that prescribed, with side arms and gloves. Proceed to the courtroom and remain outside. When you are notified that you are wanted enter the room. Then take off your cap and right hand glove, and raise your right hand above your head, palm to the front, to be sworn. After the judge advocate reads the oath, say, "I do" or "So help me God." Then sit down in the chair indicated by the judge advocate. Do not cross your legs, but sit upright. When asked, "Do you know the accused? If, so, state who he is," answer, "I do; Corporal John Jones, Co. 'B' 1st Infantry." Be sure you thoroughly understand every question before you start to reply, answering them all promptly, in a loud, distinct, deliberate voice, and confining your answers strictly to the questions asked and telling all you know.

CHAPTER XV

THE CARE, DESCRIPTION AND MANAGEMENT OF THE RIFLE

637. As the bore of the rifle is manufactured with great care in order that a high degree of accuracy may be obtained, it should be carefully cared for. What remains from smokeless powder tends to eat and wear away the bore and should, therefore, be removed as soon after firing as practicable.

The proper way of cleaning a rifle is from the breech. For this purpose the barrack cleaning rod should be used.

To clean a rifle use rags, preferably canton flannel, cut them into squares of such size that they may be easily run through the barrel. Remove the bolt from the rifle, place the muzzle on the floor and do not remove it therefrom while the cleaning rod is in the bore. Wrap a rag that has been thoroughly soaked in a saturated solution of soda and water around the point of the cleaning rod, insert it into the bore and work back and forth in the bore. Follow with dry rags until the bore is thoroughly dry, then remove the muzzle from the floor and with a small stick and a new rag, soaked in the same solution, proceed to clean the muzzle end of the bore. This should find the bore free from dirt, rust, etc. Clean again with rags dipped in oil, preferably "3 in 1," dry thoroughly and apply a thin coating of the same oil. Repeat the process of cleaning with oil daily and the bore will at all times be thoroughly clean. Five minutes work a day will accomplish this.

To clean the bolt, dismount it, clean all parts thoroughly with an oily rag, dry, and before assembling lightly oil the firing pin, the barrel of the sleeve, the striker, the well of the bolt and all cams.

The stock and hand guard should receive a light coat of raw linseed oil once a month, or after any wetting from rain, dew, etc.—this should be thoroughly rubbed in with the hand.

The chamber, magazine and other parts require very little care—wiping, drying, brushing and coating with a thin coat of oil, as in the bore, is sufficient to keep these parts clean.

Unless the rifle is to be stored away, or not used for any length of time, the use of cosmic oil should be discouraged—it is thick and sticky, which makes it hard to remove without the use of gasoline or chloroform.

Pomade is valueless in the care of the rifle; pomade is of use only in the burnishing and polishing of brasses and coppers, and even then it is not as good as "Polishine."

Never, under any circumstances, should a recruit be permitted to use emery paper on any part of his rifle—the use of the burnisher likewise should be prohibited.

In the place of emery paper or the burnisher an ordinary rubber eraser will be found very serviceable.

The Bayonet

The bayonet need not be taken apart in order to clean it. With a small stick—small enough to be used inside the cut for the scabbard

637 (contd.)

catch, hook and clearance cut—an oily rag and a rubber eraser, the bayonet can be thoroughly cleaned.

The rawhide cover of the bayonet scabbard should be washed once a month with castile soap and water, then rub a small quantity of leather dressing all over and into the leather, with a brush, sponge or rag; then wipe with a damp rag or sponge. This will remove all dirt and stains. Allow to dry and next apply a light coat of some cream paste. Wait a moment for this to dry, then polish with a clean brush or rag.

The metallic parts require nothing but an occasional wiping off with an oily rag—these parts should then be dried.

Are enlisted men allowed to take their arms apart?

No, not unless they have the permission of a commissioned officer, and even then only under proper supervision and in the manner prescribed in the descriptive pamphlet issued by the ordnance department.

(Except when repairs are needed, the following named parts should never be dismounted by the soldier, and whenever they are taken apart they should be removed only by the artificer, or some one else familiar with the handling of tools and delicate mechanisms: Bolt stop, cut off, safety lock, sleeve lock, front sight, front sight movable stud, lower band, upper band and stacking swivel screws.

Unless the screw driver is handled carefully and with some skill the screws are sure to be injured either at the head or thread.)

Is the polishing of blued and browned parts permitted?

No, and rebluing, rebrowning, putting any portion of an arm in fire, removing a receiver from a barrel, mutilating any part by firing or otherwise, and attempting to beautify or change the finish, are prohibited. However, the prohibition of attempts to beautify or change the finish of arms, is not construed as forbidding the application of raw linseed oil to the wood parts of arms. This oil is considered necessary for the preservation of the wood, and it may be used for such polishing as can be given when rubbing in one or more coats when necessary. The use of raw linseed oil only is allowed for redressing and the application for such purpose of any kind of wax or varnish, including heelball, is strictly prohibited.

Is the use of tompions[1] in small arms permitted?

No, it is prohibited by regulations.

Should pieces be unloaded before being taken to quarters or tents?

Yes, unless it is otherwise ordered. They should also be unloaded as soon as the men using them are relieved from duty.

Should a loaded or unloaded rifle or revolver ever be pointed at anyone in play?

No, sir; under no circumstances whatsover. A soldier should never point a rifle or revolver at a person unless he intends to shoot him.

NOTES

It is easier to prevent than to remove rust.

Oil to be used only to remove rust or after firing or when going out in damp weather. When occasion for its use has passed, it should be carefully wiped off, so as not to collect dust and sand.

[1] Wooden stoppers or plugs that are put in the muzzles of rifles and other arms to keep out dirt and water.

To remove rust, apply oil with rag and let it stand for awhile so as to soften rust—weapon then wiped clean with dry rag. Emery paper should never be used to remove rust.

To prevent dust and rust in bore, a good strong gun string should be frequently used.

All articles of brass to be kept brightly polished.

Never put away arms and equipment before cleaning.

Emery paper, burnisher and sand are used only on sabers, bayonets, mess kits and other bright metal. Under no circumstances should they ever be used on blued or browned metal.

Cosmic oil and emery paper may generally be gotten from the company quartermaster-sergeant. Polishine, burnisher, chamois skin, machine oil ("3-in-1") and button stick must be bought by the soldier.

The **Front Sight, Fig 42,** is secured in its slot in the front sight movable stud by the front sight pin, Fig. 43; the pin is tapering, and its small end is driven in from the right and the ends upset to prevent accidental removal.

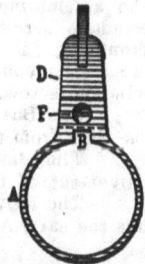

Fig. 43.

Fig. 44.   Fig. 45.

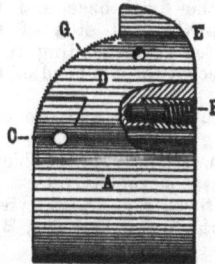

Fig. 42.

The **Front Sight Fixed Stud** and **Front Sight Movable Stud,** with front sight in place, are shown assembled, Fig. 44, right side view, and Fig. 45, front view. The front sight fixed stud, A, has a slot, B, which, bearing, on a lug on the upper side of the barrel, prevents lateral displacement of the stud; and hole, C, for the front sight stud pin, which prevents longitudinal displacement of the stud. The front sight movable stud, D, has the front sight pin hole, E. It is held by the undercut slot in the front sight stud and secured from lateral displacement by the front sight screw, F. The recess for this screw is not drilled in the movable stud until the rifle has been targeted and the correct position of the movable stud determined. The rear face, G, of both the front sight fixed stud and front sight movable stud is serrated to prevent any reflection of light from this surface interfering with the aiming.

The **Rear Sight Fixed Base, Fig. 46,** rear end view, and Fig. 47, right side view, has the holes, A and B, for the base pin and base spline,

Fig. 46.

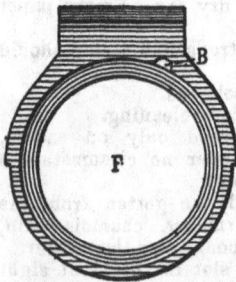

Fig. 47.

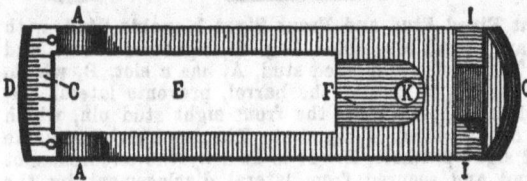

respectively, by which it is firmly secured to the barrel and lateral and longitudinal movement prevented; the undercut, D, for the tenon of the hand guard; the lightening cuts, E; the barrel hole, F; the pivot lug, G, for the movable base; the undercut, H, for the lip on the rear end of the movable base; the undercut, I, for the windage screw and the lip on the front end of the movable base; the lug, J, on the top of which are two zero marks for the wind gauge graduations; and the chamfer, K, the seat for the windage screw collar. This chamfer is carried to the rear to permit of the assembling of the fixed base and the windage screw. On the left side of the front lug the chamfer corresponding to K is merely a conical recess for the head of the windage screw.

The Base Spline locates and prevents the base from turning on the barrel.

The Base Pin, similar to the base spline, prevents longitudinal movement of the base on the barrel.

The Movable Base, Fig. 48, top view, and Fig. 49, right side view, has the ears, A, in which are the holes, B, for the joint pin, which serves

Fig. 48.

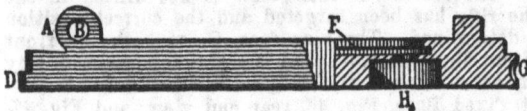

as a hinge for the leaf; the wind gauge graduations, C, each point of which corresponds to a lateral deviation of four inches for each 100 yards; the lip, D, which fits in the undercut in the rear end of the fixed base; the spring opening, E; the

Fig. 49.

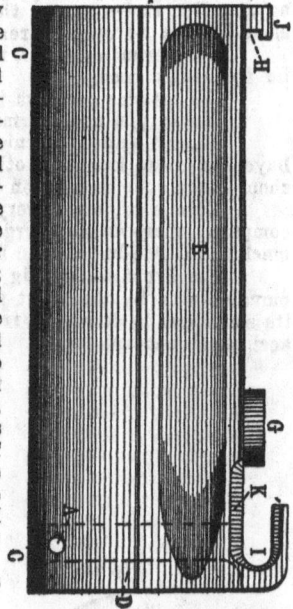

spring seat, F, which is undercut to admit the lip on the front end of the base spring, the lip, G, in which is a worm gear for the engagement of the windage screw; the pivot hole, H, for the pivot lug on the fixed base; and the shoulders, I, on which the front end of the leaf rests when down. The hole, K, is made for convenience in manufacture.

Fig. 50.

Fig. 51.

The Leaf, Fig. 50, top view (when down), and Fig. 51, right side view, has the joint, A. in which is the joint pin hole; the rib, B; the undercut. C, for the drift slide and the sighting notch, D. The free end of the base spring bears against

Fig 52. the lower end of the leaf and maintains it in either its lowered or raised position. The leaf is graduated from 100 to 2,850 yards. The lines extending across one or both branches of the leaf are 100-yard divisions, the longer of the short lines are 50-yard and the shorter lines 25-yard divisions.

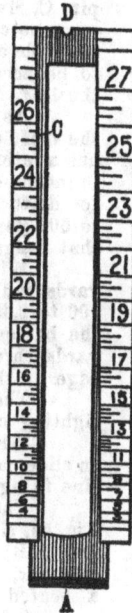

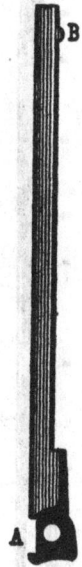

Fig. 52. Fig. 54.

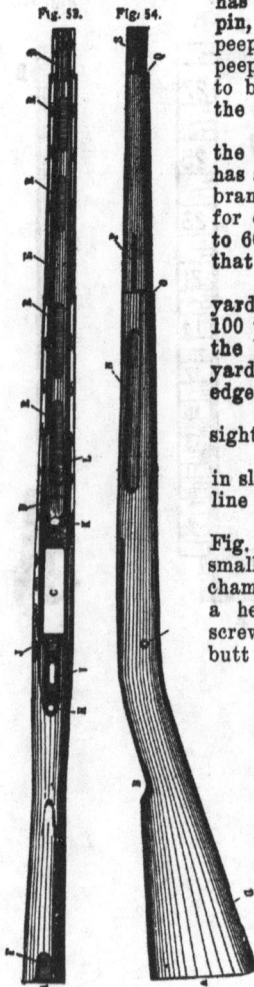

The Drift Slide, Fig. 52, top view (leaf down) has the peephole, A; the field view, B; the drift slide pin, C, riveted to the slide in manufacture; and the peep notches, D. The lines on either side of the peephole and lower peep notch enable the drift slide to be accurately set at any desired graduation on the leaf.

As the slide is moved up or down on the leaf the drift slide moves with it and at the same time has a lateral movement in the undercut between the branches of the leaf, thus automatically correcting for drift. This movement corrects for all drift up to 600 yards, but for only part of the drift beyond that range.

With the leaf up, ranges from 100 to 2,350 yards can be obtained through the peephole; from 100 to 2,450 yards through the lower peep notch at the bottom of field view; and from 1,400 to 2,750 yards through the upper peep notch in the upper edge of the drift slide.

The 2,850-yard range is obtained through the sighting notch in the upper end of the leaf.

With the leaf down and using the open notch in slide cap the sights are set for 530 yards or battle line firing for the down position of the slide.

The stock is shown in Fig. 53, top view, and Fig. 54, right side view. The parts are the butt, A; small, B; magazine well, C; barrel bed, D; air chamber, E, which reduces the charring effect of a heated barrel on the stock; small butt plate screw hole and seat for the butt plate tang, F; butt swivel plate seat, G; mortise for receiver tang lug and hole for rear guard screw, H; mortise for sear and slot for trigger, I; cut-off thumb-piece recess, J; mortise for recoil on receiver, K; bed for fixed base, L; grasping grooves, N; shoulder for lower band, O; bed for band spring, P; shoulder for upper band, Q; channels for decreasing weight, R; upper band screw hole, S; and the stock screw hole, T. The large hole in the butt is for decreasing weight, and the smaller one is a pocket for the combination oiler and thong case.

Fig. 83　　　　Fig. 84　　　637 (contd.)

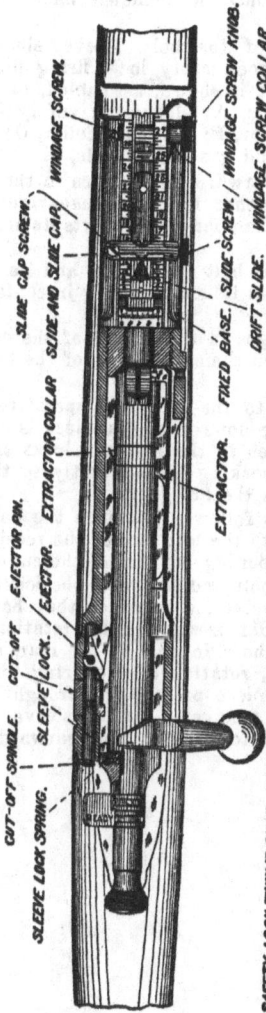

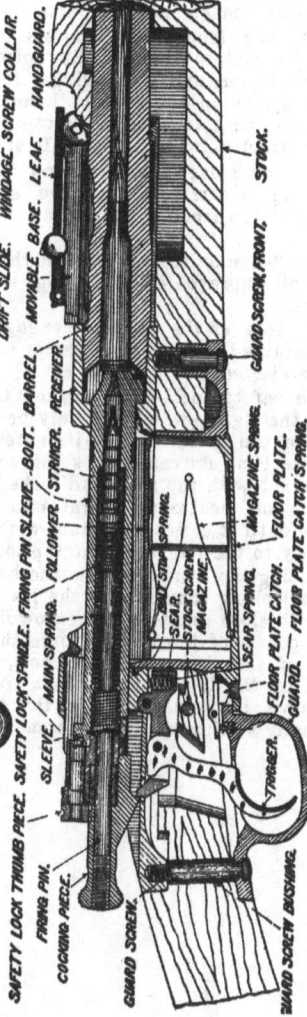

WINDAGE SCREW.

WINDAGE SCREW KNOB.

WINDAGE SCREW COLLAR.

SLIDE CAP SCREW.

SLIDE AND SLIDE CAP.

SLIDE SCREW.

FIXED BASE.

DRIFT SLIDE.

MOVABLE BASE. LEAF. HANDGUARD.

BARREL.

EXTRACTOR COLLAR

EJECTOR PIN.

CUT-OFF.

CUT-OFF SPINDLE.

SLEEVE LOCK SPRING.

SLEEVE LOCK.

EJECTOR.

EXTRACTOR.

RECEIVER.

FIRING PIN SLEEVE. BOLT.

FOLLOWER. STRIKER.

SAFETY LOCK THUMB PIECE. SAFETY LOCK SPINDLE.

SLEEVE. MAIN SPRING.

FIRING PIN.

COCKING PIECE.

SEAR STOP SPRING.

SEAR.

STOCK SCREW.

MAGAZINE.

SEAR SPRING.

FLOOR PLATE CATCH.

GUARD.

TRIGGER.

MAGAZINE SPRING.

GUARD SCREW FRONT.

STOCK.

FLOOR PLATE.

FLOOR PLATE CATCH SPRING.

GUARD SCREW.

GUARD SCREW BUSHING.

THE ASSEMBLED PARTS AND THEIR OPERATIONS

Most of the operating parts may be included under the bolt mechanism and magazine mechanism.

The Bolt Mechanism consists of the bolt, sleeve, sleeve lock, extractor, extractor collar, cocking piece, safety lock, firing pin, firing pin sleeve, striker, and mainspring. It is shown, assembled, in Fig. 82. The parts shown in the cut are handle, A; sleeve, B; safety lock, C; cocking piece, D; safety lug, E; extractor, F; extractor collar, G; locking lugs, H; extractor tongue groove, I, and gas escape hole, J.

The bolt moves backward and forward and rotates in the well of the receiver; it carries a cartridge, either from the magazine, or one placed by hand in front of it, into the chamber and supports its head when fired.

The sleeve unites the parts of the bolt mechanism, and its rotation with the bolt is prevented by the lugs on its sides coming in contact with the receiver.

The hook of the extractor engages in the groove of the cartridge case and retains the head of the latter in the countersink of the bolt until the case is ejected.

The safety lock, when turned to the left, is inoperative; when turned to the right—which can only be done when the piece is cocked—the point of the spindle enters its notch in the bolt and locks the bolt; at the same time its cam forces the cocking piece slightly to the rear, out of contact with the sear, and locks the firing pin.

The bolt mechanism operates as follows: To open the bolt, raise the handle until it comes in contact with the left side of the receiver and pull directly to the rear until the top locking lug strikes the cut-off.

Raising the handle rotates the bolt and separates the locking lugs from their locking shoulders in the receiver, with which they have been brought into close contact by the powder pressure. This rotation causes the cocking cam of the bolt to force the firing pin to the rear, drawing the point of the striker into the bolt, rotation of the firing pin being prevented by the lug on the cocking piece projecting, through the slot in the sleeve, into its groove in the receiver. As the sleeve remains longitudinally stationary with reference to the bolt, this rearward motion

Fig. 82

of the firing pin, and consequently of the striker, will start the compression of the mainspring, since the rear end of the latter bears against the front end of the barrel of the sleeve and its front end against the rear end of the firing pin sleeve.

When the bolt handle strikes the receiver, the locking lugs have been disengaged, the firing pin has been forced to the rear until the sear notch of the cocking piece has passed the sear nose, the cocking piece nose has entered the cock notch in the rear end of the bolt, the sleeve lock has engaged its notch in the bolt, and the mainspring has been almost entirely compressed.

During the rotation of the bolt a rear motion has been imparted to it by its extracting cam coming in contact with the extracting cam of the receiver, so that the cartridge case will be started from the chamber.

The bolt is then drawn directly to the rear, the parts being retained in position by the cocking piece nose remaining in the cock notch and locked by the sleeve lock engaging its notch in the bolt.

To close the bolt, push the handle forward until the extracting cam on the bolt bears against the extracting cam on the receiver, thereby unlocking the sleeve from the bolt, and turn the handle down. As the handle is turned down the cams of the locking lugs bear against the locking shoulders in the receiver, and the bolt is forced slightly forward into its closed position. As all movement of the firing pin is prevented by the sear nose engaging the sear notch of the cocking piece, this forward movement of the bolt completes the compression of the mainspring, seats the cartridge in the chamber, and, in single loading, forces the hook of the extractor into the groove of the cartridge case.

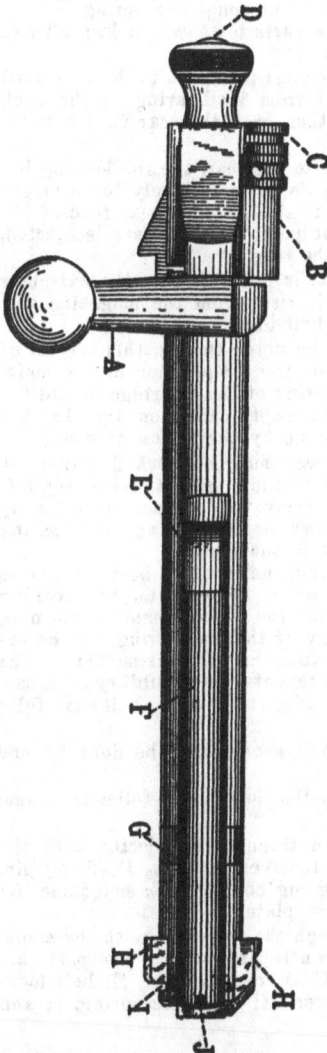

In loading from the magazine the hook of the extractor, rounded at its

[565]

637 (contd.)

lower edge, engages in the groove of the top cartridge as it rises from the magazine under the action of the follower and magazine spring.

The position then occupied by the parts is shown in Fig. 83 and Fig. 84, and the piece is ready to fire.

To pull the trigger, the finger piece must be drawn to the rear until contact with the receiver is transferred from its bearing to the heel, which gives a creep to the trigger, and then until the sear nose is withdrawn from in front of the cocking piece.

Just before the bolt is drawn fully to the rear, the top locking lug strikes the heel of the ejector, throwing its point suddenly to the right in the lug slot. As the bolt moves fully to the rear, the rear face of the cartridge case strikes against the ejector point and the case is ejected, slightly upward and to the right, from the receiver.

Double loading from the magazine is prevented by the extractor engaging the cartridge case as soon as it rises from the magazine and holding its head against the face of the bolt until ejected.

Fig. 85

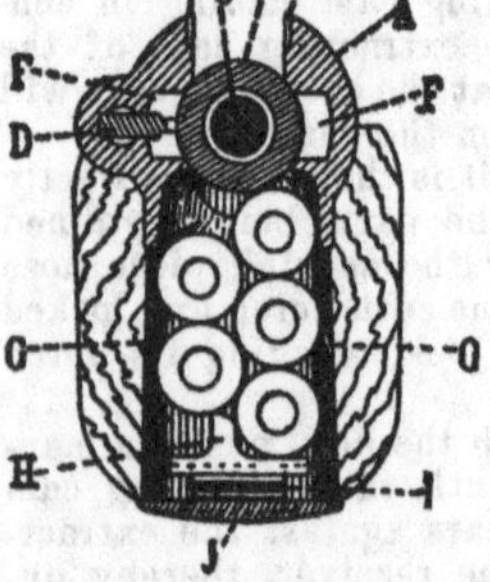

It will be noted that in this system of bolt mechanism the compression of the mainspring, the seating of the cartridge in and the starting of the empty case from the chamber are entirely done by the action of cams.

The piece may be cocked either by raising the bolt handle until it strikes the left side of the receiver and then immediately turning it down or by pulling the cocking piece directly to the rear.

In firing, unless the bolt handle is turned fully down the cam on the cocking piece will strike the cocking cam on the bolt, and the energy of the mainspring will be expended in closing the bolt, instead of on the primer; this prevents the possibility of a cartridge being fired until the bolt is fully closed.

The opening and closing of the bolt should each be done by one continuous motion.

The magazine mechanism includes the floor plate, follower, magazine spring, and cut-off.

Fig. 85 represents a cross section through the ejector with the magazine loaded. The parts shown are receiver, A; bolt, B; firing pin, C; ejector, D; clip slots, E; bolt locking lug channels, F; magazine, G; follower, H; magazine spring, I; and floor plate, J.

Fig. 86 shows a cross section through the magazine with the magazine empty, and with cut-off "on," shown in projection. The parts are receiver, A; bolt, B; firing pin, C; cut-off, D; rear lug slot, E; bolt locking lug channels, F; magazine, G; follower, H; magazine spring, I; and floor plate, J.

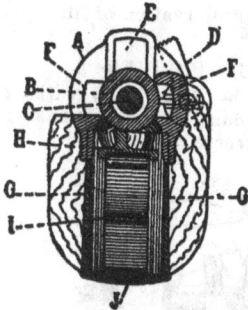

To charge the magazine, see that the cut-off is turned up showing
"on," draw the bolt fully to the rear, insert the cartridge from a clip, or from the hand, and close the bolt. To charge the magazine from a clip, place either end of a loaded clip in its seat in the receiver and, with the thumb of the right hand, press the cartridges down into the magazine until the top cartridge is caught by the right edge of the receiver. The manner in which the cartridges arrange themselves in the magazine and the position of the follower and compressed magazine spring are shown in Fig. 85. The cartridge ramp guides the bullet and cartridge case into the chamber. The magazine can be filled, if partly filled, by inserting cartridges one by one.

Fig. 86

Pushing the bolt forward, after charging the magazine, ejects the clip.

When the cut-off is turned down, the magazine is "off." The bolt can not be drawn fully back, and its front end projecting over the rear end of the upper cartridge holds it down in the magazine below the action of the bolt. The magazine mechanism then remains inoperative, and the arm can be used as a single loader, the cartridges in the magazine being held in reserve. The arm can readily be used as a single-loader with the magazine empty.

When the cut-off is turned up, the magazine is "on;" the bolt can be drawn fully to the rear, permitting the top cartridge to rise high enough to be caught by the bolt in its forward movement. As the bolt is closed this cartridge is pushed forward into the chamber, being held up during its passage by the pressure of those below. The last one in the magazine is held up by the follower, the rib on which directs it into the chamber.

In magazine fire, after the last cartridge has been fired and the bolt drawn fully to the rear, the follower rises and holds the bolt open to show that the magazine is empty.

PRECAUTIONS

If it is desired to carry the piece cocked, with a cartridge in the chamber, the bolt mechanism should be secured by turning the safety lock to the right. Under no circumstances should the firing pin be let down by hand on a cartridge in the chamber.

To obtain positive ejection, and to insure the bolt catching the top cartridge in magazine, when loading from the magazine, the bolt must be drawn fully to the rear in opening it.

When the bolt is closed, or slightly forward, the cut-off may be turned up or down, as desired. When the bolt is in its rearmost position, to pass from loading from the magazine to single loading it is necessary to force the top cartridge or followed below the reach of the bolt, to push the bolt slightly forward and to turn the cut-off down, showing 'off."

637 (contd.)

In case of a misfire it is unsafe to draw back the bolt immediately, as it may be a case of hangfire. In such cases the piece should be cocked by drawing back the cocking piece.

It is essential for the proper working and preservation of all cams that they be kept lubricated.

DISMOUNTING AND ASSEMBLING BY SOLDIER

The bolt and magazine mechanism can be dismounted without removing the stock. The latter should never be done, except for making repairs, and then only by some selected and instructed man.

Fig. 87.

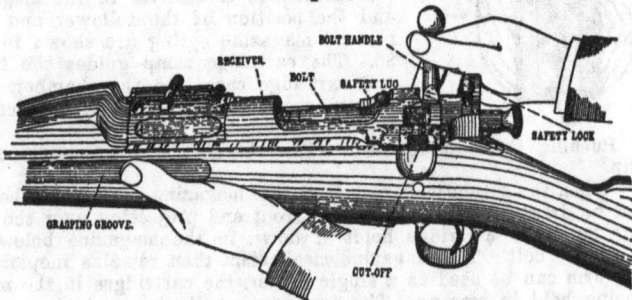

TO DISMOUNT BOLT MECHANISM

Place the cutoff at the center notch; cock the arm and turn the safety lock to a vertical position, raise the bolt handle and draw out the bolt (Fig. 87).

Fig. 88

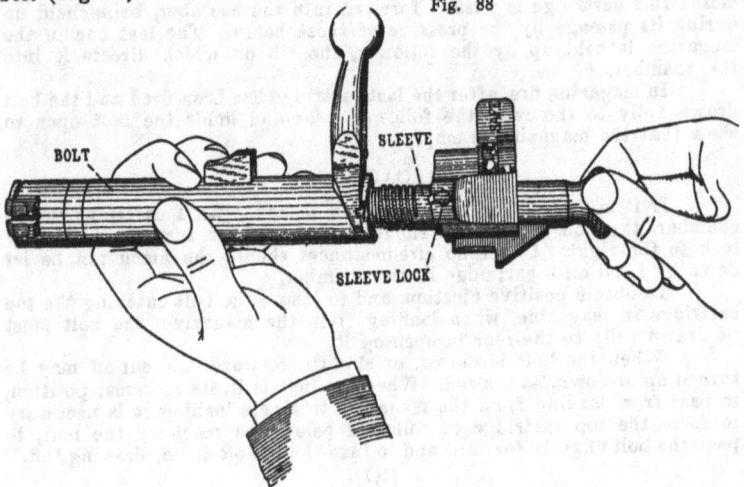

Hold bolt in left hand, press sleeve lock in with thumb of right hand to unlock sleeve from bolt, and unscrew sleeve by turning to the left (Fig. 88).

Hold sleeve between forefinger and thumb of the left hand, draw cocking piece back with middle finger and thumb of right hand, turn safety lock down to the left with the forefinger of the right hand, in order to allow the cocking piece to move forward in sleeve, thus partially relieving the tension of mainspring; with the cocking piece against the breast, draw back the firing pin sleeve with the forefinger and thumb of right hand and hold it in this position (Fig. 89) while removing the striker with the left hand; remove firing pin sleeve and mainspring; pull firing pin out of sleeve; turn the extractor to the right, forcing its tongue out of its groove in the front of the bolt, and force the extractor forward (Fig. 90) and off the bolt.

Fig. 89.

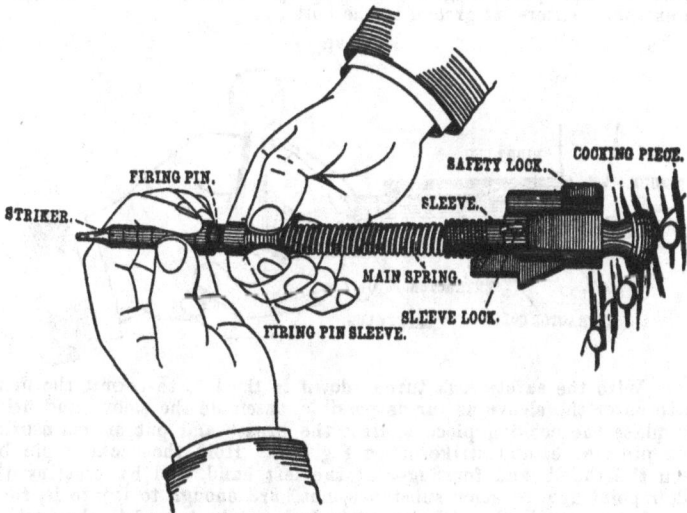

TO ASSEMBLE BOLT MECHANISM

Grasps with the left hand the rear of the bolt, handle up, and turn the extractor collar with the thumb and forefinger of the right hand until its lug is on a line with the safety lug on the bolt; take the extractor in the right hand and insert the lug on the collar in the undercuts in the extractor by pushing the extractor to the rear until its tongue comes in contact with the rim on the face of the bolt (a slight pressure with the left thumb on the top of the rear part of the extractor assists

Fig. 90.

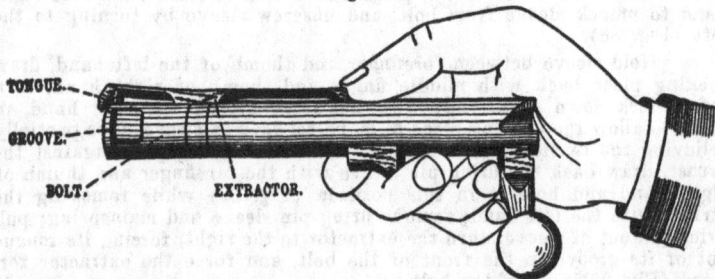

TONGUE.
GROOVE.
BOLT.   EXTRACTOR.

in this operation); turn the extractor to the right until it is over the right lug; take the bolt in the right hand and press the hook of the extractor against the butt plate (Fig. 91) or some rigid object, until the tongue on the extractor enters its groove in the bolt.

Fig. 91.

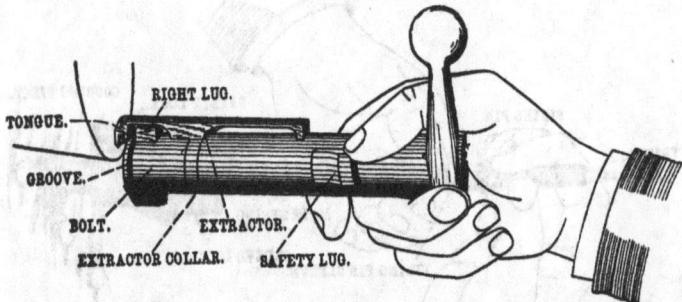

RIGHT LUG.
TONGUE.
GROOVE.
BOLT.   EXTRACTOR.
EXTRACTOR COLLAR.   SAFETY LUG.

With the safety lock turned down to the left to permit the firing pin to enter the sleeve as far as possible, assemble the sleeve and firing pin; place the cocking piece against the breast and put on mainspring firing pin sleeve, and striker (See Fig. 91). Hold the cocking pin between the thumb and forefinger of the left hand, and by pressing the striker point against some substance, not hard enough to injure it, force the cocking piece back until the safety lock can be turned to the vertical position with the right hand; insert the firing pin in the bolt and screw up the sleeve (by turning it to the right) until the sleeve lock enters its notch on the bolt.

See that the cut-off is at the center notch; hold the piece under floor plate in the fingers of the left hand, the thumb extending over the left side of the receiver; take bolt in right hand with safety lock in a vertical position and safety lug up; press rear end of follower down with left thumb and push bolt into the receiver; lower bolt handle; turn safety lock and cut-off down to the left with right hand.

TO DISMOUNT MAGAZINE MECHANISM

With the bullet end of a cartridge press on the floor plate catch (through the hole in the floor plate), at the same time drawing the bullet to the rear; this releases the floor plate.

Raise the rear end of the first limb of the magazine spring high enough to clear the lug on the floor plate and draw it out of its mortise; proceed in the same manner to remove the follower.

To assemble magazine spring and follower to floor plate, reverse operation of dismounting.

Insert the follower and magazine spring in the magazine, place the tenon on the front end of the floor plate in its recess in the magazine, then place the lug on the rear end of the floor plate in its slot in the guard, and press the rear end of the floor plate forward and inward at the same time, forcing the floor plate into its seat in the guard.

PART III*

COMPANY FIELD TRAINING

IN THE ATTACK, THE DEFENSE, THE SERVICE
OF SECURITY, THE SERVICE OF INFORMATION,
NIGHT OPERATIONS, INTRENCHMENTS, OBSTACLES,
FIELD FIRING, CAMPING, AND INDIVIDUAL COOKING.

*Part III is based on ''Company Training'', by General Haking,
British Army, which is the best book the author has ever seen on the
subject of company training.

CHAPTER I

THE COMPANY IN ATTACK

638. Importance of the Attack. Decisive results are obtained only by the offensive. Aggressiveness wins battles. (Par. 121, Field Service Regulations). Indeed, it may be said there is but one way to win battles, and that is by attacking, by going after the other fellow with hammer and tongs. The defense, the service of security, and the service of information are important only because, with an efficient commander, they are merely means that enable him to bring every possible man in the best possible condition, physically and morally, on the field of battle at a vital point, and there attack the enemy with a smashing force and determination that will drive him from the field in defeat. This is really the greatest principle of war, and it applies to squads as well as to armies.

Of course, it is easier to defend, to sit back and wait for developments, but, remember, that such tactics never have won, nor ever will win battles.

Let every officer, noncommissioned officer and private become imbued with the dominating spirit of attack, realizing that the best way to defeat the enemy, is to "go after" him, and to do so with your whole heart and soul—as if you meant it. Strike hard with the utmost speed and force and keep on striking to the limit of human endurance.

A famous general once said a soldier should know three things: First, obedience; second, obedience; third, obedience. This might well be changed to, "First, *attack;* second, *attack;* third, *attack*".

RULES AND PRINCIPLES OF ATTACK

639. Advantages of the Attack. The attackers can choose the point of attack, while the defenders must be prepared to resist at all points. The fact of advancing in spite of the defenders' fire gives the attackers the idea they are succeeding, and on the other hand it gives the defenders the idea that the other fellows are getting the better of them. Another moral advantage is that the attackers leave their dead and wounded behind them as they advance, while the casualities of the defenders usually remain in the trenches and the defenders must undergo the demoralizing ordeal of fighting amongst them.

Superiority of Fire. It is an established fact in modern warfare that it is imposible to shoot an enemy out of an entrenched position— he must be driven out with the bayonet. Now, there is only one way you can get near enough to his position to charge it and drive him out, and that is by keeping down his fire, which you can do only by gaining and maintaining what is called "Superiority of fire". We, therefore, see that "Superiority of Fire" is the key to the situation. Remember, the more effective your fire is, the less effective will that of the enemy be.

640. **Do not Open Fire Until it is Absolutely Necessary**—that is to say, continue to advance without firing as long as you can do so without ruinous losses. Remember, you must husband your ammunition as much as possible for the struggle for superiority of fire that is yet to come, and remember, too, after the attack begins the chances are you will not be able to get any more ammunition that day, except what may be gotten from the dead and wounded, and what is brought by the supports when they come up to reënforce the firing line. It is very demoralizing to the enemy to see you continue your advance on him without even returning his fire.

Make every effort by using cover or inconspicuous formations to arrive, if you possibly can, to within about 800 yards of the enemy before opening fire.

Direction of Advance. Get the direction in which you are to advance well fixed in your mind and, when operating with other companies, do not change it; for, if you do, you will interfere with the companies on your right or left. You may be tempted to change the direction of your advance so as to take advantage of a more covered approach, but, with other companies on your right and left, this will result in confusion.

Deployment. Do not deploy until it is necessary to do so. If the cover will enable you to do so, do not deploy until you get within effective rifle range of the enemy. Deployed troops are unwieldy and difficult to handle as compared with those in close order.

Who Indicates the Point or Time for Opening Fire. The major should indicate the point or time for opening fire. He may do this in his order for deployment or he may follow the firing line close enough to give the order at the proper time. If it be impracticable for him to do so, the senior officer of the battalion with the firing line selects the time for opening fire.

Assignment of the Objective. At the beginning of the attack the major assigns the objective. Unless a particular target has been assigned the company, it takes as its target that part of the general objective which lies in its front.

Protection of Flanks. Flanks must always be protected, and their protection is the duty of the commanders of all flank units down to the lowest, whether specifically enjoined in orders or not.

Close with the Enemy as soon as Possible; the longer you delay doing so, the longer will you be under his fire and consequently the greater will be your casualties.

641. Size of the Fractions that Rush. In the rushes, make the advancing fractions as large as the hostile fire and the necessity for maintaining superiority of fire, will permit. Remember, the smaller the rushing fractions are, the slower will the advance be—the longer will you be exposed to the enemy's fire and consequently the greater will your casualties be. The size of the rushing fraction will, of course, depend upon the cover available, the volume and accuracy of the hostile fire and other circumstances. It may sometimes be advisable to begin the rushes with a company and then change to half a company or platoon, and finally to a squad or file. No opportunity should be lost to increase the size of the rushing fraction.

642. Rushes to be Made Under Covering Fire. Every rush must be made under the covering fire of the adjoining fraction or fractions. As a fraction is about to rush forward the adjoining fraction or fractions must increase the rate of fire; for, when the enemy sees a fraction rushing forward he will very likely increase his fire, and we must keep it down as much as possible.

The commander of the fraction that is about to rush should not start until the remainder of the line is delivering a vigorous fire, and if necessary, he should, in case of delay, call out to the commanders of the adjoining fraction or fractions to increase their fire.

Length of Rushes. The length of the rush generally varies from 30 to 80 yards, depending upon the existence of cover, positions for firing, and the volume and accuracy of the hostile fire.

Companies to be Kept Closed on Their Centers. In order to facilitate control by the company commander, also to provide intervals on the firing line in which roënforcements may be placed, the company must be kept closed in on its center as it becomes depleted by casualties; for, squads and other units coming up from the support should take their place on the firing line in their entirety and should not be divided up and mingled with the individuals of the line.

Fixing Bayonets for the Assault. The major or senior officer in the firing line decides when bayonets shall be fixed, and gives the proper order or signal, which is repeated by all parts of the firing line. Bayonets are generally fixed before or during the last, or second last advance preceding the charge.

The Charge. The firing line having reached the position from which the charge is to be made, the major causes the ''Charge'' to be sounded, and the signal is repeated by all the musicians. The company officers lead the charge, and the skirmishers spring forward shouting, running with bayonets at charge, and closing with the enemy.

The support fixes bayonets when the firing line does.

After the Charge. The conduct of the charging troops after the charge will depend upon circumstances; they may halt and engage in bayonet combat or in pursuing fire; they may advance a short distance to obtain a field of fire, or to drive the enemy from the vicinity; they may assemble, or they may reorganize to repel a counter-attack by the enemy.

Plan for Illustrating the Application of the Rules and Principles of Attack. The application of the rules and principles of attack will be shown by taking the company through the following types of attack, which constantly occur in war:

1st. The company forming a part of a larger force that is attacking an enemy occupying a defensive position, the attacking force being compelled to advance for a considerable distance exposed only to hostile artillery fire, and subsequently to both artillery and infantry fire before the assault can be delivered. This is what may be called the stereotyped form of attack.

2nd. Same as above, excepting that the company is exposed to both artillery and infantry fire from the beginning. ("Meeting engagement".)

3rd. The advance guard attack.

4th. The company, originally on the defense, goes out and attacks a force that has been attacking it.

643. The Five Stages of the Attack. The following diagram shows the five stages of the stereotyped form of attack. This simple outline of attack should be well fixed in the mind of every man in the company:—

FIRST STAGE

(ADVANCE OF THE COMPANY UNDER ARTILLERY FIRE, BUT NOT YET NEAR ENOUGH TO THE ENEMY TO BE SUBJECTED TO HIS INFANTRY FIRE).

Leading Features:

Formations to be adopted; use of cover; pace in advancing; selection of halting-places; artillery action.

SECOND STAGE

(ADVANCE OF THE COMPANY UNDER BOTH ARTILLERY AND LONG-RANGE INFANTRY FIRE).

Leading Features:

The selection of fire-positions; the use of covering fire.

THIRD STAGE

(STRUGGLE FOR SUPERIORITY OF FIRE).

Leading Features:

Fire direction; fire control; fire discipline.

FOURTH STAGE

(ADVANCE AFTER SUPERIORITY OF FIRE HAS BEEN GAINED TO A POSITION CLOSE ENOUGH TO CHARGE THE ENEMY).

Leading Features:

Maintenance of superiority of fire; rapid advance.

FIFTH STAGE

(THE CHARGE).

Leading Features:

A vigorous, simultaneous rush with a heavy line.

When does one stage of the attack end and the other begin?

Of course, the lines of demarcation between the different stages of the attack are not sharp and well-defined, like the lines on a tennis-court, for instance, but the different stages gradually blend into one another. However, each stage has its own characteristics, its own "ear-marks", and there will be no trouble in recognizing them.

FOR EXAMPLE—

First Stage. As long as we are subjected to only artillery fire we are in the first stage.

643 (contd.)

Second Stage. When the first rifle bullets begin to whiz through the air, we then know the first stage is ending and the second commencing.

Third Stage. When we commence to see that, because of the enemy's infantry fire, we are beginning to advance more slowly, we then know that the second stage is ending and the third is beginning. And when we find that it is impossible to advance any further unless we can reduce the enemy's infantry fire, we know that the third stage has been reached, and that the struggle for superiority of fire is on.

Fourth Stage. When the enemy's infantry fire begins to subside and we find that we are able to resume the advance, we then know that we are gaining superiority of fire—that is to say, the fourth stage has been reached, and we must now advance to a fire position close enough to the enemy to enable us to charge from it.

Fifth Stage. The charge.

CHAPTER II
THE COMPANY IN ATTACK
THE FIRST STAGE OF THE ATTACK

THE ADVANCE OF THE COMPANY UNDER HOSTILE ARTILLERY FIRE, BUT NOT YET
NEAR ENOUGH TO THE ENEMY TO BE SUBJECTED TO HIS INFANTRY FIRE

644. The Situation. Let us suppose the company is assembled under cover of some hill or wood from four to five thousand yards from the enemy's position, which we are going to attack and that as soon as the company leaves this cover and begins to advance, it will be fired upon by the hostile artillery, but not by the infantry. Let us suppose, further, that we are taking part in a big battle, and consequently have other companies on our right and left and also supports and reserves in rear.

Importance of the Attack. Impress upon the company the importance of the attack, (see Par. 638.)

The Object of this Stage of the Attack. What are we trying to do when the company commences the attack? We are going to try to smash the enemy—to attack him so quickly and vigorously that we will drive him from his position in confusion and disorder. The enemy is entrenched, and experience has shown that you can't shoot an enemy out of trenches—you've got to get close in on him and drive him out with the bayonet, but before we can do this we will have a long way to go. In the beginning of our advance, which we will call the first stage of the attack, we will probably be subjected only to artillery fire, and will be too far away to use our rifles with effect. Our immediate object then, is to get over ground in our immediate front as rapidly as possible and without losing any more men than we can help.

Before the company goes into an action always tell the men:

1st. The point of attack.

2nd. The general situation—whether there are any of our troops to the front, on our flank or rear.

3rd. What part the company is to take in the fight—whether to form part of the firing line, the support or the reserve.

If the men know these things, they will be able to act more intelligently, especially in case of confusion on separation from the rest of the company.

HOW TO ACCOMPLISH OUR OBJECT

The Pace. We will suppose that there are large patches of open ground in front of us, which the enemy can see from his position, and which he can fire upon effectively with his artillery. We can not avoid these open patches, because we are taking part in an extensive attack, with other companies on our right and left, and if each company changed its original direction of advance and sought a more covered approach, confusion and disorder would result.

Let us consider the ground. Some of it will be open and exposed to the enemy's artillery fire, and some of it will be hidden from his view and merely exposed to badly-aimed artillery fire, or what is called "searching fire", but which is rarely effective. The result is that during this stage of the attack the company will be advancing a part of the time over open and exposed ground and a part of the time over ground that is hidden from the enemy's view and fire. Of course, we want to advance as rapidly as possible; for, apart from the advantage of rapidity and vigor in the attack, which alarms the enemy almost as much as our bullets and shells, the less time it takes us to cover the ground in front of us, the less time will we be under the enemy's fire. But there are two things that will hinder the rapidity of our advance: first, we must not exhaust the men at this early stage of the fight by too rapid an advance, because if we do they will not be in the best condition to continue the fight when they are close enough to the enemy to use their rifles. Secondly, the enemy's artillery is apt to stop us, because if, when we are crossing an exposed area the sky above us suddenly becomes full of bursting shells, it is the natural inclination of the ordinary human being to lie down and wait until the fire slackens. This means loss of time, and it is also just what the enemy's gunners want, as it gives them a stationary target to shoot at. As we all know, it is much easier, especially with artillery, to shoot at a stationary target than at a moving one. We, therefore, see that the company must—

1st. Halt occasionally to rest the men, and select, when possible, halting places which are not exposed to the enemy's artillery fire.

2nd. Advance as rapidly as possible over places that are exposed to fire, and resist the temptation to lie down, if suddenly it comes under a burst of artillery fire.

From this it follows that the company must move at a run over exposed ground, and walk or halt, if necessary, in order to rest the men when reaching covered ground.

Use of the Weapon. As we are at this stage of the attack too far away from the enemy's position to use our rifles with effect, to do any firing now would merely mean a waste of ammunition, every round of which, as you will see, we will need later on.

Formations. We must select a formation which is the most difficult for the enemy's artillery to hit, and which at the same time will permit the immediate use of the rifle, should the situation change and the enemy suddenly appear within rifle range.

Our Drill Regulations describe the three following formations, any one of which, depending upon circumstances, may be used during this stage of the attack:—

1. *Platoon Columns* (see Pars. 221-226), which are used when, due either to difficult ground or limited cover, there are only a few favorable routes of advance; no two platoons should march within the burst of a single sharpnel (ordinarily about 20 yards wide). Aside from the advantage of enabling the whole company to use the few favorable routes, this formation also enables the captain to maintain control over the company.

2. *Squad Columns* (see Pars. 221, 223, 225, 226), which are of value principally in facilitating the advance over rough or brush-broken ground. They afford no material advantage in securing cover. While the captain's control over the company in squad columns is somewhat less than in platoon column, still it is greater than when the company is deployed as skirmishers.

3. *A succession of thin lines*, (see Par. 227), which is used in crossing a wide stretch swept by artillery fire, or by a heavy, long-range rifle fire which can not be profitably returned. This method results in a serious (though temporary) loss of control over the company. It also takes up time. Its advantage lies in the fact that it offers a less definite target.

The two disadvantages of the skirmish line (see Par. 215) are—

1st. It offers to the hostile artillery a target that is a continuous, straight line, and consequently simplifies the question of range.

2nd. It results in a loss of control over the company. It should not, therefore, as a rule, be used until we are ready or about ready to open fire.

645. The Hostile Artillery. This, the first stage of attack, may be called the "Artillery Stage". Infantrymen should remember that the effect of artillery fire is moral rather than physical. Comparatively few of the casualties in a battle are caused by the artillery—the man who is really to be feared is the "dough boy"— he's the one who does the deadly work.

A shell contains just enough powder to burst the case and not sufficient to send the bullets flying in all directions with a velocity sufficient to damage seriously any one standing at a distance.

The velocity of the bullets in the shell depends upon the rate the shell is flying through the air at the moment it explodes, and the bullets soon lose their velocity. Should a shell, for instance, burst 200 yards in front of a soldier and one of the bullets should hit him, it would do him no serious injury, because it has not sufficient velocity left to penetrate his clothes and skin. A soldier is, therefore, practically safe 200 yards away from a shell that bursts directly in front of him. We see from this that a shell that bursts on the ground produces little or no damage—the shell must not only burst in the air, but it must burst in just exactly the right place. This is accomplished by putting a time-fuse in the shell. A mistake of only one second in the cutting of the time fuse will, at 3500 yards, make a difference of about 120 yards. The rate that the shell travels is affected materially by the density of the air. It usually takes an artilleryman from five to ten minutes to find the range—so, we need expect no serious damage, except from a chance shot, for five or ten minutes after the artillery opens up on us, unless, of course, the gunners have gotten the range before hand.

Localities to be Avoided. The enemy's artillery generally picks out certain localities where it appears likely a target will appear, and ascertains the range and determines the proper fuse cutting beforehand, so that they can open at once a heavy and effective fire on any troops that may appear in those localities. Clumps of trees, edges of woods, exits from defiles, and approaches to bridges, are amongst the most common of such localities. Any object that is fairly isolated and stands up well, such as a building, a hay stack, etc., is an execellent object for the artillery to range upon, and if they have not ascertained the range and cut the fuse beforehand, they can do so very rapidly with such range marks. Such localities should, therefore, be avoided, if possible, but if, as is usually the case, this is out of the question, then they should be passed as rapidly as possible.

646. Our Artillery. Of course, during this stage of the attack our own artillery is helping us to advance by keeping down the hostile artillery as much as possible. Our artillery has these advantages over the hostile artillery:—

1st. It has only one target (the hostile artillery) to fire at, while the hostile artillery has two (our artillery and our advancing infantry).

2nd. Our artillery has a stationary target, while the hostile artillery has a moving target.

3rd. Our artillery can disperse their guns and concentrate their fire much more than the enemy, since the latter are tied down to practically the position selected for defense, whereas we have the whole country to the front and flanks in which we can place our guns.

Conclusion. We, therefore, see that during this stage of the attack we are opposed only by the hostile artillery, over which, gun for gun, our own artillery possesses certain advantages, and consequently the enemy's artillery will probably be kept pretty busy looking after the attacking artillery.

Our object must be to cross any area exposed to the enemy's artillery as rapidly as possible, halting in places that afford cover, and taking advantage of any temporary cessation of artillery fire to rush over exceptionally exposed ground, thus avoiding a number of casualties, and thereby adding to the confidence of our men, whom we shall keep fresh, and thus bring them up to the second stage of the attack in excellent condition for the decisive fighting which is now approaching.

CHAPTER III
THE COMPANY IN ATTACK
THE SECOND STAGE OF THE ATTACK

THE ADVANCE OF THE COMPANY UNDER ARTILLERY AND LONG-RANGE INFANTRY FIRE

647. Beginning of the Stage. The beginning of the second stage of the attack is marked by the sound of the enemy's rifle bullets and the desire of our own men to return the fire.

648. Ammunition. However, everything possible must be done to have the company advance as far as we can without returning the enemy's fire; for, our fire would at this range be ineffective and would merely be a waste of ammunition, every round of which we will, as you will see, need later on. One of the few advantages of the defense is that the trenches can be filled with ammunition before the fight begins, or ammunition in unlimited quantity can be brought up during the battle. The defense can, therefore, afford to waste a certain amount of ammunition in long-range firing, but it is quite different with us whose ammunition supply is practically limited to what we have taken into the fight on our persons. However, we have a compensating moral advantage of the enemy's seeing that his fire is so ineffective that we do not even condescend to return it, and that we are coming right after him. Another moral advantage the attacker has over the defender, is that it's much more trying to remain in one place while the other fellow is hitting than it is to be hitting. This is human nature. As the fight progresses the company commander must keep himself informed as to the condition of the supply of ammunition. "Company commanders are responsible that the belts of the men in their companies are kept filled at all times, except when the ammunition is being expended in action. In the fire line the ammunition of the dead and wounded should be secured whenever practicable." (Par. 550, Infantry Drill Regulations).

"Ammunition in the bandoleers will ordinarily be expended first. *Thirty rounds in the right pocket section of the belt will be held as a reserve, to be expended only when ordered by an officer*". (Par. 551, Infantry Drill Regulations.)

"Men will never be sent back from the firing line for ammunition. Men sent forward with ammunition remain with the firing line". (Par. 552, Infantry Drill Regulations.)

According to the Field Service Regulations each man armed with a rifle carries 220 rounds of ammunition into battle—100 rounds in his belt and two bandoleers containing 60 rounds each. The bandoleers are distributed from the battalion combat wagon just before the troops go into action.

Deployment. The first sound of the enemy's bullets will be our signal to deploy as skirmishers; for, we cannot afford to advance within zone of effective rifle fire in any column formation, lest a single shot might put two or more men out of business.

Object of this Stage of the Attack. The object of this stage of the attack is to advance with the loss of as few men as possible to a fire position close enough to the enemy to enable us to use our rifles with such accuracy that we will be able to gain superiority of fire, which may be said to be three-fourths of the battle.

How to Accomplish our Object. We will, in general terms, accomplish our object by endeavoring to find resting-places under cover after long rushes to enable the men to regain their breath, by finding fire-positions from which an effective fire can be delivered, and by advancing as rapidly as is consistent with safety, without exhausting the energies of our men.

649. The Use and Selection of Fire-Positions. It does not follow we should stop and fire from any good fire-position that happens to be in front of us. The distance between fire-positions should be as great as possible without exhausting the men in rushing over the distance. The more fire-positions we occupy the longer will it take us to advance, and the more will we be subjected to the enemy's fire. And again, if an estimate could be made of the comparative casualties, it would probably be found that more men are killed and wounded while halted in fire-positions than while rushing from fire-position to fire-position. It is, therefore, very important that our platoon and squad leaders should be well trained in the selection and use of fire-positions. Whenever practicable the sights should be adjusted and the magazines filled before coming up to a fire-position, so that fire may be opened at once. For instance, if the company, platoon or squad were about to emerge from a wood or other cover, and the next fire-position were known, the sights should be adjusted and the magazines filled before leaving cover.

Good Fire-Positions. Whether or not a location is a good fire-position depends on whether it affords cover, and at the same time enables one to see the enemy, and depends, therefore, entirely upon the small features of the ground and any artificial cover that may be encountered. For example, a position just behind the crest of a hill, behind a bank of any kind, or a fold in the ground, from which the enemy can be seen, is a good fire-position.

A bank running along a road is a good fire-position, but it has the disadvantage, because of the road itself, of being a pretty good mark for the enemy's fire. If there is a bank and ditch on both sides of the road, it is better to occupy the position on the far side; for, the hostile infantry will doubtlessly fire at the road and bullets striking its hard surface will ricochet, while those striking in front of the far bank will disappear.

Bushes and Undergrowth that have no clearly defined border that makes them an easy target, are good fire-positions. While it is true they do not afford protection for fire, they conceal the attackers. If the bushes are on the side of a hill, they then make better fire-positions, as the men can then see over the bushes better without being themselves exposed.

Bad Fire-Positions. The following are bad fire-positions:—

Hedges without Cover from Fire. A hedge without cover from fire is a bad fire-position, not only because it affords a good target for both the hostile artillery and infantry, but also because it is not always an easy position to advance out of. A hedge, however, could be used to advantage if the ground in the rear rises slightly, or if the enemy's position is on high ground in front, in which case the men lying down a short distance in rear on the open ground, could see over the top of the hedge, but themselves be hidden from view.

Villages, farms, cottages, etc. As a rule any kind of a building or inclosure, such as a village, farm, or cottage, is a bad fire-position in the attack. They are easy to get into, but hard to get out of; they are often subjected to artillery fire, and the casualities are much heavier than out in the open.

Quarries and gravel pits, although possibly affording a good fire-position for a few men, are generally disastrous when occupied by a large number. While it is usually easy to jump down into such places and get temporary shelter, it is very difficult to climb out and continue the attack. When such a place is encountered, the best thing to do is for the squad or platoon that strikes it to occupy a fire-position behind it, and thus provide covering fire for the forward movement of the other units on the right and left of it.

Fire-positions that are not approximately parallel to the front of the attack are a source of trouble that often leads to a loss of direction. Such positions are generally to be found in the form of a bank, a hedge, ridge, or the border of a wood. If such a position is recognized before actual occupation, company and platoon commanders must take special steps to avoid the mistakes that are likely to occur. To begin with, the true direction of the attack should be carefully noted, and steps taken to maintain it. It is impossible to give a fixed rule for the handling of all positions that are not parallel to the front of the attack. However, the following principles are general in their application, and together with the examples given, should assist one materially in handling other cases of the same general nature:

First Principle. *Every effort must be made to prevent the occupation of a position that is enfiladed by the enemy.* The reason for this is self-evident.

Second Principle. *If necessary to occupy a fire-position that is not practically parallel to the front of the attack, occupy first that part which is nearest to the enemy.* The reason for this is that if the farther portion is occupied first, there will be a natural tendency on the part of the attack to pass the remainder of the portion and come up in line with the leading platoon or squad, whereas, if the nearest portion is occupied first, it is practically certain that the men who are holding it will stay there until other troops come up on their outer flank; and these troops, with equal certainty, will conform to the fire position already established, and thus throw out the direction of the attack.

Third Principle. *If the part of the fire-position nearest the enemy has been occupied, do not occupy the remainder at all;* or, if human nature is too strong to prevent this, then occupy it for as short a time as possible.

Example 1. *An open ridge or undulation of ground that runs diagonally left to right from our front, towards the enemy's position and falls gradually to the ordinary level some six hundred yards from that position.*

Let us suppose that the ridge lies in front of an entire battalion.

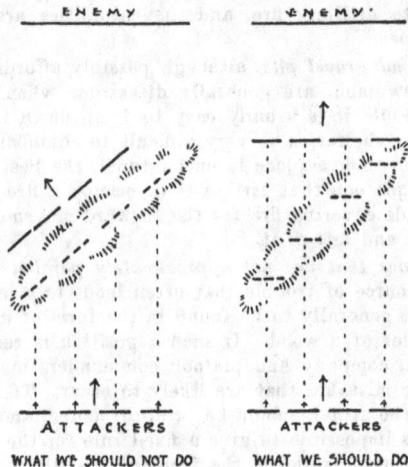

ENEMY   ENEMY

A T T A C K E R S   ATTACKERS
WHAT WE SHOULD NOT DO   WHAT WE SHOULD DO

Fig. 1.

(Fig. 1.) The company on the right, advancing ahead of the others, would go right over the end of the ridge nearest the enemy and occupy a fire-position beyond, parallel to the front of the attack. The company on the left of the first would do the same upon reaching the ridge, taking its position on the left of, and on line with, the right company. The remaining companies would follow suit in succession, holding back until the ground beyond the ridge, on their right, had been gained.

Failing this, each company would act as if it were the right company, except that platoons instead of companies would be used, the right platoon of each company pushing over the top of the ridge, and occupying beyond a position parallel to the front of the attack, the remaining platoons following suit and lining up on the left of the leading platoon.

Example 2. *A V-shaped fire-position, with the point toward the enemy,* such as a low, semicircle ridge with the circumference towards the enemy. (Fig. 2.) The effect of such a feature of the ground is the same as in the preceding case, except that the difficulties are greatly increased.

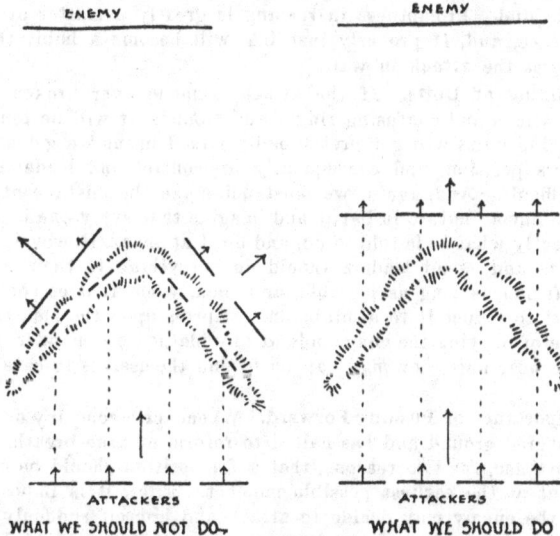

WHAT WE SHOULD NOT DO.   WHAT WE SHOULD DO

Fig. 2.

If the troops occupy such a position, they will be sure, upon leaving the position, to find themselves advancing in divergent direction. A case like this should, if possible, be treated in the same manner as the preceding example, the portion nearest the enemy being occupied first.

Example 3. *A bank along a road that runs diagonally across the front of attack.* Should be treated the same as Example 1.

Pace of Advance. The advance from one fire-position to another is usually made by rushes, but when this method becomes impracticable, any method of advance that brings the attacker closer to the enemy, such as crawling, should be employed. The length of the rushes depends on two things: 1st, the human element, which can only be ascertained by the unit commanders on the spot; 2nd, the nature of the ground and the distance to the next firing-position. The most important details to be looked after are that the men rise together, dash forward without any straggling, and, upon halting, form a fairly straight and orderly firing line. If the enemy is on the look out for rushes, and the chances are he will be, he will open fire, or increase his fire, just as soon as a unit rushes forward, or as soon as he sees that it is preparing to rush. It is, therefore, important that the adjoining unit or units should begin to deliver a covering fire just before the rush begins. Impress upon every man that the devil, in the form of a bullet, is likely to catch the men who fall behind. Promptness in rushing is greatly a matter of drill in time of peace, and, if properly instilled, will become a habit that will greatly assist the attack in war.

649a. Mixing of Units. If the attack is made over broken ground, with undulations and confusing ridges and mounds, it will be found that the advancing units will get pretty badly mixed up as we get nearer to the enemy's position, and consequently fire-control and leadership will become difficult. And, again, we must not make the mistake of closing our eyes to human nature in battle and imagine that every one is going to do just exactly what he is told to do, and do it at once. However, platoon commanders and squad leaders should do everything in their power to delay confusion as long as possible, and then, when it does come, exert every effort to reduce it to a minimum. Impress upon the men the vital importance of obeying the commands of any platoon commander or squad leader, in whose unit they may happen to find themselves in case of confusion.

Importance of Pushing Forward. Whenever a company or platoon reaches covered ground and has halted to reform or take breath, it is of vital importance, for two reasons, that a fire-position should be occupied to the front at the earliest possible moment. First, it is impossible to say when the enemy may decide to attack and himself suddenly occupy a fire-position in our immediate front—the one that we might have occupied ourselves—and thus bring the attack in that locality to a standstill; secondly, other units may be advancing over open ground on our right and left, and it will be of great assistance to them to find that a fire-position to their front has been occupied by us,—it will encourage them to pass forward.

It can be laid down as a general rule in the attack that when any part of the line reaches cover, such as a wood, for instance, a part of the line should pass on and occupy a fire-position at the far end. However, care must be taken not to press so far to the front as to become completely isolated, and run the risk of being shot into by your own men, or cut off by the enemy.

Use of Company Scouts. Cases constantly occur when platoon commanders and squad leaders must make rushes to the front without being able to see the ground which they will have to traverse, and, as a result, at the end of the rush they may find themselves in a very bad position. A case of this kind may occur when a fire-position is occupied just behind a crest from which the enemy's position can be seen, but when the ground immediately in front is covered by the top of the hill occupied by the attacker. It sometimes happens that the defense places an obstacle of some kind, under effective artillery and infantry fire, on the defenders' side of a ridge or hill, and which the attacker rushing over this ridge would not see until he was right upon it. A company scout sent on ahead would give warning of such an obstacle. It also sometimes happens that the unit rushes too far over a crest and occupies a fire-position that is exposed unnecessarily. There is no doubt that in cases of this kind, it is well to send scouts forward to select the best positions, lie down in them, and wait until the line advances. Naturally enough, in open country and in cases where the advance of a scout would mask the fire of part of the company, this plan would be impracticable.

Guarding the Flanks. Special attention is invited to the importance of the flank companies guarding their flanks, especially in closed country. Arrangements must be made to keep in communication with these flank guards, and also to see that they do not get too far ahead of the line. In a big fight our cavalry will be operating on our flanks, but they will probably be a considerable distance away, and the flanking companies must, therefore, provide for local protection.

Supports. According to our Drill Regulations a company acting alone may have a support, but if acting as part of a battalion, it has no support of its own. One or more companies of the battalion form the support for the battalion. The movements of the support as a whole and the dispatch of reënforcements from it to the firing line are controlled by the major. If at any time during the advance a company commander sees that his company has been so depleted that it can advance no farther, he should ask for support.

650. Obstacles. We will now consider the obstacles that are likely to be met during this stage of the attack, and the best means of passing them. These obstacles may be divided into three general classes.

1. *Those in which the attacker is exposed to the enemy's view and fire when he is approaching the obstacle, while getting across it, and while emerging from it on the far side.*

A stream in the open is an example of this class, and to pass such an obstacle, which is exposed to the enemy's fire throughout, it is first necessary to obtain fire superiority, if only temporarily, which means that a heavy infantry firing line must be deployed in a good fire-position in rear of the obstacle. If possible, an artillery support should assist in subduing the hostile infantry fire. Of course, the fire-position should be so chosen that the field of fire will not be masked in part or in whole, by the men when approaching or crossing the obstacle, or when deploying beyond. If it be not possible to select such a fire-position, the situation becomes most difficult, and a second tier of fire will have to be formed in rear of that immediately behind the obstacle, and, if possible, machine guns should be used extensively to assist in gaining and maintaining superiority of fire. The best formation to be used while actually crossing, depends upon the nature of the obstacle. In case of a shallow stream, for instance, which is easily fordable at all points, the squads or platoons could be sent across in line of skirmishers. On the other hand, if the passage were limited to a bridge across the stream, the command should be rushed across by successive squads or platoons, in column of files, or in column of twos, with increased intervals and distances between the men.

It is most important that a firing line be established on the far side of the obstacle at the earliest moment possible. Consequently, the leading squad or platoon should deploy immediately upon clearing the obstacle, and occupy a firing position well to the front. If it be not practicable to occupy a position well to the front at once, then a second advance should be made as soon as possible so as to reduce the distance the succeeding squads or platoons will have to run with their flanks to the enemy. The leading squad or platoon should be careful not to spread out more than is necessary; for, the more it spreads out the greater distance will the succeeding units have to run before getting on the line, to the right and left of the preceding squads. It must be distinctly understood beforehand which units are to go to the right and which to the left. If the crossing is being made by squads, the company commander may, for instance, direct that the second squad go to the right of the leading squad, the third squad to the left, the fourth squad to the right, the fifth to the left, etc. And again, the ground just beyond the obstacle might be such that it would be better for the first squad across to incline to the right (or left) upon reaching the far end of the obstacle, and the remaining squads form on its left (or right).

Of course, the thing to do in every case is to pass around the obstacle, if possible. This may often be done in case of small obstacles (those taking up only forty or fifty yards of front), but as a rule it is not possible to avoid the large ones.

2. *The second class of obstacles is those in which the attacker is protected from the enemy's view and fire while approaching the obstacle, but is partly exposed to fire when passing through it, and may be subjected to heavy fire when emerging from the obstacle.*

A stream with trees and bushes, or a thin wood along its banks might be an example of this class of obstacles. The far edge of either of these makes an excellent mark for the enemy's artillery as well as for his infantry.

It is a great mistake for a company to blunder into an obstacle without having had a reconnaissance made. The first thing to do then, is to send out a reconnoitering patrol in charge of an officer or noncommissioned officer, to ascertain whether the obstacle, or the ground just beyond, is occupied by the enemy; also, to get information regarding the best way through or over the obstacle, and whether there is a good position in which the line can deploy and form preparatory to continuing the attack. Great care must be taken to see that the company does not come under a heavy fire near the far edge in unsuitable formation, and we must also see that the company is not exposed unnecessarily to the enemy's fire before everything is ready for an immediate advance.

3. *Those obstacles in which the attacker is protected from the enemy's view and fire in approaching and crossing the obstacle, but may be subjected to a heavy fire when emerging from it.*

A thick wood is an example of an obstacle of this class. The following points should be borne in mind:—

Be careful not to lose your direction; connect with the companies on your right and left, and maintain the general alignment.

Take every possible precaution to prevent the enemy from learning that you have reached the obstacle, and especially that you are about to emerge on the far side. If he knows this, he will more than probably be waiting for you and will greet you with a heavy, well-directed fire just as soon as you appear on his side of the obstacle.

Do not mistake any of your own troops for the enemy.

Before reaching the far side of the obstacle ascertain by means of scouts or patrols whether it will be exposed to fire; also, if possible, locate a fire-position 50 or 100 yards beyond the obstacle.

Arrange to rush out of the obstacle with as much of the company as can be conveniently handled at one time. Do not emerge in driblets, and don't make the common mistake of forming a firing line along the edge of the wood, thus occupying a well-defined line that stands out as an excellent mark for the enemy's fire.

Cornfields. It is thought the best way of crossing a cornfield, is by a series of rushes in squad columns, or some other column of files. A line of skirmishers in a field of high corn is difficult to control; and, furthermore such a formation is conducive to skulking, men on the ground failing to rise and advance at the command and remaining undiscovered.

Marshes. If impossible to go around wet, marshy ground, there is but one thing to do. Go right through it, lying down in the mud and water when ordered, and rising and advancing when the command is given. The formations to be adopted in crossing a marsh would be determined by the same general principles that apply in getting over level dry ground, with such modifications as might be made necessary by deep pools, difficult mud holes, etc. The main point to impress upon the men is that they must not be afraid of mud and water—they are much less dangerous than bullets.

650a End of Stage. The culminating point of this stage of the attack is the establishment of a fire-position close enough to the enemy to enable us to gain superiority of fire, and by "superiority of fire" we mean that our fire must be so accurate, heavy and deadly that most of the enemy will be keeping their heads under cover, and, consequently we will be able to advance right up to them without a good part of the company getting put out of business.

How near must this fire-position be to the enemy?

The distance of this fire-position from the enemy will depend upon the effectiveness of the enemy's fire and nature of the ground in front of his position, which will determine the cover afforded the attack, the availability of fire-positions and the field of view afforded the enemy. Of course, we will want to get as close as possible to the enemy before beginning the final struggle for fire-superiority—the closer the better. If the ground to the front of the enemy is open, his fire and field of fire are good, and the attackers' fire-positions are poor, it will probably be impossible to get any nearer than 800 yards or more without first gaining superiority of fire. On the other hand, if the ground in front of the enemy is broken, and affords cover and fire-positions to the attacker, it may be possible to get as near as one or two hundred yards before beginning the final struggle for fire superiority. Remember, we must husband our energy and our ammunition for this struggle for fire superiority.

CHAPTER IV

THE COMPANY IN ATTACK

THE THIRD STAGE OF THE ATTACK

THE STRUGGLE FOR SUPERIORITY OF FIRE OVER THE ENEMY

651. Situation. We have now reached a fire-position, say, within five or six hundred yards of the enemy, beyond which it is impossible to advance without reducing the hostile fire. In other words, our advance has been stopped. This stage of the attack is, necessarily, a stationary operation.

Object of this Stage of the Attack. The object of this stage of the attack is to gain superiority of fire so that we can advance to a position so near to the enemy that we will be able to charge him, and every effort must be made to gain superiority of fire as soon as possible; for, if it is not gained within a reasonable time, the energies of the firing line will become exhausted and the attack die out—hence, the vital importance of bringing this stage of the attack to a head as early as possible.

How near should this position be? It should be as near as possible—not over 800 yards—for, the shorter the distance the shorter will be the time that the attacker will be subjected to fire, and the less exhausted will he be upon reaching the defender's position, and, consequently, the better able will he be to cope with the enemy in the bayonet combat that is likely to take place. While this fire-position should be as near the enemy as possible, the company commander should not make the mistake, in case of open ground, of pushing too close to the hostile position without first gaining superiority of fire. If he does, he will probably find himself at the mercy of the enemy, who will have better cover, and will also, because of the open nature of the country, be able to prevent the arrival of reënforcements. On the other hand, if the fire-position is established too far back, the chances are that neither side will gain superiority of fire. The following principles should guide the company commander:

1st. If the cover is good both for the firing line and the reënforcements, advance as far as you can, whether or not you have gained superiority of fire.

2nd. Whether the ground is broken or very open, the fire-position must in every case be near enough to insure a decision as regards superiority of fire.

3rd. The location of the fire-position must be such that it will be possible to send up reënforcements without exposing them to fire too long on open ground.

651 (contd.)

How to accomplish our Object. In order to accomplish our object it will be necessary for every one to put his shoulder to the wheel with the determination "to do or die". Remember, that this is really the crucial stage of the fight—it is the test that is going to decide whether we are going to advance, and drive the enemy out of his position, or whether he is going to stop us, and, if so, probably drive us back. Our salvation, our success, depends upon *effective fire*—a heavy, deadly fire—in order to produce which:—

1st. The soldier must shoot accurately;

2nd. The fire must be well directed by the company commander, and properly controlled by the platoon commanders and squad leaders;

3rd. There must be cooperation, *teamwork*, between the different companies of the firing line and between the different platoons and the different squads of the same company.

This stage of the attack should bring home to every man the importance of being able to shoot well, and doing his level best in time of peace to become a fair shot, if not a good one; a good one, if not an excellent one. The knowledge of being able to hit what you shoot at gives you confidence in yourself. If you know that the men on your right and left can shoot well, it will give you confidence in your company, and when the hour of action comes you will find this confidence to be a wonderful bracer—a great tonic—a big courage producer. This is also the stage of the fight that brings home to us the great importance and value of proper training in field firing.

Reenforcements. When the battalion commander sees that the firing line has been halted and can advance no further without superiority of fire, he will, of course, send up reënforcements. The Drill Regulations prescribe that reënforcements shall take their places on the flanks, so as not to mix up the units, but experience has shown that in practice the men intermingle very much with those already on the firing line, and, as a result, there is considerable confusion. The men must be taught that they must at once place themselves under the orders of the corporals whose squads they happen to join. There will always be lots for the arriving spare officers and noncommissioned officers to do in the way of assisting in the fire-control, encouraging the men, etc. When the company commander sees the reënforcements approaching he should start a vigorous covering fire.

Artillery. Of course, during the struggle for superiority of fire, the company will probably be exposed to the enemy's artillery fire. However, our artillery will make every endeavor to help us gain superiority of fire, and, as we have seen (Par. 646), it has certain advantages over the hostile artillery.

Counter-Attack. During this stage of the attack we must be on the look out for a counter-attack by the defense. Generally a counter-attack is delivered by the reserve of the defense, on one of the flanks of the attack, but it may be made on the front of the attacking line. However, whether made on our flank or front, a counter-attack should be met with vigor—we should move right out and go after the attackers, and not make the mistake of remaining in our fire-position while they attack us, thus ourselves assuming a defensive attitude.

651a. *How will we know when we have gained superiority of fire and can resume the advance?* Upon occupying the fire-position beyond which we can not go without gaining superiority of fire, something like the following will probably occur: Both sides will open up with a general fusilade, which will gradually subside; the burst of fire from one side is answered by a burst of about the same volume from the other, and this continues for some time, until we find that when one side opens fire, the other answers with a heavier fire that reduces or actually silences the opponent's fire. This is the first sign that superiority of fire is being gained. For a time this superiority of fire may be gained by one side and then by the other, and as time goes on we find that one side gains superiority of fire oftener than the other. Let us suppose that the attack is in the ascendant. Finally we find that as soon as the enemy opens fire there is a terrific burst of fire from the attack, and the fire of the defense at once slackens or ceases altogether. We then know that we are approaching the end of this stage, and we must, therefore, begin to make preparations to advance to the position from which we are going to charge.

CHAPTER V

THE COMPANY IN ATTACK

THE FOURTH STAGE OF THE ATTACK

THE ADVANCE AFTER SUPERIORITY OF FIRE HAS BEEN GAINED TO A POSITION CLOSE
ENOUGH TO CHARGE

652. Situation. We have gained superiority of fire and are now ready
to advance to a position close enough to the enemy to assault.

The Object of this Stage of the Attack. As just stated, the
object of this stage of the attack is to advance to a position close enough
to the enemy to assault. The distance of this position from the enemy
will, as previously explained, depend upon various conditions. "It may
be from 25 to 400 yards". (Par. 465, Infantry Drill Regulations.)

How to accomplish our Object. In order to accomplish our
object we must maintain our superiority of fire, and cover the rest of the
ground with the least possible delay. Impress upon every one—officers,
noncommissioned officers and privates—the vital importance of moving
forward as soon as fire superiority has been gained, and of covering the
ground with the least delay practicable. Delay will only result in increased casualties, and success now no longer depends entirely on fire
effect, but it depends on the assault, which must de delivered as soon as
possible.

Who gives the Order to Advance? It is not necessary to wait
for an order from the battalion commander to advance. As soon as any
company commander sees that he has gained fire superiority he should
at once commence the forward movement on his own responsibility. An
advance started even by a squad has often set an example that was at
once followed by other parts of the firing line, and that resulted in the
prompt building up of a fire position considerably nearer the enemy's
position. Cases might arise when it would be better for the original
firing line not to advance from this position, but, instead, to have the
support come from the rear, pass through the line and take up the advance
from that point, under the covering fire of the firing line. Where the
firing line could fire over the heads of the support, would be a case where
this might be done.

Covering Fire. Even though there may be no doubt about our
having gained superiority of fire, we must not attempt to continue the
advance without covering fire. There are two types of covering fire:
(1) when the fire is delivered by troops in rear over the heads of those advancing, and (2) when the fire is delivered by units to cover the advance
of other units to their right and left. The covering fire delivered by a

unit that has already advanced, to support the advance of another unit in rear, is, of course, only a modification of the second type mentioned.

The overhead fire delivered by troops in rear is by far the best kind of covering fire, because it can be furnished by support or reserve companies, and this leaves the firing line free to move forward, further assisted by the second type of covering fire which it can provide for itself. Of course, the first type of covering fire can be used only when it is safe to shoot over the heads of the troops in front. If it is possible to fire over the heads of advancing squads and platoons, we might do one of three things:

1st. Have the original firing line remain in it position and furnish a covering fire for the support, which is sent forward through the firing line to assault.

Advantages. The men of the support being fresher, will doubtless possess more energy, especially for the assault, which is to follow soon, and when passing through the original firing line, they may take a part of it with them, thus increasing the strength of the assault. Again, the original firing line would probably be able to furnish a more efficient covering fire than the troops in rear, because they know the range and are familiar with the points to aim at.

Disadvantage. The possible disadvantage is that the original firing line might be short of ammunition, and ammunition would be required for covering fire more than for the assaulting line. However, this disadvantage could be overcome by having the support drop part of its ammunition as it passes through.

2nd. Have the original firing line advance under cover of the support in rear.

Disadvantage. The original firing line might not have enough energy left to carry out the assault.

3rd. Reënforce the firing line as heavily as possible, and then have it advance under it own covering fire. In case any part of the advancing line is checked, then the company commander must open a rapid fire with the rest of the company and thus at once restore superiority of fire, and immediately rush the greater part of the company to the front.

Formation. By this stage of the fight the firing line will have been well reënforced, so that the men will probably be as close together as is compatible with the proper use of their rifles. The advances must be made by rushes by squads and platoons. The length of the rushes will depend on the nature of the ground, but, remember, you must get over the remaining ground as rapidly as possible—we must get up to the assaulting point at the earliest moment possible.

Artillery. Our artillery will be assisting us with rapid bursts of fire, of which we should take full advantage to advance.

653. Obstacles. Any natural obstacles that may be encountered during this stage of the attack will be passed as explained in the second stage of the attack. They will present very little more difficulty, if any, for, although we are now much nearer the enemy, we must remember that we have gained superiority of fire. The artificial obstacles that we may encounter are the ones that will give us trouble, because the enemy places them where we are apt to come up on them unexpectedly. As a rule, however, such obstacles will not be very extensive; for, it takes time and material to make them, and extensive obstacles in front of a defensive position render a counter-attack very difficult, if not impossible. The same general principles that apply to the passage of natural obstacles are applicable to the crossing of artificial ones.

CHAPTER VI

THE COMPANY IN ATTACK

THE FIFTH STAGE OF THE ATTACK

THE CHARGE

654. Situation. We have at last reached a position which is near enough for us to charge the enemy.

The Object of this Stage of the Attack. The object of this stage of the attack is to close in on the enemy with fixed bayonets and drive him from his position in confusion and disorder.

How to accomplish our Object. The charging line must be properly built up before the beginning of the charge, which must be delivered with the utmost vigor, and without any restraint whatsoever on the ardor of the charging troops by an attempt to maintain alignment. The charge should be made simultaneously by all the units participating. Confidence in their ability to use the bayonet in combat is a great stimulus to assaulting troops. Impress upon your men the importance of becoming proficient in time of peace in the use of the bayonet. The charge is usually immediately preceded by clip fire.

When to fix Bayonets. The major or senior officer in the firing line determines when bayonets shall be fixed, and gives the proper command or signal. It is repeated by all parts of the firing line. Bayonets must be fixed at once, but in such a way that there will be no marked pause in the firing. A good plan is to have the even numbers fix bayonets first and then the odd numbers, the odd numbers increasing their rate of fire while the even numbers are fixing bayonets, and the even numbers increasing their rate of fire while the odd numbers are fixing their bayonets. The support also fixes bayonets. Bayonets will be fixed generally before or during the last, or second last, advance preceding the charge.

When to Charge. Upon reaching the position from which the assault will be made, build up the line as rapidly as possible with the units arriving from the last fire-position, and before charging be sure to see that enough troops are on·hand to make it a success. However, do not have too dense a mass; for, then the men will be in one another's way. Also, see whether the adjoining companies have yet built up their lines and are ready to charge. Reserves joining the firing line now will give the charge a strong impetus. It is impossible to give any fixed rule as to just when the charge should be started. "The psychological moment of the charge cannot be determined far in advance. The tactical instinct of the responsible officer must decide". (Par. 464, Infantry Drill

654 (contd.)

Regulations.) "The commander of the attacking line should indicate his approval, or give the order before the charge is made. Subordinate commanders, usually battalion commanders, whose troops are ready to charge signal that fact to the commander". (Par. 466, Infantry Drill Regulations.) However, history shows cases where a corporal or a drummer boy sprang forward at the "psychological moment", and was followed by the rest of the firing line in a charge that completely routed the enemy. "Subject to orders from higher authority, the major determines the point from which the charge is to be made. The firing line having arrived at that point and being in readiness, the major causes the charge to be sounded. The signal is repeated by the musicians of all parts of the line. The company officers lead the charge. The skirmishers spring forward shouting, run with bayonets at charge, and close with the enemy". (Par. 319, Infantry Drill Regulations.)

A charge to be successful must above all things have cohesion—the men must start together, keep together, and fight together—they must charge with vigor and determination. The charge must be started promptly, when ordered whether the men are one or one thousand yards from the enemy—the distance has nothing to do with it so far as the men are concerned.

Conduct after the Charge. The further conduct of the charging troops will depend upon circumstances; they may halt and engage in bayonet combat, or in pursuing fire; they may advance a short distance to obtain a field of fire or to drive the enemy from the vicinity; they may assemble or reorganize, etc. If the enemy vacates his position every effort should be made to open fire at once on the retreating mass, reorganization of the attacking troops being of secondary importance to the infliction of further losses upon the enemy and to the increase of his confusion. (Par. 319, Infantry Drill Regulations.)

Our Artillery. Our Artillery will be on the look out for the charge, and, about this time, will increase their range so as to burst their shells just beyond the enemy's position, so as to check the possible arrival of hostile supports or reserves. A premature charge by a part of the line should be avoided, but if begun, the other parts of the line should join at once, if there is any prospect of success. Under exceptional circumstances a part of the line may be compelled to charge without authority from the rear. The intention to do so should be signalled to the rear. (Par. 470, Infantry Drill Regulations.)

Supports and Reserves. At the signal for the charge the nearby supports and reserves rush forward. (Par. 466, Infantry Drill Regulations.)

Counter-Attack. We must not forget that even at this stage of the fight the enemy may have enough energy left to leave his position and attack us. However, such an attack must be met with vigor—we should go right after the enemy at once. To assume a defensive attitude at this stage of the attack would be suicidal. "If the attack receives a temporary set back and it is intended to strengthen and continue it, officers will make every effort to stop the rearward movement, and will reëstablish the firing line in a covered position as close as possible to the enemy." (Par. 474, Infantry Drill Regulations.) "If the attack must be abandoned, the rearward movement should continue with promptness until the troops reach a feature of the terrain that facilitates the task of checking and reorganizing them. The point selected should be so far to the rear as to prevent interference by the enemy before the troops are ready to resist. The withdrawal of the attacking troops should be covered by the artillery and by the reserves, if any are available." Par. 475, Infantry Drill Regulations.)

Conclusion. It must not, of course, be supposed every attack, or even the majority of them, will have five separate stages, and will be conducted just as we have conducted this one; for, such will not be the case. Every attack has its own individual characteristics and must be handled according to the nature of the ground, and the tactical and other conditions involved. However, the stereotyped form of attack through which we have just taken you gives a very good, general idea of what an attack is like, and the general basic principles presented are applicable to any attack.

CHAPTER VII

THE COMPANY IN ATTACK

THE COUNTER-ATTACK

655. Classes of Counter-Attacks. There are two general classes of counter-attacks—what may be called the "general counter-attack", and the "local counter-attack". The general counter-attack is usually made by launching the reserve against one of the enemy's flanks when his attack is in full progress; by making a frontal attack with the firing line and supports after repulsing the enemy's attack and demoralizing him with pursuing fire; or, by the troops in rear of the firing line, when the enemy has reached the defensive position, and is in disorder. However, as the general counter-attacks are made by large bodies of troops, under a higher commander and involve higher tactics, their principles do not directly concern company commanders, and, therefore, will not be discussed here.

Local Counter-Attacks. The local counter-attack is the one in which the company is directly interested. Such an attack might be made to drive the enemy from an important position he has gained in our immediate front and from which he is doing us considerable damage, or to halt his advance during the third or fourth stage of the attack, or it might be made to block the enemy's charge. For instance, let us suppose the enemy has gained superiority of fire and is moving up to a position from which to assault. What are we going to do? Are we going to turn tail and run, await the assault in our trenches, or shall we go out of our trenches and assault the attackers? Is it better to sit in our own trenches and wait until the hostile troops come pouring over the parapet, shouting and yelling and digging their bayonets into us, or is it better to rush forward over the parapet ourselves, do the shouting and yelling ourselves, and do a little bayonet digging on our own hook?

When to deliver the Counter-Attack. There would be no object in delivering a local counter-attack during the first two stages of the attack. It is only when the attack has been stopped temporarily by the fire of the defense and must gain superiority of fire before it can advance further, that the necessity for delivering a local counter-attack would arise. Naturally enough, some parts of the defender's position will be stronger than others; in the strong parts, where the defense has a good field of fire, the attack will be at a disadvantage in the struggle for superiority of fire and may never gain it sufficiently to assemble for the assault. This is not the part of the defender's position where the counter-attack is likely to be needed—it is likely to be needed in one of the weak parts of the defender's position, where, almost invariably owing to cover, the attacker will be able to get much closer. If the attacking line is allowed to remain in its fire-position and obtain superiority of fire, there is every chance that this part of the defender's line will be defeated. There are two ways of preventing this: One is for the defenders, while the struggle for superiority of fire is in the balance, to gain temporary superiority of fire and then attack the enemy, advancing in a thick line, covering the ground with as few halts as possible and charging the enemy with the bayonet. The other alternative, but a far more risky one for the safety of the position, is to wait until the attack has gained superiority of fire and is assembling for the assault. This form of counter-attack would be resorted to only when previous efforts as suggested above have failed. In this case the company would charge with bayonet straight over the trenches.

In those parts of the defender's line where the main defense will depend upon the success of local counter-attacks, each company should be assigned an objective before the counter-attack is commenced. Also, a good fire-position in front of the general line of defense should be selected before hand, and even improved, so that when the blow has been delivered the companies will not find themselves lying out in a very exposed position where it is impossible for them to use their rifles with effect. Local counter-attacks have often failed to produce any permanent effect, either because they have gone too far, or because they have reached a position where they were unduly exposed to the enemy's fire; the enemy has then attacked and driven back the counter-attack, and practically followed it into the main position.

When deciding upon the objective of a counter-attack, so far as it can be done by studying the ground in front of the position, a company commander should bear in mind—

(1) The position which the enemy may reach before being brought to a standstill by the fire of the defense, and beyond which he cannot advance without first obtaining superiority of fire.

(2) The line, perhaps in dead ground, where the attack, if it gains superiority of fire, is likely to assembly preparatory to the assault.

(3) A suitable fire-position, always in advance of the last-mentioned line, which the counter-attack can reach, and which can be supported later by fire from the main position, if possible.

By not aiming at too much it is probable that the blow will be more permanently effective, it will be much easier to deliver, the losses will be greatly reduced, and if superiority of fire is obtained at once over the enemy in front, often his leading troops will be driven back by the counter-attack. A further advance with a view to defeating the enemy and driving him back from this part of the battlefield can be easily initiated.

655 (contd.)

The most important point to remember is that when everything else has failed, and the enemy, having gained superiority of fire, is assembling close in front of the trench for the final assault, he must be charged before he has time to charge himself. Furthermore, if one company commander sees another company on his right or left issuing from their trenches to charge, he must do the same with his company, whatever the situation in his immediate front may be. The only exception will be when the attackers in front of his company have not been able to gain superiority of fire, in which case the company would remain in its trenches and turn its fire upon the enemy in front of the companies delivering the counter-attack.

CHAPTER VIII

THE COMPANY IN ATTACK

MEETING ENGAGEMENT, IN WHICH THE COMPANY IS EX-POSED TO BOTH ARTILLERY AND INFANTRY FIRE FROM THE BEGINNING

656. Meeting Engagement. "Meeting engagement" is the name given to an action when the opposing forces meet and commence to fight before either side has had time to make much preparation for the attack or defense. Such an engagement is characterized by the necessity for hasty reconnaissance, or the almost total absence of reconnaissance; by the necessity of rapid deployment, often under fire; and usually the absence of trenches or other artifical cover. It is, therefore, evident that the first and second stages of the attack described in the previous chapters either disappear entirely or are greatly curtailed.

Battlefield. The first thing to realize is that, as a rule, the ground has not been selected by either side as being especially suitable for either attack or defense, and, owing to the lack of time, it will contain few, if any, artificial obstacles; also, the entrenching, if any, will be hasty and limited. However, as a rule, there will be several important tactical features, such as streams, ridges, woods, hills, etc., the possession of one or more of which in particular will be an advantage to the side that holds it. The feature may not be very marked of itself, but it will be very important as compared with the rest of the features of the ground. Such a position will generally assist either the attack or defense. The thing to do, of course, is, if possible, to get possession of the feature without delay.

Necessity to Attack. We see from the above that the conditions of a meeting engagement give the attack advantages in addition to those mentioned in previous chapters, and, also, that the conditions of such an engagement make it necessary to assume the offensive at once and attack with energy, vigor, and determination, so as to gain ground to the front and throw the enemy on the defensive at all costs. Everything must, therefore, be done to gain superiority of fire just as soon as possible. Hence the importance of deploying and maintaining a strong firing line from the first.

Final Stages. Having gained superiority of fire in the initial stage of the encounter, the final stages, including the maintenance of fire superiority, the advance to within charging distance, and the charge, would be carried out in a manner similar to that described in the stereotyped form of attack, except that, the ground not having been especially selected for defensive action and being free of artificial obstacles, the last stages of the attack would not be so difficult.

[605]

CHAPTER IX

THE COMPANY IN ATTACK

ADVANCE GUARD-ACTION

657. Observation of Country While Marching. When the company is acting as an advance guard, as the troops are advancing through the country the captain must be constantly examining the ground in his vicinity, and planning in his own mind what he would do in case the enemy should be suddenly met in that particular neighborhood. All other officers, as well as the noncommissioned officers must also constantly observe the country as the column advances.

Action preceding meeting Engagements. A meeting engagement is usually preceded by an action between the opposing advance guards, or between the opposing advance guard of one side and the outpost of the other. An advance guard moving in the direction of the enemy may meet a hostile advance guard of a body of troops sent forward to meet it, or it may meet an outpost line covering the enemy's main body. In either case the advance guard should attack and defeat the enemy's advance detachments before they can be reënforced from the rear.

Company Forming Part of Advance Guard Support

Meeting Hostile Advance Guard. Let us suppose the company forms a part of the support of the advance guard. It will generally meet a detachment of the enemy that is hardly stronger than itself, and which should in every case be attacked immediately, with all possible vigor. Not far in rear of this hostile detachment we know more troops are following, and we must hit a good, strong blow before any of those troops can come up from the rear. The fact that the ground may be poor for attack, but good for defense, must not tempt us to assume a defensive attitude. Start right out after the enemy and strike him hard before he has time to realize what is happening. The question of committing the advance guard to a serious engagement without instructions from the commander of the main body, need not worry the commander of the support. We know we have in our immediate front an enemy that cannot be very strong for the moment, and if we can defeat him and drive him back on the rest of his advance guard, and perhaps even on the main body, the troops in his rear may be thrown on the defensive—such is the tremendous moral power of attack and initial success. Even if our attack on the support is brought to a standstill, we will have at least thrown the enemy on the defensive, and also will have more than likely caused him to greatly exaggerate our strength.

If both sides should assume the offensive, the one that develops the most effective fire first and makes the best use of the ground will throw the other side on the defensive. We, therefore, see the vital importance of gaining superiority of fire as soon as possible and losing no time in gaining a position close enough to assault.

Protection of Flanks. The importance of protecting our flanks in advance guard action cannot be over estimated, and we should practice it diligently during times of peace.

Meeting Hostile Outpost. If the company encounters a hostile outpost, we will find the enemy already occupying a defensive position. In view of the fact that his reënforcements would not be actually marching up in his rear, as in the case of an advance guard, a little more time would elapse before they could come up; also, his position would be well chosen and undoubtedly his supports will have occupied the best positions in the vicinity. Therefore, the enemy having already assumed the defensive, the speed and immediate application of all the force available to throw him on the defense would not be required. Our object would be to push forward, establish a fire-position in the best place available, and endeavor to gain superiority of fire over the defense. As a rule, this should not be very difficult, because we would probably be stronger than the outpost in our immediate front, and by such action we would disclose the enemy's position and to some extent his strength, and thus assist materially the operations of the rest of the advance guard when it comes up to the firing line. On the other hand, should the company when it meets the hostile outpost, content itself with merely holding its ground and adopting a purely defensive attitude until the arrival of the rest of the advance guard, the enemy would probably be encouraged and we would not be able to get much, if any, information about him. In fact, it might not be at all clear what was in front of the advance guard, which, perhaps, was being held back by only a few dismounted men. It may be remarked here that it is a general rule always to make an effort to develop the strength and position of the enemy before planning your main attack, so that you may know what you are up against, and in order to accomplish this purpose it is sometimes necessary to send out combat patrols to draw the enemy's fire.

Company Forming Part of the Advance Guard Reserve

Meeting Hostile Advance Guard. Let us now suppose that the company forms a part of the reserve instead of the support of the advance guard. While a vigorous offensive is of the greatest value at all times, it is never more so than when in action against a hostile advance guard. As we form a part of the reserve, the plan of attack, of course, will have been decided upon before we come into action, and our main object will be to reach a fire-position close enough to gain superiority of fire with the least possible delay. It should be borne in mind that the enemy may be reënforced at any time by troops in his rear, so that we must take advantage of every condition to push forward within assaulting distance.

Protection of Flanks. We have already mentioned the importance of flank protection in advance guard action. If the company happens to be on the flank, two or three squads should be placed on the outer flank to protect the company, especially from mounted troops. This flank guard under certain conditions could be used to provide covering fire for the company. Scouts should be sent out beyond the flank.

Company as Support to Artillery

658. When artillery forms part of the advance guard it is sometimes necessary to detach a company to protect the guns. In such case, you should first ascertain the direction from which the enemy would be most likely to attack the artillery and then make your arrangements to meet the attack. Of course, the enemy could not attack over the ground that is being used by the advance guard; consequently, the possible directions are from the two flanks and the rear. Should the general situation, or the nature of the terrain make an attack from one or more of these directions impossible, it would, naturally enough, simplify the task of the infantry escort.

If an attack were expected from one flank only, our main object would be to prevent the enemy from occupying any locality from which they could fire upon the flanks of the guns or against the wagons in rear. It is generally impossible for the company to occupy all such localities. However, if there is a position that commands the others, it should be occupied by the company, provided the enemy cannot get between the company and the artillery without being exposed to the fire of the former. The company commander should send out scouts or patrols to all dead ground in the vicinity to give warning of a hostile advance. The company must not occupy a position too close to the artillery, because it might interfere with the working of the guns, and it would also probably suffer casualties from any hostile artillery fire that might be directed against our guns.

The company commander should arrange with the artillery commander to be informed at the earliest possible moment of any change in the position of the guns, because the infantry marches much slower than the artillery, and is apt to be left behind, especially when scouts or detachments must be called in. The company commander, in case the company is to follow the artillery to a new position, should always ascertain from the artillery commander the exact location of the position, and, if possible, the best road leading thereto.

When the enemy may attack from either flank or from the rear, arrangements similar to those already described to deal with each eventuality, must be made. In a case like this it would be best to keep the greater part of the company together in the most important position, from which, perhaps, two or three possible lines of approach could be commanded. Should the enemy succeed in occupying a position from which their rifle fire would interfere with the service of our guns, the company, must promptly attack the hostile position and carry it by assault.

CHAPTER X

THE COMPANY IN ATTACK

GENERAL RULES AND PRINCIPLES

659. Necessity for defensive Action. As we have previously shown, the attack is the dominating spirit of war—it is the only spirit that wins battles. However, there are times when the defense is necessary—indeed, when there could be no successful attack without a preceding or simultaneous defense.

660. Classes of Defense. There may be said to be two general classes of defense: 1st, the so-called *passive defense*, when a certain position is to be held only for a short time because of the vastly superior strength of the enemy—for example, the action of a rear guard that is merely delaying the enemy so that the main body can get away, or the action of an outpost that is driven in by the enemy.

2nd, the so-called *active defense*, in which the spirit of attack dominates, and which is only resorted to as a stepping stone, as a means, to the attack. In other words, we start out with the intention, the determination, of attacking the enemy, but we first delay his advance, make his attacking us as difficult as possible, cause him to lose more men than we do, break down as much as we can his power of attack, and finally attack him ourselves in turn. This is the usual form of defense, whose very soul should be the spirit of attack, but which, however, should never be resorted to unless the conditions make it absolutely necessary. Let us again repeat that the attack is the only thing that wins battles, and if compelled to assume the defensive we must not maintain that attitude any longer than is necessary, but must assume the offensive just as soon as we possibly can.

661. Advantages of the Defense. The defense has these advantages over the attack:

1. A larger amount of ammunition can be made available.

2. The men undergo comparatively little fatigue and consequently can shoot better.

3. The defense, usually having good cover, will suffer fewer losses than the attackers, and the use of smokeless powder makes it difficult for the attackers to locate the actual position of the defenders.

4. The advancing attackers must sooner or later offer a good target.

Disadvantages of the Defense. 1. The attacker can choose his point of attack, while the defender must be prepared against it at all points.

2. The defender must generally fight amongst his dead and wounded, which is demoralizing.

3. Seeing the attacker continuing to advance in spite of the defender's fire has a bad moral effect on the defender.

662. Requisites of a good defensive Position. The requisites to be sought in a good defensive position are:

1. A good field of fire to the front and flanks, to distance of 600 to 800 yards or more.

2. Effective cover and concealment for the firing line as well as for the supports and reserves.

3. Flanks that are naturally secure, or that can be made so by the use of reserves.

4. Extent of ground suitable to the size of the force that is to occupy it.

5. Good communications throughout the position—that is, between different parts of the firing line, between the firing line and the supports, and between the supports and the reserves.

6. A good line of retreat.

7. The position should be one that the enemy can not avoid, but must attack or give up his mission.

Of course, a position having all these advantages will rarely, if ever, be found. The one should be taken which conforms closest to the description.

Two of the most important requisites of a good defensive position, viz: cover for the men and a good field of fire, are conflicting; for, as a rule, the farther back we get on the top of a hill, the better will the cover be, but, on the other hand, the farther back we get the more "dead space" will there be in front—that is to say, the poorer will the field

of fire be. In selecting our fire position we must, therefore, balance these conflicting requirements and strike a compromise between the two.

Straight Lines, best Form of Defense. Of course, the simplest and most effective form of defense is for the fire trench or trenches to be constructed in a straight line. However, with a command of any size this will rarely be possible; for, we will find that, owing to the features of the ground, a straight line of any length will provide variable cover for the defense and a very ununiform field of fire to the front—in some parts the cover and field of fire will be excellent, and in others it will be very poor. We must, therefore, bear in mind that in selecting our fire-positions, the position of the defense as a whole must be considered.

663. Salients. By salients we mean hills, spurs, woods, buildings, etc., that jut out from the general line of defense, and thus form a projecting object. Salients are always a weakness in a line, not only because it is easier for the enemy to concentrate a converging fire, especially artillery, on a small, projecting locality like a salient, which can often be enfiladed, but, generally, the number of men required to occupy a salient is out of proportion to the amount of the defenders' front covered by the salient. Hence, whenever possible, salients must be avoided and the space to their front and sides covered by the fire of troops occupying parts of the line adjoining the salients. But, if this can not be done, and the occupation of a salient can not be avoided, special pains must be taken to provide overhead and other cover (bomb-proofs and loop-holes), and to guard against being enfiladed by the enemy.

664. Avoidance of the Skyline. Generally when we speak of the "Skyline" in conversation, we mean the highest part of a hill or ridge—the "natural crest"—, and, roughly speaking, the skyline and the natural crest are often practically the same. However, strictly speaking such is not the case; the exact position of the skyline is variable—it becomes nearer as one approaches the top of the hill and gets farther away as one goes away from the top of the hill. For instance, if standing at "A" the skyline would be at "C", and if standing at "B" it would be at "D". The thing to do then, is to ascertain the point beyond which the enemy will probably be unable to advance without gaining superiority of fire—that is to say, the position where he will make his fight for superiority of fire—and then fix the skyline with reference to that point. As we all know, objects seen on the skyline loom up as prominent targets. Furthermore, it is evident the occupation of the skyline as a fire position would result in lots of "dead space" in our front, and, again, it is a known fact that men will usually fire at this skyline rather than at any other part of a hill, if they cannot see the defenders' trenches. We should, therefore, if possible, avoid the skyline, and occupy a fire-position in front of it. However, there may be occasions when the occupation of the skyline cannot be avoided. For instance, in the case of a plateau, with sudden and steep slopes down to the valleys below. Because of the vast amount of time and labor involved in cutting down the sides of the slope before getting to the trench proper, the construction of trenches on the slopes would be impossible—also, their concealment would be very difficult. When such a skyline is occupied bushes should be placed along it—they will prevent the enemy from seeing the men when they rise to fire.

If possible, before selecting a defensive position, go forward several hundred yards and examine it for the direction in which the enemy will approach.

665. Cover, Trenches, and Obstacles. The natural cover of the position should be fully utilized, and, in addition, it should be strengthened by trenches and obstacles, if the time permits.

The best protection is afforded by deep, narrow, inconspicuous trenches. If little time is available, as much as practicable must be done. That the trenches may not be needed should not cause their construction to be omitted, and the fact that they have been constructed should not influence the action of a commander, if conditions are found to be other than expected.

The fire trenches should be well supplied with ammunition.

Supports. When natural cover is not available for the supports, they are placed close at hand in cover trenches.

Dummy Trenches frequently deceive the enemy and cause the hostile artillery as well as infantry to waste time and ammunition and to divert their fire.

666. Advance Posts. As a rule, the occupation of positions in front of the general line of defense should be avoided, as they tend to disperse the power of the defender. However, there are times when it is desirable to occupy such positions in order to delay the enemy, deflect his course, make him deploy sooner, or for some other purpose. In such case, the number of troops used for the purpose should be no larger than is absolutely necessary, and care must be taken not to let them get cut off, special provision being made for their retreat. The commander of an advance post should always be given definite and explicit instructions as to just what he is to do and how long he is to occupy the position.

Occupation of the Trenches. Unless the difficulty of moving the troops into the trenches be great, most of them should be held in rear until the infantry attack begins. The position itself would be occupied by a small garrison only, with the necessary outguards or patrols in front.

667. Use of Bayonet. Fire alone cannot be depended upon to stop the attack. The troops must be determined to resort to the bayonet, if necessary.

668. Night Attack. If a night attack or close approach of the enemy is expected, patrols or outposts should be thrown out in front, troops in a prepared position should strengthen the outguards and firing line, and as many obstacles as possible should be constructed. Supports and reserves should move close to the firing line and should, with the firing line, keep their bayonets fixed. If practicable, the front should be illuminated, preferably from the flanks.

Only short range fire is of any value in resisting night attacks. *The bayonet is the chief reliance.*

669. Buildings, Farm Inclosures, Etc., should not be occupied by the defense, when it can be avoided, especially if they are to be exposed to artillery fire.

Buildings and farm inclosures are bad fire-positions, not only because they make excellent targets for the artillery, but also because the men in them are in a very confined place, usually crowded together, thus presenting a vulnerable target. If a building exposed to artillery fire must be defended, it can usually be done better from the outside than from the inside. The field of fire from the outside is generally better, and it is easier and quicker to dig good trenches, even with head cover, than it is to loop-hole and otherwise prepare a building for defense. Again, the effect of the hostile artillery will not be so demoralizing outside of a house as inside of it, and the defenders will not be driven out by the building catching fire. However, if a building is not exposed to artillery fire and it also affords a good field of fire, there is no doubt that it affords a good means of resistance, as it is very difficult to assault a building, and by the use of sand bags and other material effective protection can be provided against infantry fire. In fighting inside a building the very strictest discipline must be maintained and every man made to do his full duty; for, there is no doubt that many men who go into a building take no part in its defense. Men must be assigned to various rooms, and squad leaders and the one in general charge, must frequently visit the various parts of the building to see that every one is doing his duty. Precautions must be taken to guard against fire. In preparing a building for defense, improve the field of fire in the vicinity as much as possible; provide water and heaps of earth in the rooms; break and remove the glass in all the windows; doors and windows on the ground floor that are accessible must be blocked up and loop-holed; arrange for means of communication throughout the building and for means of retreat. Ordinary furniture, chairs, tables, cupboards, bedsteads, etc., make good obstacles. Boxes. chests, trunks, sacks, mattresses, bags and pillow cases, when filled with earth, afford protection against rifle fire.

However, it is generally safer to avoid buildings, and if they must be defended, then to defend them from the outside.

670. Edge of a Wood. If exposed to hostile artillery fire, the edge of a wood is not a good defensive position, because, as explained in the attack, it affords a good target for the enemy's artillery, and it also affords a good target for the hostile infantry. However, experience in war shows that such a position is not a bad fire position if not exposed to artillery fire, especially if the defenders happen to have some artillery to assist them. The trenches should be placed as far back as possible inside the wood, so as to escape the heavy fire that is sure to be directed against the edge of the wood. However, care must be taken not to locate the trenches so far back that the field of fire will be obstructed.

The principal advantage of such a position is that the supports can be brought up and the wounded removed from the trenches under the cover of the woods.

671. Clearing in a Wood. A clearing in a wood, especially if it be two or three hundred yards wide and the farther side be exposed to the defenders' fire, makes a good defensive position. This is not due so much to the fact that the defense has good cover and a good field of fire as it is to the fact that it is only with great difficulty that the attack can make an organized advance out of the wood. However, such a position is better suited to a passive defense, as it is about as difficult for the defender to deliver a counter-attack from his edge of the wood as it is for the attacker to move forward from his side, but the defender has a slight advantage over the attacker in this respect, in that he can prepare beforehand paths for purposes of communication, and also in that he will not have suffered the casualties and been subjected to the confusion which the attacker has experienced while passing through the wood.

If necessary to make a clearing in a wood, it must be done by cutting down trees and clearing out the undergrowth. Of course, as a rule, it will not be possible to cut down all the trees and the largest ones will have to be left standing. However, the spaces between the trees left standing must be assigned to various units, so that our fire will be sure to be distributed along the entire front. Any logs that may be left should lie at right angles to the defenders' front so as not to afford any cover for the enemy. The enemy's side of the clearing should be cut in the shape of a W, with the pointed angles towards the defenders. This will cause the attackers to crowd into the pointed angles and to form along the edges of the angles, thus affording vulnerable targets, exposed to an enfilading fire from the defender. This saw-tooth effect on the attackers' side, in order to reduce the amount of labor, would be produced as follows: After the clearing has been made, cut indentations, (A, B, C, D,) in the woods, a few yards apart. Place

the limbs and other material from these indentations at W, X, Y, Z, in the form of angles. If the time and labor permit, and if we have a couple of machine-guns, so that one can be placed at each flank of the clearing, our position can be made almost impregnable by cutting two rides running like a V, with the point towards the enemy, so that the machine-guns can shoot down the rides. Before being able to reach the edge of the clearing the enemy would have to cross the V-shaped paths and would be mowed down by the machine guns.

In making a clearing always begin on the defenders' side, and as the work progresses, keep the width of the clearing about the same throughout its entire length; for, if compelled to stop suddenly before the desired width is reached, it is better to have a narrower clearing of uniform width than one with some points of the far side near the defenders' position and others far away.

When necessary to occupy a fire-position in a wood, with no clearing and without time to make one, whether or not the position is to be entrenched, always occupy the highest ground available. There is no theoretical reason for this, but experience shows that in wood fighting the severest struggles almost invariably take place on the highest ground, and success depends upon ability to hold such ground. It may be that this is due to the fact that in nearly all wood fighting the bayonet is a great factor, and it can be used with greater effect in rushing down hill than in charging up hill.

672. **Clearing Field of Fire.** If there is not a good field of fire to the front, we must improve the field as much as possible by clearing away all obstructing objects. However, in clearing the ground close to the trench, we must be very careful not to create what will appear to the enemy like a straight line, because it will give him a good idea of the location of the trench, and probably unnecessarily expose the defenders to view. Bushes and scrub that do not interfere with the field of fire and which, when seen from the front, do not present a clearly defined line, should not be cut down, as they make it difficult for the enemy to locate the trenches or see the defenders, when they put up their heads to fire.

A Wood. The method of clearing a wood was explained above.

Crops. Crops that cannot be burned may be trampled down by having troops march over them in close order.

Streams running across the front within effective rifle range should be cleared of all obstructions to view, so that the defense will be able to fire on both banks and on the approaches. However, a few trees or bushes should be left here and there along the banks as aiming marks to assist the squad leaders and platoon commanders in directing and distributing the fire of their men.

Ranges. The ranges to various points in front and on the flanks must be ascertained in advance.

Obstacles. The extent to which artificial obstacles shall be used will depend upon the necessity therefor as determined by the nature of the ground, the extent to which it is intended to defend the position, and the time and material available. (See "Obstacles", Par. 706.)

673. **Defense of Bridges and other Defiles.** As a rule, a bridge, causeway, ford, or other open defile, can be defended better from the near side— that is to say, from the defenders' side—by bringing a cross-fire upon the

defile the enemy is attempting to pass. Everything possible must be done to prevent him from forming a fighting front on the defenders' side.

In defending mountain passes or other defiles whose flanks are not open, such as village streets, it is usually better to dispute the passage itself, inch by inch, and prepare to receive the enemy with a strong, effective cross-fire should he succeed in reaching the open space at the defenders' end of the defile.

674. Defense of Villages. To prepare a village for defense:—

1. Construct trenches controlling the principal avenues of approach, which should, if practicable, be mined; station sharpshooters and expert riflemen in the belfries of churches and other commanding places, and construct barricades across the streets where heavy fighting is likely to occur.

2. Divide the village into sections, with well-defined lines of communication, each section being held by a separate tactical unit, which will provide for the care of the wounded.

3. Prepare the buildings for defense, provide for food, water and ammunition. (Buildings prepared for occupation must not be occupied if subjected to artillery fire.)

Should the attackers penetrate the outer line of defense, the defenders must stubbornly contest every inch of the ground, fighting from buildings, barricades, and trenches.

675. Defense against Cavalry. Infantry, unless taken by surprise or demoralized, need fear nothing from cavalry. All you need to do is not to lose your head, shoot straight and aim low. The kneeling position is usually the best. If attacked by a cavalryman with a drawn saber, try to get on the near side of the horse. The rider cannot use his saber effectively against a dismounted man on his left.

CHAPTER XI

THE COMPANY IN DEFENSE

USUAL TYPES OF DEFENSE

676. Having discussed the general rules and principles of defense, we will now take the company through the usual types of defense, viz:

1. The ordinary type of defense, where there are many companies assisted by artillery, occupying a position for the purpose of fighting entirely on the defensive; or, for the purpose of fighting on the defensive, with the view of taking the offensive later.

2. An advance guard is suddenly thrown on the defensive by the action of the enemy.

3. An attack is suddenly thrown on the defensive by the action of the enemy.

4. An advance guard that has routed the enemy and captured the position is in turn attacked by the enemy and thrown on the defensive.

5. An attack that has routed the enemy and captured the position, is, in turn, attacked by the enemy and thrown on the defensive.

6. A rear guard covering the withdrawal of a force from the battlefield.

7. Defense at night.

8. Defense of a position by an outpost.

Many Companies, Assisted by Artillery, Occupying A Position For the Purpose of Fighting Entirely on The Defensive; Or, For The Purpose of Fighting on the Defensive With the View of Taking the Offensive Later

It may be assumed that in this type of defense there will be sufficient time to dig fire trenches, construct obstacles, and clear the foreground.

The location of the fire trenches will, to a great extent, depend upon the amount of time available; for, a fire trench that is well located when the foreground is cleared, or some obstacle constructed, may be very badly placed if there is not enough time to do more than to prepare hasty entrenchments. When allotted the amount of front he is to defend, the company commander should, therefore, ascertain, if possible, about how much time he will have to strengthen his position.

The first thing for the company commander to decide, is whether he will begin work by digging his fire trench, clearing the foreground or constructing obstacles, and it is impossible to lay down any definite rule, except it may be said that, generally, artificial obstacles are luxuries, and as such would be the last to receive attention. However, it is not always easy to determine which should be done first—the fire trenches dug, or foreground cleared.

For example, if the position should be taken up in the afternoon and our outposts were well out to the front, it is not likely that the enemy would make a decisive attack before the following day, even though our advance guard may have been in action against the enemy's advanced troops. In such a case it would be better to employ a few men to work on the position of the fire trenches and construct a certain amount of cover, while the rest of the company cleared the foreground, as it is comparatively easy to dig trenches at night, but it is not so easy to site them, or to clear the foreground in darkness.

If the position selected for the fire trench is very exposed to the enemy's fire, which would mean that the ground in front must be fairly open, it would be best to begin work digging trenches with the majority of the company and let the remainder clear away the most serious obstacles.

If the position selected for the fire trench is naturally strong and provides more cover for the defenders than the attackers could possibly get within five or six hundred yards of the position, it would then be better to improve the field of fire by clearing the foreground, before digging any trenches, especially if there happened to be considerable obstruction to our fire, and view.

Should the section of the line allotted to the company run through a wood, we should first make a clearing, then construct obstacles on the enemy's side of the clearing and dig the fire trench last.

The one salient, important fact we must always bear in mind is this: *The defender must prevent the attacker from gaining superiority of fire.* In some cases the defender may accomplish this object better by digging his fire trenches first and in other cases by clearing the foreground first, depending upon circumstances, and in deciding which shall be done first, the company commander must use his judgment and common sense. With regard to obstacles it may be said that they can only delay the attack, but will not prevent it from gaining superiority of fire. It is most important, therefore, that we should first determine as nearly as possible the position that the attacker can reach but can not pass without gaining superiority of fire, and then plan accordingly with regard to locating the fire trench, clearing the foreground, and constructing obstacles, so that we may be in the best possible circumstances to prevent the enemy from subduing our fire.

An Advance Guard Is Suddenly Thrown on the Defensive by The Action of the Enemy

In this case the first thing to do is to occupy as soon as possible the best fire position available in the immediate neighborhood. It is all the better if this position should happen to be ahead—the very fact of the company advancing, even though a few yards, will help to improve the morale of the men. However, if it is out of the question to advance, and it is extremely probable that the position we occupy cannot be held without soon losing superiority of fire, then a fire-position must be occupied in rear, but we must not fall back a single foot more than is absolutely necessary. We must remember that reënforcements are close behind us, and that any wavering or hasty retreat on our part may easily lead to disaster. Every inch of ground must be stubbornly contested, and immediately upon reaching a good fire position, we must defend it to the last. Let the company commander and everyone else bear in mind that the safety of the whole of the main body in rear and the success of the subsequent battle may depend upon the vigorous, determined action of the advanced guard.

An Attack is Suddenly Thrown on The Defensive By The Action of The Enemy

If the attacking force should be suddenly thrown on the defensive by the enemy, the first object of the company commander should be to resume the offensive just as soon as he possibly can, seizing the very first opportunity to renew the attack, and then, at the earliest possible moment gain superiority of fire, advance to a position close enough to charge, and charge. Naturally enough, the ground will not be the same in front of all the companies, and those that are favored by the terrain should push forward and thus relieve the strain on those that are not so fortunate in that respect. The momentum of the companies that thus press forward will carry along other companies less fortunately situated. We should remember that the longer the company remains on the defensive the harder will it be to have it resume the attack. As in the case of the advance guard, we should not fall back, if it can possibly be avoided; for, if we do, other companies will probably follow suit—retreating is very infectious, and it works on the cumulative principle of a snow-ball rolling down the side of a hill. Indeed, it is generally better to continue occupying an indifferent fire-position than to seek a better one farther back. We must also remember that reënforcements will be coming up as rapidly as possible to help us, and their arrival should be the signal for a forward move. An advance of only a few yards will often change the whole moral aspect of the situation.

An Advance Guard That Has Routed The Enemy And Captured the Position, Is In Turn Attacked By The Enemy And Thrown On The Defensive

The situation is this: Our Advance guard has defeated the leading troops of the enemy, and is in possession of the ground; the leading troops of the enemy are being heavily reënforced with the view of retaking the ground they lost. How should we meet the situation? To begin with, our attack having been successful, not only is the morale of our troops very high, but, also, the advance guard commander will be able to make more elaborate arrangements than would be possible in any ordinary meeting engagement.

When the advance guard drove the leading troops of the enemy from their position it did all that was required of it for the time being; it is now merely called upon to retain the ground it has taken from the enemy.

Assuming that we made our attack as we should, it was a somewhat rapid operation with a very strong firing line from the beginning—hence, it is not likely that the companies are badly mixed up.

The first thing for the company commander to do is to reorganize his company, and get the squads and platoons under their proper commanders. He must then choose and occupy the best fire position available. Let us suppose the position where we now find ourselves is bad for defense and that any further attack would involve the advance guard in a premature action against the enemy's main body. It would be necessary for some of the companies to remain in the position gained and act as an outpost, while the rest of the companies would be withdrawn to a better defensive position shortly in rear. Of course, it would be quite impossible to improve any of the foreground that is exposed to the enemy's fire. However, other parts might be cleared. Also, as a rule, we would have to confine our entrenchments to such as the men can make with their entrenching tools while lying down. We, therefore, see we must so choose our fire position as to make the most of the ground as it exists, and, consequently, the chances are it will be most difficult to insure even a fairly straight line of defense. However, we should by all means avoid pronounced salients in our position. (See ''Salients'', Par. 663.)

If we are compelled to occupy a bad or indifferent fire-position, we must then make the best of the situation by assuming a very active defense with counter-attacks. Remember, fire-positions are bad, as a rule, because the terrain in front is such that it is easy for the attacker to advance under cover, gain superiority of fire, assemble close to the defenders and charge,—and these are the very advantages the defender will have when he assumes the offensive and delivers a counter-attack over the same ground; the tables being turned, the enemy will have a bad fire-position, and it will be easy for the defender to advance under cover, gain superiority of fire, assemble close to the enemy, and charge.

Any fire-position that may be selected must be improved by intrenching as much as the enemy's fire will permit. By means of scouts and patrols sent out to the front and flanks, we must get all the information we can about the enemy—his exact position, whether he is intrenching, etc.

After our previous fight during the attack of the advance guard, the men will be badly in need of water, and the company commander should arrange as soon as possible for them to get water.

There is one thing that every one (officers, noncommissioned officers and privates) must understand: *The safety of our main body and probably the success of the approaching battle will depend upon the defense of the position we are now occupying, and which we secured at the sacrifice of many casualties. It must, therefore, be held at all cost.*

An Attack That Has Routed The Enemy And Captured The Position, Is In Turn Attacked By The Enemy And Thrown On The Defensive

Although this situation is, in the main, but a modification of the one we have just discussed, it is easier to handle; for, the main body being involved, there are more troops available for the purpose, and the troops are even in higher spirits than in the preceding case.

The first thing to do is to push forward with the men who have just charged, and secure a good fire-position on the enemy's side. We will find our companies badly disorganized and mixed up with men of other companies and of other regiments. As we push forward the men, under the direction of officers and noncommissioned officers, should be grouping themselves into squads, platoons and companies, corporals taking charge of squads, platoon leaders grouping two or three squads together under their command, and captains assuming control of several of these groups. A fire-position must be established on the enemy's side of the locality, because he must be pursued immediately by fire to prevent his reforming; we must be prepared to repulse without delay, by fire and assault, any hostile counter-attack that may be attempted; ground must be gained to the front so that our artillery can come up in safety and assist us in the fire pursuit of the enemy, and in resisting any counter-attack; and time must be gained to enable our troops in rear to reform and prepare for further offensive operations, either by way of continuing the attack or pressing the pursuit.

Having secured a good fire-position, we must next fix aiming marks and take ranges, and complete the reorganization of the companies, getting the officers, noncommissioned officers and privates of the same regiment together in squads, platoons and companies.

The replenishment of the ammunition will be looked after by the proper field and staff officers.

Let us all bear in mind that whatever the circumstances may be, it is the duty of each and every one of us in the front line to advance just as soon as the enemy is driven back, and occupy a fire-position that will secure the ground we have captured.

A Rear Guard Covering The Withdrawal of a Force From The Battlefield

This type of defense is different from the others we have so far discussed, in that it is strictly passive, and special attention must be given to the successful withdrawal of the company as soon as it has accomplished its object.

The object of the operation is to give the main body time to reorganize and retire in good order. The procedure followed by a rear guard is quite simple. The companies are deployed on a wide front with few supports and no reserves. When the enemy has been made to deploy and attack us, or has been compelled to work around our flank, and we have occupied our position as long as is safe, a part of the rear guard retires under the covering fire of the other part, and occupies another position in rear, from which it can cover the retirement of the other part. Whether the right of the line should retire first, to be followed later by the left, or whether the center should withdraw first, or whether both flanks should withdraw first, leaving the center to follow later— these are matters to be determined by the nature of the ground and the comparative pressure of the attack that is brought against different parts of the line. The decision as to which company should retire first will rest with the officer in local command of the companies, but in the absence of specific orders, a company that is defending the only locality from which the line of retreat is visible, or that is holding a position commanding part of the rest of the line, would be last to retire. As a rule, before retiring, an officer or well trained noncommissioned officer should be sent back to ascertain the general line of defense in the next position in rear, and to reconnoiter suitable fire-positions, thus enabling the company commander to take his new position without delay. In case of broken country and considerable distance between the companies. scouts should be posted on our flanks to let us know when the adjoining companies begin to withdraw. The actual withdrawal must be made in fighting formation, and as rapidly as possible, so as to increase the distance between the enemy and ourselves. Also, rapid withdrawal and prompt occupation of the next position, will enable us to delay the enemy more than we could by a running fight.

In selecting fire-positions we must remember that in addition to the conflicting requirements of a good field of fire and good cover for the defenders, the position must also permit the company to withdraw easily. In other words, the fire-position must permit of such easy with-drawal that when the company falls back it will be able to gain cover in rear before the enemy can occupy the position that the company has just vacated, and this is something we want to remember; for, it makes unsuitable certain fire-positions which, under other circumstances, would be very desirable. For example, the foot of a slope in open country with a bare hillside in rear, although affording a fine field of fire and splendid cover for the defenders, would not be selected because of the difficulty of withdrawing over the hill and the casualties that would doubtless result. In such a case as this, the top of the hill, even with lots of dead ground immediately in front of it, would be a much more suitable posi-tion. The principle to work on is this: Our object is to delay the enemy in his advance, bring him to a standstill as far as possible from our position, and force him to obtain superiority of fire before getting any closer; we have no intention of standing a charge and consequently dead ground in our immediate front is immaterial; for, we will have retired before the enemy reaches that dead space. What we must do, then, is to select fire-positions that command all ground from about two hundred to eight hundred yards in front of the locality we are defending.

Care must be taken, especially in open country, to protect our flanks against hostile cavalry.

There is a great deal of ammunition expended in rear guard action, and the ammunition supply must receive the attention of the company commander.

Defense At Night

677. Level open ground is the best kind of ground over which to make a night attack; for, it is easy to traverse in the darkness. However, a defensive position with that kind of ground in front of it will have by day a good field of fire and will require comparatively few troops to defend it. We, therefore, see that, as a rule, those parts of the defensive line that are strongest by day are the most liable to attack by night. Of course, there are some exceptions to this, because, for example, a morass, vineyard, or other natural obstacle in front of a portion of the line, may make that part of the ground as impassable by night as by day.

Obstacles are useful for delaying or breaking up night attacks, especially if their location is unknown to the enemy and there are troops stationed close to prevent their destruction.

The best way to guard against night attacks is to have the front of the position well patrolled, and to throw out outposts, protected by artificial obstacles. If the outposts are forced to retire, the enemy's advance will be discovered and a night attack can be made upon him. In connection with the use of obstacles for night work, it may be remarked that they should not be visible by day. The company must, therefore, be able to construct obstacles in the darkness, and should be trained to do so in time of peace.

Experience has shown that, taken all in all, the bayonet is the most important and reliable weapon we have for night fighting. However, the bayonet cannot be used effectively behind entrenchments—its great power and value is when in the hands of well-trained men who are prepared to charge, without even firing a shot, as soon as the enemy has been located. If the enemy can attack at night, there is no reason why we cannot do the same, even though we may be on the defensive for the time being.

The whole situation may, therefore, be summed up in these few words: When acting on the defensive at night the whole position must be watched by sentries and patrols; the most probable avenues of approach must be obstructed by obstacles, with outposts behind them, ready to delay the enemy and prevent the destruction of the obstacles; and as soon as the enemy has been located, if advancing to attack us, companies must be sent out to attack him with the bayonet—from which we see that by far the most important part of defense by night is an attack or counter-attack.

Defense Of A Position By An Outpost

This type of defense is treated in detail in the following chapter, "The Company on Outpost".

CHAPTER XII
THE COMPANY ON OUTPOST
ESTABLISHING THE OUTPOST

678. We will now apply some of the general principles of outposts (see Par. 514) to a company taking up its position on the line of outposts.

Let us suppose that our battalion has been detailed for outpost duty.

In order to understand more fully the duties and functions of the company commander, we will first consider what the major does. To begin with, he and the battalion will have been detailed for outpost duty before the march was completed, and he will have been told, amongst other things, what is known of the enemy and also what is known of other bodies of our own troops, where the main body will halt, the general position to be occupied by the outpost, and what the commander intends doing in case of attack.

The major verbally designates, say, two companies, as the reserve, and the other two companies, including our own, as the support. He places the senior officer of the reserve companies in command of the reserve and tells him where he is to go, and he indicates the general line the outpost is to occupy and assigns the amount of front each of the other companies is to cover. The limits of the sector so assigned should be marked by some distincitive features, such as trees, buildings, woods, streams, etc., as it is important that each company should know the exact limits of its frontage. He tells the company commanders what he knows of the enemy and of our own troops so far as they affect the outposts, he indicates the line of resistance and how much resistance is to be afforded in case of attack, states whether intrenchments and obstacles are to be constructed, gives instructions about lighting fires and cooking, and states where he can be found.

Upon receiving his orders from the major, the company commander, *with a proper covering detachment,* moves to the locality allotted him and as he arrives upon the ground he is to occupy, he sends out, as temporary security, patrols or skirmishers, or both, a short distance in front of the general position the outguards will occupy, holding the rest of the company back under cover. If practicable, the company commander should precede the company and make a rapid examination of the ground. He then sends out *observation groups,* varying in size from four men to a platoon, generally a squad, to watch the country in the direction of the enemy. These groups constitute the *outguards,* and are just sufficient in number to cover the front of the supports, and to connect where necessary with the outguards of adjoining supports.

The company commander next selects a defensive position on the general line of resistance, from which not only can he command the approaches, but where he can also give assistance to the adjoining supports; he then gives instructions in regard to the intrenchments and obstacles, after which he makes a more careful reconnaissance of the section assigned him; corrects the position of the outguards, if necessary; gives them instructions as to their duties in case of attack or when strangers approach their posts; tells them the number (if any) of their post, the number of the outguard and support and the numbers of the adjoining outguards and supports; points out lines of retreat in case they are compelled to fall back to the support, cautioning the men not to mask the fire of the support; he tells them the names of all villages, rivers, etc., in view, and the places to which the wagon roads and the railroads lead; selects, if necessary, places for additional posts to be occupied at night and during fog; sees that suitable connections are made between him and the adjoining outguards, and between his support and the adjoining supports; and questions subordinate commanders to test their grasp of the situation and knowledge of their duties, and on returning to the support he sends a report with a *sketch* to the outpost commander, showing the dispositions made.

After the line of observation has been established, the support stacks arms and the men are permitted to remove their equipments, except cartridge belts. One or more sentinels are posted over these supports, and they guard the property and watch for signals from the outguards. Fires are concealed as much as possible and the messing is done by reliefs. Mounted messengers ordinarily do not unsaddle; they rest, water and feed as directed.

After the major has received reports from both company commanders, he will himself visit the outguards and supports and make such changes as he may deem necessary, immediately after which he will submit to the commander of the troops a written report, accompanied by a combined sketch showing the positions of the different parts of the outpost. The major might begin his inspection of the line of outguards before receiving the reports of the company commanders.

In training and instructing the company in outpost work, it is always best to send out a few patrols and scouts an hour or two in advance, with definite instructions as to what they are to do, and have them operate against the company as hostile scouts and patrols. If the rest of the company know that patrols and scouts are operating in their front, and will try to work their way through the outpost line, they will naturally take a keener interest in their work. Exercises of this kind create a feeling of rivalry between the scouts and patrols, who, on the one hand, are trying to work their way through the line of outposts, and the outguards and patrols, who, on the other hand, are trying to prevent them from so doing. It makes the work much more *human*.

CHAPTER XIII

THE COMPANY IN SCOUTING AND PATROLLING

679. The general principles of patrolling are explained in Par. 455; so we need not repeat them here.

Many of the principles of scouting are, in reality, nothing but the fundamentals of patrolling, and the main function of scouting, *reconnoitering*, is also the function of a certain class of patrols. So, we see that scouting and patrolling are inseparably connected, and the importance of training the members of the company in the principles of scouting is, therefore, evident.

680. Requisites of a good Scout. A man, to make a good scout, should possess the following qualifications:—

Have good eyesight and hearing;

Be active, intelligent and resourceful;

Be confident and plucky;

Be healthy and strong;

Be able to swim, signal, read a map, make a rough sketch, and, of course, read and write.

681. Eyesight and Hearing. To be able to use the eye and the ear quickly and accurately is one of the first principles of successful scouting. Quickness and accuracy of sight and hearing are to a great extent a matter of training and practice. The savage, for instance, almost invariably has quick eyesight and good hearing, simply from continual practice.

Get into the habit of seeing, *observing*, things—your eyesight must never be resting, but must be continually glancing around, in every direction, and *seeing* different objects. As you walk along through the country get into the habit of noticing hoof-prints, wheel-ruts, etc., and observing the trees, houses, streams, animals, men, etc., that you pass.

Practice looking at distant objects and discovering objects in the distance. On seeing distant signs, do not jump at a conclusion as to what they are, but watch and study them carefully first.

Get into the habit of listening for sounds and of distinguishing by what different sounds are made.

682. Finding your Way in a strange Country. The principal means of finding one's way in a strange country are by map reading, asking the way, the points of the compass and land marks.

Map Reading. This, of course, presupposes the possession of a map. The subject of map reading is explained in Pars. 539-555.

Asking the Way. In civilized countries one has no trouble in finding his way by asking, provided, of course, he speaks the language. If in a foreign country, learn as soon as you can the equivalent of such

---

The best book on scouting that the author has even seen, is Baden-Powell's "Aids to Scouting", which was consulted in the preparation of this chapter.

expressions as "What is the way to ......?" "Where is ......?" "What is the name of this place?", and a few other phrases of a similar nature. Remember, however, that the natives may sometimes deceive you in their answers.

Points of the Compass. A compass is, of course, the best, quickest and simplest way of determining the directions, except in localities where there is much iron, in which case it becomes very unreliable.

For determining the points of the compass by means of the North Star and the face of a watch, see Par. 689.

The points of the compass can also be ascertained by facing the sun in the morning and spreading out your arms straight from the body. Before you is east; behind you, west; to your right, south; to your left, north.

The points of the compass can be determined by noting the limbs and bark of trees. The bark on the north side of trees is thicker and rougher than that on the south side, and moss is most generally found near the roots on the north side. The limbs and branches are generally longer on the south side of trees, while the branches on the north are usually knotty, twisted and drooped. The tops of pine trees dip or trend to the north.

683. Lost. In connection with finding your way through strange country, it may be said, should you find you have lost your way, do not lose your head. Keep cool—try not to let your brains get into your feet. By this we mean don't run around and make things worse, and play yourself out. First of all, sit down and think; cool off. Then climb a tree, or hill, and endeavor to locate some familiar object you passed, so as to retrace your steps. If it gets dark and you are not in hostile territory, build a good big fire. The chances are you have been missed by your comrades and if they see the fire, they will conclude you are there and will send out for you. Also, if not in hostile territory, distress signals may be given by firing your rifle, but don't waste all your ammunition.

If you find a stream, follow it; it will generally lead somewhere— where civilization exists.

The tendency of people who are lost is to travel in a circle uselessly.

Remember this important rule: *Always notice the direction of the compass when you start out, and what changes of direction you make afterwards.*

684. Landmarks. Landmarks or prominent features of any kind are a great assistance in finding one's way in a strange country. In starting out, always notice the hills, conspicuous trees, high buildings, towers, rivers, etc. For example, if starting out on a reconnaissance you see directly to the north of you a mountain, it will act as a guide without your having to refer to your compass or the sun. If you should start from near a church, the steeple will serve as a guide or landmark when you start to make your way back.

When you pass a conspicuous object, like a broken gate, a strangely shaped rock, etc., try to remember it, so that should you desire to return that way, you can do so by following the chain of landmarks. On passing such landmarks always see what they look like from the other side; for, that will be the side from which you will first see them upon the return trip.

The secret of never getting lost is to note carefully the original direction in which you start, and after that to note carefully all landmarks. Get in the habit of doing this in time of peace—it will then become second nature for you to do it in time of war.

It may sometimes be necessary, especially in difficult country, such as when travelling through a forest, and over broken mountains and ravines, for you to make your own landmarks for finding your way back by "blazing" (cutting pieces of bark from the trees), breaking small branches off bushes, piling up stones, making a line across a crossroad or path you did not follow, etc.

685. Concealment and Dodging. Both in scouting and patrolling it must be remembered not only that it is important you should get information, but it is also fully as important that the enemy should not know you have the information—hence, the necessity of hiding yourself. And remember, too, if you keep yourself hidden, not only will you probably be able to see twice as much of what the enemy is doing, but it may also save you from being captured, wounded or killed.

Should you find the enemy has seen you, it is often advisable to pretend that you have not seen him, or that you have other men with you by signalling to imaginary comrades.

As far as possible, keep under cover by travelling along hedges, banks, low ground, etc. If moving over open country, make your way as quickly as possible from one clump of trees or bushes to another; or, from rocks, hollows or such other cover as may exist, to other cover. As soon as you reach new cover, look around and examine your surroundings carefully.

Do not have about you anything that glistens, and at night be careful not to wear anything that jingles or rattles. And remember that at night a lighted match can be seen as far as 900 yards and a lighted cigarette nearly 300 yards. In looking through a bush or over the top of a hill, break off a leafy branch and hold it in front of your face.

In selecting a tree, tower or top of a house or other look-out place from which to observe the enemy from concealment, always plan beforehand how you would make your escape, if discovered and pursued.

A place with more than one avenue of escape should be selected, so that if cut off in one direction you can escape from the other. For example, should the enemy reach the foot of a tower in which you are, you would be completely cut off, while if he reached a house on whose roof you happened to be, you would have several avenues of escape.

Although trees make excellent look-out places, they must, for the same reasons as towers, be used with caution. In this connection it may be remarked unless one sees foot marks leading to a tree, men are apt not to look up in trees for the enemy—hence, be careful not to leave foot marks. When in a tree, either stand close against the trunk, or lie along a large branch, so that your body will look like a part of the trunk or branch.

In using a hill as a look-out place, do not make the common mistake of showing yourself on the skyline. Reach the top of the hill slowly and gradually by crouching down and crawling, and raise your head above the crest by inches. In leaving, lower your head gradually and crawl away by degrees, as any quick or sudden movement on the sky-line is likely to attract attention. And, remember, just because you don't happen to seen the enemy that is no sign that he is not about. At maneuvers and in exercises soldiers continually make the mistake of exposing themselves on the skyline.

At night confine yourself as much as possible to low ground, ditches, etc. This will keep you down in the dark and will enable you, in turn, to see outlined against the higher ground any enemy that may approach you.

At night especially, but also during the day, the enemy will expect you along roads and paths, as it is easier to travel along roads and paths than across country and they also serve as good guides in finding your way. As a rule, it is best to use the road until it brings you near the enemy and then leave it and travel across country. You will thus be able better to avoid the outposts and patrols that will surely be watching the roads.

Practice in time of peace the art of concealing yourself and observing passers-by. Conceal yourself near some frequented road and imagine the people traveling over it are enemies whose. numbers you wish to count and whose conversation you wish to overhear. Select a spot where they are not likely to look for you, and which has one or more avenues of escape; choose a position with a background that matches your clothes in color; keep quiet, skin your eyes; stretch your ears.

A mounted scout should always have wire cutters when operating in a country where there are wire fences.

686. Tracking. By "tracking" we mean following up footmarks. The same as the huntsman tracks his game so should we learn how to track the enemy. One of the first things to learn in tracking is the pace at which the man or horse was traveling when the track was made.

A horse walking makes pairs of footmarks, each hind foot being close to the impression of the forefoot. At a trot the tracks are similar, but the pairs of footmarks are farther apart and deeper, the toe especially being more deeply indented than at the walk. At a canter there are two single footmarks and then a pair. At a gallop the footmarks are single and deeply indented. As a rule, the hind feet are longer and narrower than the forefeet.

In the case of a man walking, the whole flat of the foot comes equally on the ground, the footmarks usually about 30 inches apart. If running, the toes are more deeply indented in the ground, and the footmarks are considerably farther apart than when walking. Note the difference between footmarks made by soldier's shoes and civilian's shoes, and those made by men and those made by women and children.

Study the difference between the tracks by a gun, a carriage, an escort wagon, an automobile, a bicycle, etc., and the direction in which they were going.

In addition to being able to determine the pace of tracks, it is most important that you should be able to tell how old they are. However, ability to do this with any degree of accuracy, requires a vast amount of practice. A great deal depends on the kind and the state of the ground and the weather. For example, if on a dry, windy day you follow a certain track over varying ground, you will find that on light sandy soil, for instance, it will look old in a very short time, because any damp earth that may have been kicked up from under the surface will dry very quickly to the same color as the rest of the surface, and the edge of the footmark will soon be rounded off by the breeze blowing over the dry dust. The same track in damp ground will look much fresher, and in damp clay, in the shade of trees, a track which may be a day old will look quite fresh.

The following are clues to the age of tracks: Spots of rain having fallen on them since they were made, if, of course, you know when the rain fell; the crossing of other tracks over the original ones; the freshness or coldness of the droppings of horses and other animals (due allowance being made for the effect of the sun, rain, etc.), and, in the case of grass that has been trodden down, the extent to which it has since dried or withered.

Having learned to distinguish the pace and age of tracks, the next thing to do is to learn how to follow them over all kinds of ground. This is a most difficult accomplishment and one that requires a vast amount of practice to attain even fair proficiency.

In tracking where it is difficult to see the track, such as on hard **ground, or in the grass**, note the direction of the last foot-print that you can see, then look on ahead of you a few yards, say, 20 or 30, in the same direction, and, in grass, you will probably see the blades bent or trodden, and, on ground, you will probably see stones displaced or scratched—or some other small sign which otherwise would not be noticed. These indistinct signs, seen one behind the other, give a track that can be followed with comparative ease.

If you should lose the track, try to find it again by placing your handkerchief, hat, or other object on the last footmark you noticed, and then work around it in a wide circle, with a radius of, say, 30, 50, or 100 yards, choosing the most favorable ground, soft ground, if possible. If with a patrol, only one or two men should try to find the onward track; for, if everyone starts in to find it, the chances are the track will be obliterated with their footmarks. In trying to find the continuation of a track this way, always place yourself in the enemy's position, look around the country, imagine what you would have done, and then move out in that direction and look for his tracks in soft ground.

Practice

In order to learn the appearance of tracks, get a suitable piece of soft ground, and across this have a man walk and then run, and have a horse walk, trot, canter and gallop. The next day make similar tracks alongside the first ones and then notice the difference between the two. Also, make tracks on ordinary ground, grass, sand, etc., and practice following them up. Finally, practice tracking men sent out for the purpose. The work will probably be very difficult, even disheartening at first, but you will gradually improve, if you persevere.

Above all things, get into the habit of seeing any tracks that may be on the ground. When out walking, when going through exercises at maneuvers, and at other times, always notice what tracks are on the ground before you, and study them.

The following exercises in scouting and patrolling afford excellent practice and training:—

687. The Mouse and Cat Contest. 1. A section of country three or four miles square, with well-defined limits, is selected. The boundaries are made known to all contestants and anyone going outside of them will be disqualified.

2. Two patrols of eight men each are sent out as "mice". They occupy any positions they may wish within the boundaries named, and conceal themselves to watch for hostile patrols.

3. Half an hour later two other squads, wearing white bands around their hats, or having other distinguishing marks, are sent out as "cats" to locate, if possible, and report upon the position of the "mice".

4. An hour is fixed when the exercise shall end, and if within the given time the "cats" have not discovered the "mice", the "mice" win.

5. The "cats" will write reports of any "mice" patrols they may see.

Rules

1. An umpire (officer or noncommissioned officer) goes with each patrol and his decisions as to capture and other matters are the with each patrol and his decisions as to capture and other matters are the orders of the company commander. The umpires must take every possible precaution to conceal themselves so as not to reveal the position of the patrols with which they are.

Each umpire will carry a watch, all watches being set with that of the company commander before the exercise commences.

2. Any "cat" patrol coming within 50 yards of a "mouse" patrol, without seeing the "mice," is considered captured.

3. When the time is up, the umpires will bring in the patrols and report to the company commander.

688. Flag-Stealing Contest. 1. A section of country of suitable size, with well-defined limits, is selected, the boundaries being made known to the contestants.

2. The contestants are divided into two forces of about 20 men each, and each side will establish three Cossack posts along a general line designated by the company commander, the two positions being selected facing each other and being a suitable distance apart. The men not forming part of the Cossack posts will be used as reconnoitering patrols.

3. About three-quarters of a mile in rear of the center of each line of outposts four flags will be planted, in line, about 30 yards apart.

4. The scouts and patrols of each force will try to locate the outposts of the other force, and then to work their way around or between them, steal the flags and bring them back to their own side. They will endeavor to prevent the enemy from doing the same.

5. One scout or patrol will not carry away more than one flag at a time, and will have to return to their side safely with the flag before they can come back and capture another.

6. Scouts may work singly or in pairs. Any scout or patrol coming within 80 yards of a stronger hostile party, or Cossack post, will be considered as captured, if seen by the enemy, and if carrying a captured flag at the time, the flag will not count as having been captured. Of course, if a scout or patrol can pass within 80 yards of the enemy without being discovered, it may do so.

7. An umpire (officer or noncommissioned officer) will be with each Cossack post, each patrol, and at the position of the flags.

8. The hour when the exercise ends will be designated in advance and at that hour the umpires will bring in the Cossack posts and patrols. The same requirements regarding watches obtains as in the Mouse and Cat Contest.

9. At the conclusion of the contest the commander of each side will hand in to the company commander all sketches and reports made by his men.

10. Points will be awarded as follows:

Each flag captured, 5.

For each sketch and hostile report of the position of a Cossack post, 3.

For each report of movements of a hostile patrol, 2.

The side getting the greatest number of points will win.

11. Umpires may penalize the contestants for a violation of the rules.

The same contest may be carried out at night, substituting lighted Japanese lanterns for the flags.

CHAPTER XIV*
NIGHT OPERATIONS

689. Importance. Because of the long range and great accuracy of modern fire-arms there has been in recent years a marked increase in the practice of night operations. During the Russo-Japanese War both sides frequently resorted to night attacks, especially to cover ground that could not, because of hostile fire, be traversed by day. There is no doubt that in future wars night operations will be of common occurrence, not only for massing troops under cover of darkness in favorable positions for further action, but also for the actual assault of positions.

GENERAL PRINCIPLES

(From Infantry Drill Regulations. The paragraph numbers follow the paragraph.)

The Attack. To enable large forces to gain ground toward the enemy, it may sometimes be cheaper and quicker in the end to move well forward and to deploy at night. In such case the area in which the deployment is to be made should, if practicable, be occupied by covering troops before dark.

The deployment will be made with great difficulty unless the ground has been studied by daylight. The deployment gains little unless it establishes the firing line well within effective range of the enemy's main position. (450).

Attack of Fortifications. If the enemy is strongly fortified and time permits, it may be advisable to wait and approach the charging point under cover of darkness. The necessary reconnaissance and arrangements should be made before dark. If the charge is not to be made at once, the troops intrench the advanced position, using sand bags if necessary. Before daylight the foreground should be cleared of obstacles. (482).

If the distance is short and other conditions are favorable, the charge may be made without fire preparation. If made, it should be launched with spirit and suddenness at the break of day. (483).

Deployment for Defense. If a night attack or close approach by the enemy is expected, troops in a prepared position should strengthen the outguards and firing line and construct as numerous and effective obstacles as possible. Supports and local reserves should move close to the firing line and should, with the firing line, keep bayonets fixed. If practicable, the front should be illuminated, preferably from the flanks of the section. (509).

---

*Many of the hints in this chapter were gathered from "Night Operations for Infantry", by Colonel Dawkins, British Army, and "Night Movements", translated from the Japanese by Lieut. Burnett, U. S. Army.

Only short range fire is of any value in resisting night attacks. The bayonet is the chief reliance. (510).

NIGHT OPERATIONS

By employing night operations troops make use of the cover of darkness to minimize losses from hostile fire or to escape observation. Night operations may also be necessary for the purpose of gaining time. Control is difficult and confusion is frequently unavoidable.

It may be necessary to take advantage of darkness in order to assault from a point gained during the day, or to approach a point from which a daylight assault is to be made, or to effect both the approach and the assault. (558).

Offensive and defensive night operations should be practiced frequently in order that troops may learn to cover ground in the dark and arrive at a destination quietly and in good order, and in order to train officers in the necessary preparation and reconnaissance.

Only simple and well-appointed formations should be employed.

Troops should be thoroughly trained in the necessary details— e. g., night patrolling, night marching, and communication at night. (559).

The ground to be traversed should be studied by daylight and, if practicable, at night. It should be cleared of hostile detachments before dark, and, if practicable, should be occupied by covering troops.

Orders must be formulated with great care and clearness. Each unit must be given a definite objective and direction, and care must be exercised to avoid collision between units.

Whenever contact with the enemy is anticipated, a distinctive badge should be worn by all. (560).

Preparations must be made with secrecy. When the movement is started, and not until then, the officers and men should be acquainted with the general design, the composition of the whole force, and should be given such additional information as will insure coöperation and eliminate mistakes.

During the movement every precaution must be taken to keep secret the fact that troops are abroad.

Unfriendly guides must frequently be impressed. These should be secured against escape, outcry, or deception.

Fire action should be avoided in offensive operations. In general, pieces should not be loaded. Men must be trained to rely upon the bayonet and to use it aggressively. (561.)

Long night marches should be made only over well-defined routes. March discipline must be rigidly enforced. The troops should be marched in as compact a formation as practicable, with the usual covering detachments. Advance and rear guard distances should be greatly reduced. They are shortest when the mission is an offensive one. The connecting files are numerous. (562).

A night advance made with a view to making an attack by day usually terminates with the hasty construction of intrenchments in the dark. Such an advance should be timed so as to allow an hour or more of darkness for intrenching.

An advance that it to terminate in an assault at the break of day should be timed so that the troops will not arrive long before the assault is to be made; otherwise the advantage of partial surprise will be lost and the enemy will be allowed to reenforce the threatened point. (563)

The night attack is ordinarily confined to small forces, or to minor engagements in a general battle, or to seizure of positions occupied by covering or advanced detachments. Decisive results are not often obtained.

Poorly disciplined and untrained troops are unfit for night attacks or for night operations demanding the exercise of skill and care.

Troops attacking at night can advance close to the enemy in compact formations and without suffering loss from hostile artillery or infantry fire The defender is ignorant of the strength or direction of the attack.

A force which makes a vigorous bayonet charge in the dark will often threw a much larger force into disorder. (564).

Reconnaissance should be made to ascertain the position and strength of the enemy and to study the terrain to be traversed. Officers who are to participate in the attack should conduct this reconnaissance. Reconnaissance at night is especially valuable. Features that are distinguishable at night should be carefully noted, and their distances from the enemy, from the starting point of the troops, and from other important points should be made known.

Preparations should have in view as complete a surprise as possible. An attack once begun must be carried to its conclusion, even if the surprise is not as complete as was planned or anticipated. (565).

The time of night at which the attack should be made depends upon the object sought. If a decisive attack is intended, it will generally yield the best results if made just before daylight. If the object is merely to gain an intrenched position for further operations, an earlier hour is necessary in order that the position gained may be intrenched under cover of darkness. (566).

The formation for attack must be simple. It should be carefully effected and the troops verified at a safe distance from the enemy. The attacking troops should be formed in compact lines and with strong supports at short distances. The reserve should be far enough in rear to avoid being drawn into the action until the commander so desires. Bayonets are fixed, but pieces are not loaded.

Darkness causes fire to be wild and ineffective. The attacking troops should march steadily on the enemy without firing, but should be prepared and determined to fight vigorously with the bayonet.

In advancing to the attack the aim should be to get as close as possible to the enemy before being discovered, then to trust to the bayonet.

If the assault is successful, preparations must be made at once to repel a counter-attack. (567).

On the defense, preparations to resist night attacks should be made by daylight, whenever such attacks are to be feared.

Obstacles placed in front of a defensive position are especially valuable to the defense at night. Many forms of obstacles which would give an attacker little concern in the daytime become serious hindrances at night.

After dark the foreground should be illuminated whenever practicable and strong patrols should be pushed to the front.

When it is learned that the enemy is approaching, the trenches are manned and the supports moved close to the firing line.

Supports fix bayonets, but do not load. Whenever practicable and necessary they should be used for counter-attacks, preferably against a hostile flank.

The defender should open fire as soon as results may be expected. This fire may avert or postpone the bayonet combat, and it warns all supporting troops. It is not likely that fire alone can stop the attack. The defender must be resolved to fight with the bayonet.

Ordinarily fire will not be effective at ranges exceeding 50 yards.

A white rag around the muzzle of the rifle will assist in sighting the piece when the front sight is not visible. (568).

TRAINING OF THE COMPANY

Night movements are amongst the most difficult operations of war, and, therefore require the most careful, painstaking and thorough training and instruction of troops in all matters pertaining thereto. The history of night fighting shows that in most cases defeat is due to disorganization through panic. It is said that in daylight the moral is to the physical as three is to one. That being the case, it is hard to say what the ratio is at night, when a general atmosphere of mystery, uncertainty and fear of surprise envelops the operations, and, of necessity affects the nerves of the men. The vital importance, therefore, of accustoming troops as much as we can in peace to the conditions that will obtain in night fighting, cannot be overestimated. The following outline shows the subjects in which individual and collective instruction and training should be given:

INDIVIDUAL TRAINING

General. The first thing to be done is to accustom the soldier to darkness and to teach him to overcome the nervousness which is natural to the average man in darkness.

689 (contd.)

The best way to do this is to begin by training him in the use of his powers of vision and hearing under conditions of darkness, which are strange to him. The company should be divided into squads for this instruction.

Vision. Take several men to ground with which they are familiar. Have them notice the different appearance which objects present at night, when viewed in different degrees of light and shade; the comparative visibility of men under different conditions of dress, background, etc.; the ease with which bright objects are seen; the difference between the visibility of men standing on a skyline and those standing on a slope. Post the men in pairs at intervals along a line which the instructors will endeavor to cross without being seen. The instructors should cross from both sides, so as to compel observation in both directions. Have a man (later, several) walk away from the rest of the men and when he is about to disappear from view, halt him, and estimate the distance. Send a man (later, several) outside the field of vision, to advance on the rest of the men. Halt him when he enters the field of vision and estimate the distance. Send a number of men outside the limit of vision and then let them advance on the rest of the men, using cover and seeing how near they can approach unobserved.

Hearing. Place a number of men a few yards apart and make them guess what a noise is caused by, and its approximate position. The rattle of a meat can, the movement of a patrol, the working of the bolt of a rifle, the throwing down of accoutrements, low talking, etc., may be utilized. Take special pains to impress upon the men the penetrating power of the human voice, and the necessity of preserving absolute silence in night operations. Have blank cartridges fired and teach the men to judge their direction and approximate distance away.

Finding Bearings. Show the men how to determine the points of the compass from the North Star. The Big Dipper constellation looks like this:

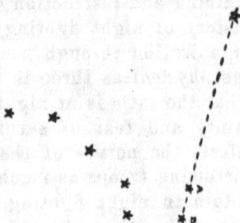

The North Star is on the prolongation of a line joining the two "pointing" stars, and at above five times the distance between the two stars. At another time have those same men individually locate the North Star. Using this star as a guide, practice the men moving in different directions, by such commands as, "Smith, move southeast". "Jones, move northwest", etc.

To test a man's ability to keep a given direction when moving in the darkness, choose a spot from which no prominent land marks are visible, advance toward it accompanied by a man, from a distance not less than 200 paces. While advancing the soldier must take his bearings. On arriving at the spot chosen the instructor will turn the soldier around rapidly two or three times and then have him continue to advance in the same direction as before. No prominent landmarks should be visible from the starting point.

Moving in the Dark. Form four or five men in line with about one pace interval, the instructor being on one of the flanks. Place some clearly visible mark, such as a lantern, for the instructor to march on. Impress upon the men the importance of lifting their feet up high and bringing them to the ground quietly and firmly, and of keeping in touch with the guide and conforming to his movements without sound or signal. The pace should be slow and frequent halts should be made to test the promptness of the men in halting and advancing together. As the line advances, each man will in turn take his place on the flank and act as guide. The light on which the men are marching should be hidden from view at intervals, in order to test the ability of the men to maintain the original direction. Later on, the number of men in a line may be increased considerably. The rougher the ground, the darker the night and the longer the line, the slower must the pace be and the more frequent the halts. After passing an obstacle men instinctively line up parallel to it, and consequently if the obstacle does not lay at right angle to the line of advance, the direction will be lost; so, be sure to guard against this.

Night Fencing. Practice the men in charging in the dark against a white cloth or the dummy figure of a man. In the beginning have the figure in a fixed place, but later have the soldier charge seeking the figure, and not knowing just exactly where it is beforehand.

Night Entrenching. It is frequently necessary in time of war to dig trenches at night in front of the enemy, and while this work is easy in the moonlight, it is very difficult in the dark. Bear in mind the following points:

1. The tendency is to make the trench too narrow; hence, guard against this.

2. Be careful not to throw the earth too far or too near.

3. Do not strike your neighbor's tools in working.

4. Do not use the pick unless necessary, because it makes considerable noise.

5. Do not scrape the tools together in order to get off the dirt; use a chip of wood or the toe of the shoe.

6. Make as little noise as possible in digging and handling your tools.

7. If discovered by the enemy's searchlights, do not become excited or confused; simply lie down.

8. If attacked by the enemy, do not get rattled and throw your tools away—put them in some fixed place where they can be found again.

Equipment. At first the men should be taken out without arms, but later on they should be trained to work in full equipment. Teach every man what parts of his equipment are likely to make a noise under special circumstances, such as lying down, rising, crossing obstacles, etc., and instruct him how to guard against it. Bayonets should always be fixed, but in order to avoid accidents the scabbard should be left on them.

From the beginning of the training continually impress upon the men that it is absolutely criminal to fire without orders during a night operation and that the bayonet is the only weapon he can use with advantage to himself and safety to his comrades.

Night Firing. As a rule men fire too high in the dark. They must, therefore, be cautioned not to raise the rifle above the horizontal, or incline the upper part of the body to the rear. When the firing is stopped be sure to turn on the safety-lock. Experience during the Russo-Japanese War taught the Japanese the kneeling position is the most suitable for horizontal firing. The following method, to be conducted in daytime, may be employed in training the soldier to hold his rifle parallel to the ground while firing in the dark:—Have each soldier, kneeling, close his eyes and bring his rifle to the position of aim, barrel parallel to the ground. With the rifle in this position, let him open his eyes and examine it. Then have this done by squad, by command. When they become proficient in this movement, have them close their eyes and while the eyes are closed, put up a target and have them practice horizontal firing, opening their eyes each time after pulling the trigger and then examining the position of the piece.

COLLECTIVE TRAINING

At first practice squads, then the platoons and later the company in simple movements, such as squads right and left, right and left oblique, etc., gradually leading up to more complicated ones in close and extended order, such as right and left front into line, advancing in platoon and squad columns, charging the enemy, etc. As far as possible the movements should be executed by simple prearranged signals from the unit commanders. The signals, which must not be visible to the enemy, may be made with a white handkerchief or a white flag, if the night be not too dark; with an electric flashlight, a dark lantern or luminous disk. The light of the flashlight or lantern must be screened, so it cannot be seen by the enemy. The following signals are suggested:

To advance: Raise vertically the lantern or other object with which the signal is made.

To halt: Lower and raise the object several times.

To lie down: Bring the object down near the ground.

To form squad columns: Move the object several times to the right and left.

To form platoon columns: Describe several circles.

As skirmishers: Move the object front to rear several times.

Night Marches. In acting as an advance guard to a column, the company would send out a point a few yards ahead, which would be followed by the rest of the company. Three or four scouts should be sent out a hundred yards or so ahead of the point. They should advance at a quick pace, keeping in the shadow on the side of the road, being constantly on the alert, using their ears even more than their eyes. They will halt to listen at cross-roads and suspicious places, and move on again when they hear the company approaching. Should the enemy be discovered, one of the scouts will return to warn the advance guard—the others will conceal themselves and watch. Under no circumstances must the scouts ever fire, unless it be for the purpose of warning the company and there is no other way of doing so. The following is suggested as a good formation for a company acting as advance guard at night:

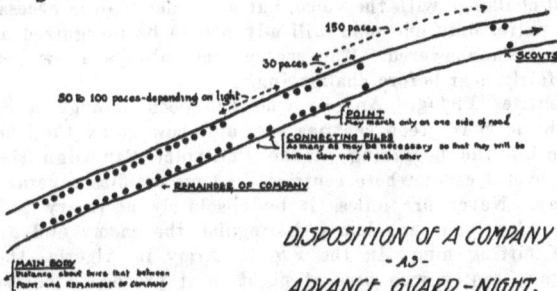

150 paces

30 paces

80 to 100 paces—depending on light

POINT
May march only on one side of road

CONNECTING FILES
As many as may be necessary so that they will be within clear view of each other

SCOUTS

← REMAINDER OF COMPANY

MAIN BODY
Distance about twice that between Point and Remainder of Company

DISPOSITION OF A COMPANY
- AS -
ADVANCE GUARD - NIGHT.

A company marching alone would move in the same formation as when acting as advance guard, except that it would protect its rear with a few scouts. Of course, the nature of the country and proximity and activity of the enemy, will determine the best formation to be used, but whatever the formation may be, always remember to cover well your front, rear and flanks, with scouts, whose distance away will vary with the light and nature of the country. *Don't forget that protection in rear is very important.*

The men must be warned against firing, smoking, talking, striking matches, making noise, etc. They should also be informed of the object in view, direction of the enemy, etc.

In night marches the rests should not exceed five minutes; otherwise, many men will fall asleep.

OUTPOSTS

Careful training in outpost duty at night is very harassing, but, in view of its importance, should not be neglected. This instruction should be given with the greatest thoroughness, strictness and attention to detail.

690. Sentries Challenging. In challenging sentries must be careful to avoid any noise that would disclose their position. In fact, challenging by voice should be reduced to a minimum by arranging a system of signals by which the officers of the day, patrols, etc., can be recognized. The following signals, any one of which may be decided upon, which would be made first by the sentry and then answered by the approaching party, are suggested: Clap the hands together twice; strike the ground twice with the butt of the rifle; strike the butt of the rifle twice with the hand; whistle softly twice. The replying signal would be the same as the sentry's signal, except that in case of the use of the butt of the rifle, an officer would reply by striking twice on his revolver holster. After repeating the signal once, if it is not answered, the sentry will challenge with the voice, but no louder than is necessary. In case of a patrol only one man will advance to be recognized after the signal has been answered. The sentry must always allow persons to approach fairly near before challenging.

691. Sentries Firing. Anyone who has been through a campaign knows how nervous green sentries are, and how quick they are about firing. During the beginning of the Philippine Campaign the author heard of several cases where sentries fired on fire-flies several hundred yards away. Never fire unless it be absolutely necessary to give an alarm, or unless you can clearly distinguish the enemy and are fairly certain of hitting him. In the French Army in Algeria, there is a rule that any sentry who fires at night must produce a corpse, or be able to show by blood marks that he hit the person fired at. If he can do neither, he is punished for giving a false alarm.

Marking of Route from Outguards to Supports. The route from the support to the outguards, and from pickets to their sentries, should, if necessary, be clearly marked with scraps of paper, green sticks with the bark peeled off, or in any other suitable way.

Readiness for Action. The supports should always be ready for action. The men must sleep with their rifles beside them and in such places that they will be able to fall in promptly in case of attack. Some men have a way of sleeping with their blankets over their heads. This should not be allowed—the ears must always be uncovered. The commander, or the second in command, with several men, should remain awake. When the commander lies down he should do so near the sentry, which is always posted over the support.

GENERAL

Connections. It is of the greatest importance that proper connection be maintained between the different parts of a command engaged in night operations. It is astonishing with what facility units go astray and how difficult it is for them to find their way back where they belong.

Preparation. It matters not what the nature of the night operation may be, the most careful preparation is necessary. Success often depends upon the care and thoroughness with which the plans are made.

All possible eventualities should be thought of and provided for as far as practicable. The first thing to do is to get as much information as possible about the ground to be covered and the position of the enemy, and care must be taken to see that the information is accurate. Reconnaissance must be made by night as well as by day; for, ground looks very different at night from what it does during the day.

CHAPTER XV
FIELD ORDERS OF ENLISTED MEN

692. The men of the company should learn the substance of the following orders:

Platoon Leaders
693.   **In Battle**

My field orders are:—

1. In battle I will do all I can to preserve the integrity of squads; I will designate new squad leaders to replace those disabled, and will organize new squads when necessary.

2. On the firing line I will carry out faithfully and thoroughly the duties stated in paragraph 289.

3. In battle, in camp, on the march , and at all other times, I will see that the guides, squad leaders and privates around me carry out their field orders, and that they comply with the known wishes and desires of the company commander, and I will myself obey and carry out so much of the field orders of the privates as apply to me.

694.   Guides

My field orders are:—

1. I will endeavor, by assisting officers and platoon leaders and otherwise, to preserve the integrity of squads.

2. On the firing line I will carry out faithfully and thoroughly the duties stated in paragraph 289.

3. In battle, in camp, on the march, and at all other times, I will see that the squad leaders and privates around me carry out their field orders and that the known wishes and desires of the company commander are complied with, and I will myself obey and carry out so much of the field orders of the privates as apply to me.

695.   Squad Leaders

My field orders are:—

1. In battle I will do all I can to preserve the integrity of my squad.

2. On the firing line I will carry out faithfully and thoroughly the duties stated in paragraph 289.

3. In battle, in camp, on the march, and at all other times, I will see that the privates around me carry out their field orders and that they comply with the known wishes of the company commander, and I will myself obey and carry out so much of the field orders of the privates as apply to me.

696.   Musicians

My field orders are:—

1. Whenever the company is deployed, I will at once join the captain and remain with him until further orders.

2. On the firing line I will carry out faithfully and thoroughly the duties stated in paragraph 235, and in case the company forms part of a battalion, I will be on the constant look out for orders and signals from the battalion commander.

3. When the order to charge is sounded I will at once repeat it.

4. I will obey and carry out as much of the field orders of a private as apply to me, and I will at all times faithfully comply with the known wishes and desires of the company commander.

697.   **Privates**

IN BATTLE

My battle orders are:—

1. I will not straggle, nor will I under any circumstances skulk, but, at the command to advance, I will always do so *at once*.

2. In advancing by rushes, or any other way, I will always endeavor to be the first man to start the advance. I am aware of the fact that in advancing by rushes the last men to reach the new position are exposed to the enemy's fire that much longer, and, consequently, are more apt to get hit than the others.

3. I will not endeavor to carry any wounded to the rear. That is the business of the litter-bearers. My business is to remain on the firing line and help with my rifle.

4. I will not fail to change my sight when new ranges are announced, nor will I forget to change my sight when advancing by rushes, whether or not the new range is announced.

5. I will never lose an opportunity to replenish my ammunition supply from the belts of the dead and wounded. The time may come before the fight is over when ammunition will be worth a hundred times its weight in gold.

6. I will use a rest for my rifle whenever I can. It will improve my shooting.

7. I will obey at once all the commands and orders of my squad leader, and platoon commander.

8. In case of surprise, excitement or confusion, I will at once listen for the orders of my officers and noncommissioned officers, and I will obey them immediately and implicitly.

9. I will take advantage of cover, but, if by so doing, I cannot see the enemy, I will then get where I can see him; for, it is much more important that I should be able to see the enemy so as to shoot at him, than it is for me to conceal myself from his sight. I will always take special pains to avoid the sky-line (the tops of hills and ridges); for, a man on the sky-line looms up as a clear, distinct target.

10. When on the firing line, I will be on the lookout for signals and orders from my squad leaders; I will exercise proper care in setting my sights and delivering my fire; I will aim deliberately; I will observe the

enemy carefully, increase my fire when the target is favorable and cease firing when the enemy disappears; I will not neglect a target because it is not very distinct; I will not waste my ammunition, but will be economical with it.

(*Sharpshooters and expert riflemen.*)   I will be on the lookout for the enemy's officers and will fire at every one I see.

11.   I will use the ammunition in bandoleers first.   I will keep thirty (30) rounds in the right pocket section of my belt as a reserve to be used only when ordered by an officer.   (Par. 551, Infantry Drill Regulations).

12.   When reënforcing the firing line I will find out at once the range and target from the men already there.

13.   I will at all times make every possible effort not to get separated from my squad, but should I unavoidably become separated, I will immediately try to rejoin it.   Should I fail in this, I will then join the nearest squad and put myself under the orders of its leader.   Should I not be able to do this, and thus find myself without a leader for the time being, I will not lose my head, but will go on fighting on my own hook, remembering that the only way for us to win the battle is for each and every man to fight, **fight, FIGHT** for all that he is worth. To give up fighting will only make it just that much easier for the enemy to kill me and my comrades.

698.   NIGHT OPERATIONS

My orders in night operations are:—

1.   I will not talk or make other noise, but will preserve absolute silence.   Nor will I smoke or strike matches, because the light might be seen by the enemy.

2.   I will be constantly on the lookout for signals and orders from my officers and noncommissioned officers, and I will obey at once all signals and orders.

3.   If ordered to fire, I will be sure to hold my piece parallel to the ground, so as not to shoot high.

4.   Under no circumstances will I ever fire in a night movement unless ordered by a superior, or unless I am placed in a position where I must fire in order to give the alarm.

699.   OUTPOST ORDERS

My orders as a sentry on outpost are:—

1.   The number of my post (if any) is No......, of Outguard No. ...... Outguard No...... is on my right, and No......., on my left.

2.   The support of this outguard is located.....(define location), and if compelled to fall back, I will retreat......(state line of retreat).

3. (If any). There are——advance detachments in front of me, located as follows.....(give location), and.....friendly patrols are operating in my front.

4. Should any friendly patrol attempt to cross the outpost line near me without telling me who they are, where they are going, about how long they will be out and how they will return, I will halt them and get this information before allowing them to proceed.

5. The enemy is (or is supposed to be) ..... (define location as accurately as possible), and if he approaches us, he will very likely do so by way of ....(state direction from which the enemy is expected).

6. In case the enemy approaches, I will.....(state fully and specifically what you would do.)

7. I know the names of all the villages, streams and other prominent features in sight and also where the roads lead. The village over yonder (pointing) is .....; this road (pointing) leads to.....; that high mountain is called....., etc.

8. I will keep a constant watch to the front and flanks and will pay special attention to unusual or suspicious noises or occurrences.

9. If I see any indications of the enemy, I will at once notify the outguard commander. In case of great and immediate danger or in case of attack, I will give the alarm by firing my piece rapidly.

10. Officers, noncommissioned officers and detachments that I recognize as parts of the outposts, and officers that I know have authority to do so, will be allowed to pass in and out of the outpost line. I will detain all others and notify the commander of the outguard.

11. I will fire upon individuals or detachments who fail to halt, or, otherwise disobey me after a *second* warning, or sooner, if they attempt to attack or escape.

12. I will halt deserters approaching, order them to lay down their arms and notify the commander of the outguard. I will order deserters pursued by the enemy to drop their arms and will at once notify the commander of the outguard. Should deserters fail to lay down their arms after a *second* warning, I will fire upon them.

13. I will halt bearers of flags of truce, cause them to face about, and will notify the commander of the outguard.

14. I will salute only when I address, or am addressed by officers.

15. In case of doubt as to what to do, I will call for the commander of the outguard.

16. At night I will allow persons to approach fairly close before challenging, and I will challenge in a low voice so as not to reveal my position to the enemy.

17. I will never fire at night unless I can clearly see the enemy and I am sure I can hit him, or unless it be absolutely necessary to fire in order to give the alarm.

700. ADVANCE AND REAR GUARDS

My orders when acting as connecting file on advance or rear guard duty, are:—

1. I will be on the constant lookout for signals, which I will always transmit at once.

2. I will always take distance from, and keep in sight of, the body in my rear.

3. When the column halts, if I am not already in a position where I can see both bodies between which I am acting as connecting file, I will, if possible, place myself in such position, and will keep on the constant lookout for signals from both bodies. Should I receive a signal from one body when I cannot see the other, I will at once run to a position from which I can see the other and repeat the signal.

701. PATROLLING

My orders when patrolling are:—

I will take special pains to remain concealed as much as possible; for, when seeking information about the enemy, it is often almost as important not to be seen as it is to see.

702. MESSAGES

1. When I am given a verbal message by a noncommissioned officer to carry, I will always repeat the message to the noncommissioned officer before leaving, to see that I understand it, and, as I am leaving, I will go over the message several times in my own mind. If given a message by an officer, and not directed by him to repeat it, I will before leaving ask, for instance, "May I repeat the message so as to be sure that I understand it?", and, as I am leaving, I will go over the message several times in my own mind.

2. I will always conceal in my shoe or elsewhere any written message I may be carrying, and, if captured, will try to destroy the message the very first chance I get.

703. ON THE MARCH

My orders on the march are:—

1. I will fill my canteen before the march begins.

2. I will not leave ranks to get water or for any other purpose without permission of my company commander, and during halts I will not leave the immediate vicinity of the company without permission.

3. Should I wish to relieve myself when the company halts, I will do so as soon as the halt is made and not wait until it is nearly over.

4. I will at all times keep my proper place in column.

5. I will not nibble food while actually marching.

6. I will not sit on damp ground during halts.

7. I will not enter yards, orchards, or gardens, during halts, nor will I ever enter a house unless invited to do so by the occupants.

8. When the command is given to fall in after a halt, I will fall in promptly.

704. IN CAMP

My orders in camp are:—

1. Upon first reaching camp I will not leave camp until I find out from the First Sergeant what the orders are about the men leaving camp.

2. I will not introduce liquor into camp.

3. In camp and on the march I will take good care of my feet and look after my health according to the instruction received in garrison.

GENERAL

1. In battle, in camp, on the march, and at all other times, I will comply faithfully with all the known wishes and desires of the company commander.

2. If I see I am going to be captured, I will, if possible, throw away the bolt of my rifle and should I have field glasses in my possession I will break the glass.

3. Should I be taken prisoner, I will not, under any circumstances give any information about our troops, and should I be compelled to answer questions, I will give misleading answers. Nor will I talk with any of our men about our own troops, what we were doing when captured, etc., because the chances are the enemy or some of their spies will overhear my conversation. I will take advantage of the first opportunity to make my escape and get back to our troops with all the information that I can get about the enemy.

CHAPTER XVI

INTRENCHMENTS

705. Lying Trench.—When intrenching under fire, cover is first se-cured in the lying position, each man scooping out a depression for his body and throwing the earth to the front. Such a trench affords limited protection against rifle fire and less against shrapnel. Soldiers should be taught to construct such trenches as rapidly as possible, avoiding all neatness, which takes time, having in view only the rapid construction of a row of pits.

Sitting Trench.—If time permits, the orginal excavation may be enlarged and deepened until it is possible to assume a sitting position, with the legs crossed and the shoulder to the parapet. In such a posi-tion a man presents a smaller target to shrapnel bullets than in the lying position and can fire more comfortably and with less exposure than in the kneeling position.

Standing Trench.—From the sitting trench the excavation may be continued until a standing position is possible.

Classification of Trenches:—

Trenches may be classified into *fire trenches, cover trenches,* and *communicating trenches,* the first named being occupied by the firing line, the second by the supports, and the last by troops passing between the first and second.

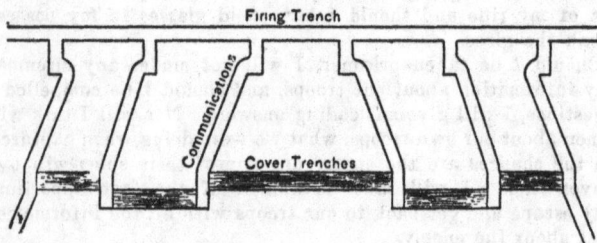

Fire Trenches:—

The lessons learned from the Russo-Japanese War have resulted in the discarding of the old-fashioned wide and shallow kneeling trenches. The simplest form of *fire trench* is deep and narrow and has a flat concealed parapet, as shown in Fig. 1. In ordinary soil, and on a basis of two reliefs and tasks of 5 feet, it can be constructed in about two hours with intrenching tools. This trench affords fair cover for troops subjected to artillery fire, but not actually firing.

When it is probable that time will permit elaboration, the simple trench should be planned with a view to developing it ultimately into more complete forms, as shown in Figs 2 and 3.

In very difficult soil, if the time is short, it may be necessary to dig a wider, shallower trench with a higher parapet.

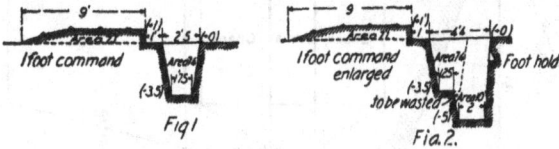

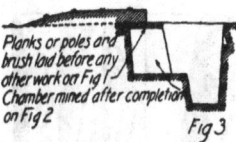

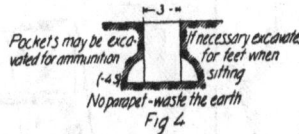

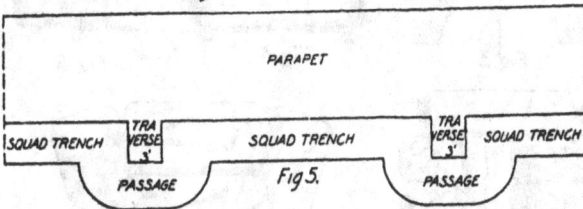

Area 21   1 foot command   1'+ 2.5'+(-0)   Area 14 w 2.5'   (-3.5)

Fig 1

9   1 foot command enlarged   Area 14 w 2.5'   Foot hold   (-3.5)   to be wasted 2'   (-5')

Fig. 2.

Planks or poles and brush laid before any other work on Fig 1 Chamber mined after completion on Fig 2   *Fig 3*

3   Pockets may be excavated for ammunition   (-4.5)   No parapet - waste the earth   If necessary excavate for feet when sitting   *Fig 4*

PARAPET

SQUAD TRENCH   TRAVERSE 3'   SQUAD TRENCH   TRAVERSE 3'   SQUAD TRENCH

PASSAGE   *Fig 5.*   PASSAGE

A Trench without Parapet.—Where the excavated earth is easily removed, a fire trench without parapet, as shown in Fig. 4, may be the one best suited to the soil and other conditions affecting the choice of profile. The enemy's infantry, as well as his artillery, will generally have great difficulty in seeing this trench.

It must be remembered that the type profiles given are not at all rigid and that they should be modified to suit the ground.

Head cover is the term applied to any horizontal cover which may be provided above the plane of fire. It is obtained by notching or loopholing the top of the parapet so that the bottoms of the notches or loopholes are in the desired plane of fire. The extra height of parapet may be 12 to 18 inches and the loopholes may be 3 to 3½ feet center to center.

Head cover is of limited utility. It increases the visibility of the parapet and restricts the field of fire. At close range the loopholes serve as aiming points to steady the enemy's fire and may do more harm than good at longer ranges. This is especially the case if the enemy can see any light through the loophole. He waits for the light to be obscured, when he fires, knowing there is a man's head behind the loophole. A background must be provided or a removable screen arranged so that

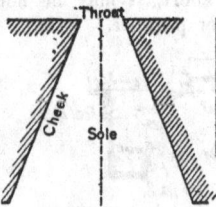

Fig. 1

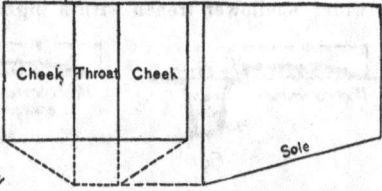

Fig. 2

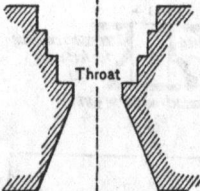

Fig. 3

Fig. 4

Fig. 5

Fig. 6

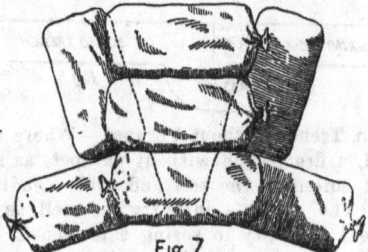

Fig. 7

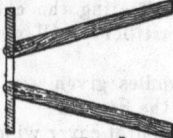

Fig. 8

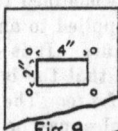

Fig. 9

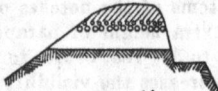

Fig. 10

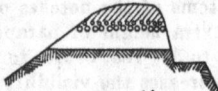

Fig. 11

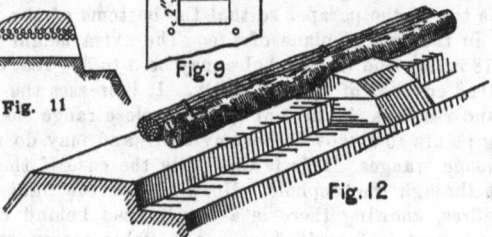

Fig. 12

there will be no difference in the appearance of the loophole whether a man is looking through it or not. Head cover is advantageous only when the conditions of the foreground are such that the enemy can not get close up.

Notches and **loopholes**, Figs. 1-3, are alike in all respects, except that the latter have a roof or top and the former have not. The bottom, also called **floor** or **sole**, is a part of the original superior slope. The sides, sometimes called **cheeks**, are vertical or nearly so. The plan depends upon local conditions. There is always a narrow part, called the **throat**, which is just large enough to take the rifle and permit sighting. From the throat the sides diverge at an angle, called the **splay**, which depends upon the field of fire necessary.

The position of the throat may vary. If on the outside, it is less conspicuous but more easily obstructed by injury to the parapet and more difficult to use, since in changing aim laterally the man must move around a pivot in the plane of the throat. If the material of which the loophole is constructed presents hard surfaces, the throat should be outside, notwithstanding the disadvantages of that position, or else the sides must be stepped as in Fig. 3. In some cases it may be best to adopt a compromise position and put the throat in the middle, Fig. 3. Figs. 4 to 7 show details and dimensions of a loophole of sand bags.

A serviceable form of loophole consists of a pyramidal box of plank with a steel plate spiked across the small end and pierced for fire. Fig. 8 shows a section of such a construction. It is commonly known as the **hopper loophole**. The plate should be $\frac{3}{8}$ in. thick, if of special steel; or $\frac{1}{2}$ in., if ordinary metal. Fig. 9 shows the opening used by the Japanese in Manchuria and Fig. 10 that used by the Russians.

The construction of a notch requires only the introduction of some available rigid material to form the sides; by adding a cover the notch becomes a loophole. Various methods of supporting earth will be described under "Revetments." Where the fire involves a wide lateral and small vertical angle, loopholes may take the form of a long slit. Such a form will result from laying logs or fascines lengthwise on the parapet, supported at intervals by sods or other material, Fig. 12, or small poles covered with earth may be used, Fig. 11.

Overhead cover. This usually consists of a raised platform of some kind covered with earth. It is frequently combined with horizontal cover in a single structure, which protects the top and exposed side. The supporting platform will almost always be of wood and may vary from brushwood or light poles to heavy timbers and plank. It is better, especially with brush or poles, to place a layer of sods, grass down, or straw, or grain sacks over the platform before putting on the earth, to prevent the latter from sifting through.

The thickness of overhead cover depends upon the class of fire against which protection is desired, and is sometimes limited by the vertical space available, since it must afford headroom beneath, and generally should not project above the nearest natural or artificial horizonal cover. For splinter proofs a layer of earth 6 to 8 ins. thick on a support of brush or poles strong enough to hold it up will suffice if the structure is horizontal. If the front is higher than the rear, less thickness is necessary; if the rear is higher than the front, more is required. For bombproofs a minimum thickness of 6 ins. of timber and 3 ft. of earth is necessary against field and siege guns, or 12 ins. timber and 6 ft. of earth against the howitzers and mortars of a heavy siege train.

In determining the **area of overhead cover** to be provided, allow 6 sq. ft. per man for occupancy while on duty only, or 12 sq. ft. per man for continuous occupancy not of long duration. For long occupation 18 to 20 sq. ft. per man should be provided.

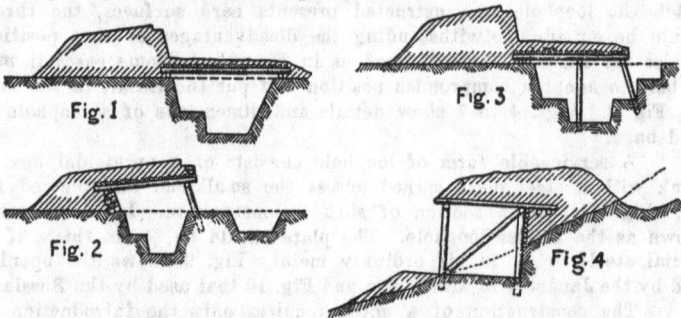

Types of overhead cover.

Lookouts.—To enable the garrison of a trench to get the greatest amount of comfort and rest, a *lookout* should be constructed and a sentinel stationed therein.

The simplest form would consist of two sandbags placed on the parapet and splayed so as to give the required view, and carefully concealed.

Better forms may be constructed, with one side resting on the berm by using short uprights with overhead cover, a slit on all sides being provided for observation.

Location.—There are two things to be considered in locating trenches: (1) The tactical situation, and (2) the nature of the ground. The first consideration requires that the trenches be so located as to give the best field of fire. Locating near the base of hills possesses the advantage of horizontal fire, but, as a rule, it is difficult to support trenches so located and to retreat therefrom in case of necessity. While location near the crest of hills—on the "military crest"—does not possess the advantage of horizontal fire, it is easier to support trenches so located and to retreat therefrom. Depending upon circumstances, there are times when it will be better to intrench near the base of hills and there are other times when it will be better to intrench on the "military crest", which is always in front of the natural crest. The construction of trenches along the "military crest" does not give any "dead space"—that is, any space to the front that can not be reached by the fire of the men in the trenches.

Whether we should construct our trenches on high or low ground is a matter that should always be carefully considered under the particular conditions that happen to exist at that particular time, and the matter may be summarized as follows:

The advantages of the high ground are:

1. We can generally see better what is going on to our front and flanks; and the men have a feeling of security that they do not enjoy on low ground.

2. We can usually reënforce the firing line better and the dead and wounded can be removed more easily.

3. The line of retreat is better.

The disadvantages are:—

1. The plunging fire of a high position is not as effective as a sweeping fire of a low one.

2. It is not as easy to conceal our position.

The advantages of low ground, are:—

1. The low, sweeping fire that we get, especially when the ground in front is fairly flat and the view over the greater part of it is uninterrupted, is the most effective kind of fire.

2. As a rule it is easier to conceal trenches on low ground, especially from artillery fire.

3. If our trenches are on low ground, our artillery will be able to find good positions on the hill behind us without interfering with the infantry defense.

The disadvantages are:—

1. As a rule it will be more difficult to reënforce the firing line and to remove the dead and wounded from the trenches.

2. On a low position there will usually be an increase of dead space in our front.

3. The average soldier acting on the defensive dreads that the enemy may turn his flank, and this feeling is much more pronounced on low ground than on high ground. Should the enemy succeed in getting a footing on our flank with our trenches on top of the hill, it would be bad enough, but it would certainly be far worse if he got a footing on top

705 (contd.)

of the hill, on the flank and rear, with our company on low ground in front. We, therefore, see there are things to be said for and against both high and low ground, and the most that can be said without examining a particular piece of ground is: Our natural inclination is to select high ground, but, as a rule, this choice will reduce our fire effect, and if there is a covered approach to our fire trenches and very little dead ground in front of it, with an extensive field of fire, there is no doubt the lower ground is better. However, if these conditions do not exist to a considerable degree, the moral advantage of the higher ground must be given great weight, especially in a close country.

With regard to the nature of the ground, trenches should, if practicable, be so located as to avoid stony ground, because of the difficult work entailed and of the danger of flying fragments, should the parapet be struck by an artillery projectile.

To locate the trace of the trenches, lie on the ground at intervals and select the best field of fire consistent with the requirements of the situation.

Trenches should be laid out in company lengths, if possible, and adjoining trenches should afford each other mutual support. The flanks and important gaps in the line should be protected by fire trenches echeloned in rear.

Clearing the Foreground.—Time permitting, it is very important that the ground in front of the trenches should be cleared of brushwood, high grass and everything else that might screen the enemy.

Concealment of Trenches.—The location of the trenches should be disguised by covering the side toward the enemy with grass, branches, leaves, etc.

Obstacles.—It is sometimes desirable to place obstacles in front of trenches, so as to obstruct the advance of the enemy, break up his formation and detain him under the fire of the men in the trenches. See "Obstacles", Par. 706.

Cover Trenches.—Where natural cover is not available for the support, each fire trench should have artificial cover in rear for its support,—either a *cover trench* of its own or one in comman with an adjoining fire trench.

The *cover trench* is simple and rectangular in profile. Concealment is most important, but when impossible, the trench should have substantial overhead cover. It is generally concealed by the contour of the ground or by natural features.

Cover trenches should be made as comfortable as possible. It will often be advisable to make them extensive enough to provide cooking and resting facilities for the garrisons of the corresponding fire trenches.

CHAPTER XVII

OBSTACLES

(From Engineer Department Manual on Field Fortifications.)

706. Object. The main object in placing obstacles in front of defensive positions is to stop the enemy's advance, or to delay him while under the defenders' fire, thus causing him to lose more men than the defenders and making his attack as difficult as possible.

Location. Obstacles must be so located that they will be exposed to the defender's fire, both artillery and infantry, and so that they will not obstruct the counter-attack on the part of the defender. They should also be invisible from the direction of the enemy's approach, should be difficult to destroy and should afford no screen or cover to the enemy.

Kinds of Obstacles. The following are the most common kinds of obstacles:—

707. Abatis consisting of trees lying parallel to each other with the branches pointing in the general direction of approach and interlaced. All leaves and small twigs should be removed and the stiff ends of branches pointed.

Abatis on open ground is most conveniently made of branches about 15 feet long. The branches are staked or tied down and the butts anchored by covering them with earth. Barbed wire may be interlaced among the branches. Successive rows are placed, the branches of one extending over the trunks of the one in front, so as to make the abatis 5 feet high and as wide as desired. It is better to place the abatis in a natural depression or a ditch, for concealment and protection from fire. If exposed to artillery, an abatis must be protected either as above or else by raising a glacis in front of it. Fig. I shows a typical form of abatis with a glacis in front. An abatis formed by felling trees toward the enemy, leaving the butt hanging to the stump, the branches prepared as before, is called a **slashing**, Fig. 2. It gives cover, and should be well flanked.

708. A **Palisade** is a man-tight fence of posts. Round poles 4 to 6 inches in diameter at the large end are best. If the sticks run 5 to 8 inches, they may be split. If defended from the rear, palisades give some shelter from fire and the openings should be made as large as possible without letting men through. If defended from the flank, they may be closer, say 3 to 4 inches apart. The top should be pointed. A strand or two of barbed wire run along the top and stapled to each post is a valuable addition.

Palisading is best made up in panels of 6 or 8 feet length, connected by a waling piece, preferably of plank, otherwise of split stuff. If the tops are free, two wales should be used, both underground. If the tops are connected by wires, one will do.

[659]

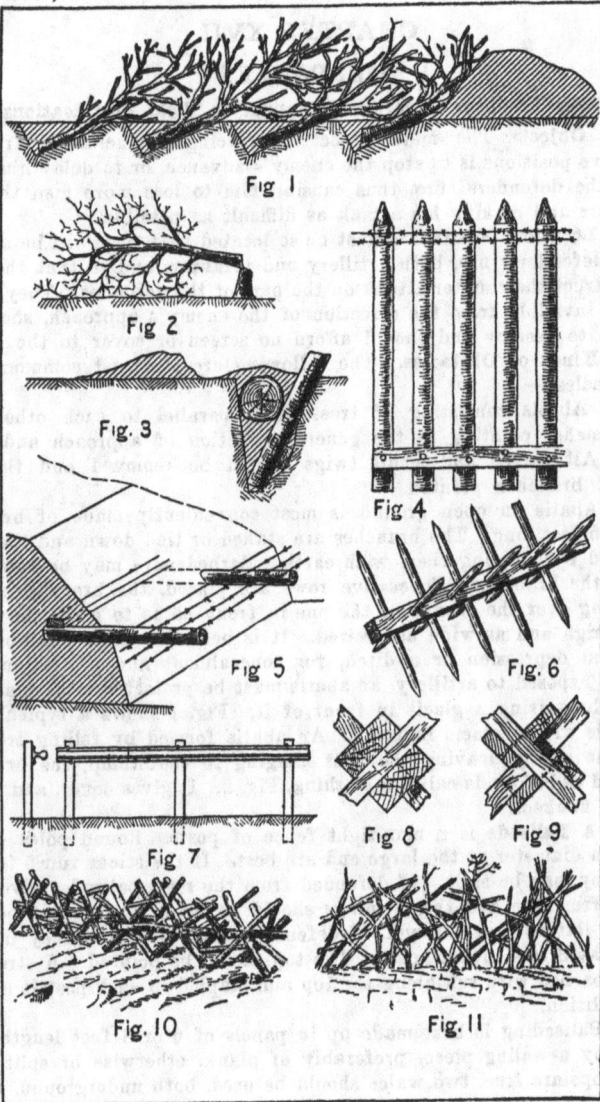

Fig 1

Fig 2

Fig 3

Fig 4

Fig 5

Fig. 6

Fig 7

Fig 8

Fig 9

Fig. 10

Fig. 11

Palisades should be planted to incline slightly to the front. As little earth should be disturbed in digging as possible, and one side of the trench should be kept in the desired plane of the palisade. If stones can be had to fit between the posts and the top of the trench, they will increase the stiffness of the structure and save time in ramming, or a small log may be laid in the trench along the outside of the posts. Figs. 3 and 4 show the construction and placing of palisades.

709. A **Fraise** is a palisade horizontal, or nearly so, projecting from the scarp or counterscarp. A modern and better form consists of supports at 3 or 4 feet interval, connected by barbed wire, forming a horizontal wire fence. Fig. 5

710. **Cheveaux de frise** are obstacles of the form shown in Fig. 6. They are usually made in sections of manageable length chained together at the ends. They are most useful in closing roads or other narrow passages, as they can be quickly opened for friendly troops. The lances may be of iron instead of wood and rectangular instead of round; the axial beam may be solid or composite. Figs. 8 and 9 show methods of constructing cheveaux de frise with dimension stuff.

A **formidable obstacle against cavalry** consists of railroad ties planted at intervals of 10 feet with the tops $4\frac{1}{2}$ feet above the ground, and connected by a line of rails spiked securely to each, Fig. 7. The rail ends should be connected by fish plates and bolted, with the ends of the bolts riveted down on the ends.

Figs. 10 and 11 show forms of heavy obstacles employed in Manchuria by the Russians and Japanese, respectively. The former is composed of timber trestles, made in rear and carried out at night. The latter appears to have been planted in place.

711. A **wire entanglement** is composed of stakes driven in the ground and connected by wire, barbed is the best, passing horizontally or diagonally, or both. The stakes are roughly in rectangular or quincunx order, but slight irregularities, both of position and height should be introduced.

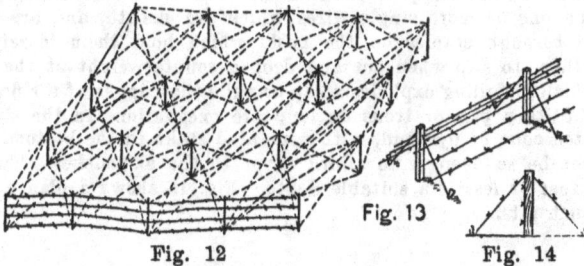

Fig. 13

Fig. 12   Fig. 14

In the **high entanglement** the stakes average 4 feet from the ground, and the wiring is horizontal and diagonal, Fig. 12.

The **low wire entanglement** has stakes averaging 18 inches above the ground and the wire is horizontal only. This form is especially effective if concealed in high grass. In both kinds the wires should be wound around the stakes and stapled and passed loosely from one stake to the next. When two or more wires cross they should be tied together. Barbed wire is more difficult to string but better when done. The most practicable form results from the use of barbed wire for the horizontal strands and smooth wire for the rest.

This is the most generally useful of all obstacles because of the rapidity of construction, the difficulty of removal, the comparatively slight injury from artillery fire, and its independence of local material supplies.

Time and materials.—One man can make 10 sq. yds. of low and 3 sq. yds. of high entanglement per hour. The low form requires 10 feet of wire per sq. yd. and the high 30 feet. No. 14 is a suitable size. The smooth wire runs 58.9 ft. to the lb. A 100-lb. coil will make 600 sq. yds. of low or 200 sq. yds. of high entanglement. If barbed wire is used, the weight will be about 2½ times as much.

712. Wire fence.—An ordinary barbed-wire fence is a considerable obstacle if well swept by fire. It becomes more formidable if a ditch is dug on one or both sides to obstruct the passage of wheels after the fence has been cut. The fence is much more difficult to get through if provided with an apron on one or both sides, inclined at an angle of about 45°, as indicated in Figs. 13 and 14. This form was much used in South Africa for connecting lines between blockhouses. When used in this way the lines of fence may be 300 to 600 yds. long, in plan like a worm fence, with the blockhouse at the reëntrant angles. Fixed rests for rifles, giving them the proper aim to enfilade the fence, were prepared at the blockhouses for use at night.

Such a fence may be arranged in many ways to give an automatic alarm either mechanically or electrically. The mechanical forms mostly depend on one or more single wires which are smooth, and are tightly stretched through staples on the posts which hold them loosely, permitting them to slip when cut and drop a counterweight at the blockhouse, which in falling explodes a cap or pulls the trigger of a rifle.

713. Military pits or trous de loup are excavations in the shape of an inverted cone or pyramid, with a pointed stake in the bottom. They should not be so deep as to afford cover to the skirmisher. Two and one-half feet or less is a suitable depth. Fig. 15 shows a plan and section of such pits.

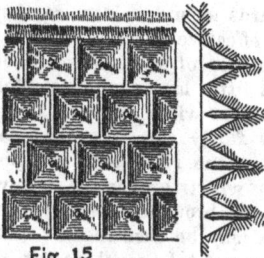

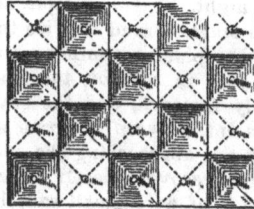

Fig 15   Fig. 16

They are usually dug in 3 or 5 rows and the earth thrown to the front to form a glacis. The rear row is dug first and then the next in front, and so on, so that no earth is cast over the finished pits.

An excellent arrangement is to dig the pits in a checkerboard plan, leaving alternate squares and placing a stake in each of them to form a wire entanglement, Fig. 16. One man can make 5 pits on a 2-hour relief.

Miscellanous barricades.—Anything rigid in form and movable may be used to give cover from view and fire and to obstruct the advance of an assailant. Boxes, bales and sacks of goods, furniture, books, etc., have been so used. The principles above stated for other obstacles should be followed, so far as the character of the materials will permit. The rest ingenuity must supply. Such devices are usually called **barricades** and are useful in blocking the streets of towns and cities.

714. **Inundations.**—Backing up the water of a stream so that it overflows a considerable area forms a good obstacle even though of fordable depth. If shallow, the difficulty of fording may be increased by irregular holes or ditches dug before the water comes up or by driving stakes or making entanglements. Fords have frequently been obstructed by ordinary harrows laid on the bottom with the teeth up.

The unusual natural conditions necessary to a successful inundation and the extent and character of the work required to construct the dams make this defense of exceptional use. It may be attempted with advantage when the drainage of a considerable flat area passes through a restricted opening, as a natural gorge, a culvert, or a bridge.

Open cribs filled with stones, or tighter ones with gravel or earth may form the basis of the obstruction to the flow of water. The usual method of tightening cracks or spaces between cribs is by throwing in earth or alternate layers of straw, hay, grass, earth, or sacks of clay. Unless the flow is enough to allow considerable leakage, the operation will not be practicable with field resources.

When the local conditions permit water to be run into the ditch of a parapet it should always be done.

715

715. **Obstacles in Front of Outguards** should be low so they cannot be seen at night. A very simple and effective obstacle can be made by fastening a single strand of wire to the top of stout stakes about a foot high, and then placing another wire a little higher and parallel to, and about one yard in rear of, the first. The wires must be drawn tight, and securely fastened, and the stakes fairly close together, so that if the wire is cut between any two stakes the remainder will not be cut loose. Any one approaching the enemy will trip over the first wire, and before he can recover himself he will be brought down by the second. In the absence of wire, small sapplings may be used instead. Of course, they are not as good as wire, but it does not take much to trip up a man in the dark.

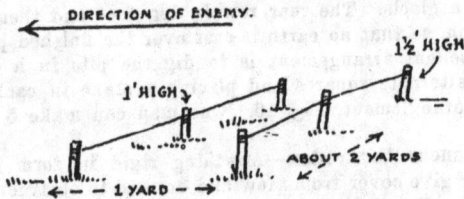

CHAPTER XVIII

FIELD FIRING*

716. Definition. By "Field firing" (called "Combat practice" in the Small-Arms Firing Manual) we mean collective firing at targets which simulate the appearance of an enemy under conditions approaching those found in war, and the application of this class of fire to tactical exercises.

717. Object. While it is important that soldiers should be good individual shots, individual shooting is not everything. The maximum effect of fire can be gotten only by instructed and disciplined troops under a commander capable of directing their fire properly. The fire of the company may be likened to spraying water from a hose, and as the fireman can shift his stream of water from one point to the other with certainty, being able to direct and control it with promptness and accuracy, so should the company commander be able to switch the cone of fire of his company from one target to another, having it at all times under direction and control. In other words, as the pliable, manageable hose responds to the will of the fireman, so should the company be so trained and instructed that it will respond to the will of the company commander on the firing line, in the midst of the noise and confusion of battle. No one except a man who has been in battle can realize how great are the noise and confusion, and how necessary and important are coöperation, teamwork, discipline, and communication, in order for a company commander to control and direct the fire of the company—there must be absolute coöperation, teamwork, and communication between all parts of the company—between the captain and the platoon leaders, the platoon leaders and the squad leaders, and the squad leaders and the members of their squads. Each and every man must know and do his part and endeavor all he can to keep in touch with and help the others. Now, the foundation of teamwork and coöperation, is communication—communication between the company commander and the men on the firing line—the means by which, the medium *through* which, he will make known his will to the men on the firing line. As stated before, because of the noise and confusion on the firing line this is no easy matter. The ideal way would be for the company commander to control the company by communicating direct with every man on the firing line, as graphically shown on the following page:

---

*This chapter is based on the Infantry Drill Regulations and the musketry instruction bulletins issued in 1913 by the Headquarters, Fifth Brigade, Second Division, at Galveston, Texas

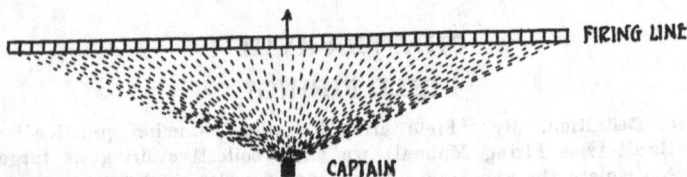

FIRING LINE

CAPTAIN

However, in the noise and confusion of battle it would be utterly impossible for all the men to hear the captain's voice. Experience shows that from 20 to 35 rifles are as many as one leader can control. The captain, must, therefore, control the company through the platoon commanders—that it is say, he *actually directs* the fire and the platoon commanders, assisted by the squad leaders, *actually control it.* In other words, the captain communicates with the men on the firing line, he makes his will known to them, through his platoon commanders, as graphically shown in this diagram:

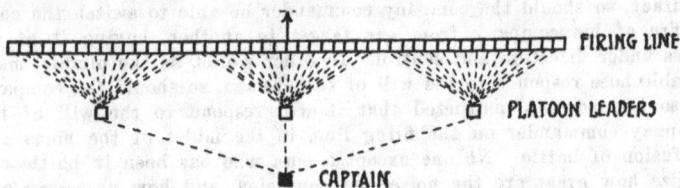

FIRING LINE

PLATOON LEADERS

CAPTAIN

However, in order for our system of communication to be successful, each and every man, as stated above, must know and do his part and endeavor all he can to help the others. If this is done, then the different parts and elements of the company will dove-tail and fit into one another, resulting in a complete, homogeneous whole, in the form of an efficient, pliable, manageable instrument in the hands of the company commander. And this is the object, the result, sought by practice and instruction in field firing, and which will be obtained if the captain, the platoon leaders, the squad leaders, the file closers, the musicians, and the privates, will perform the following duties and functions:

718. *The Captain.* (Fire Direction).

The captain *directs* the fire of the company or of designated platoons. He designates the target, and, when practicable, allots a part of the target to each platoon. Before beginning the fire action he determines the range, announces the sight setting, and indicates the class of fire to be employed and the time to open fire. Thereafter, he observes the fire effect, corrects material errors in sight setting, prevents exhaustion of the ammunition supply, and causes the distribution of such extra ammunition as may be received from the rear. (I. D. R. 249.)

Having indicated clearly what he desires the platoon leaders to do, the captain avoids interfering, except to correct serious errors or omissions. (I. D. R. 240).

719. *The Platoon Leaders.* **(Fire control).**

In combat the platoon is the *fire unit.* (I. D. R. 250).

Each platoon leader puts into execution the commands or dir-rections of the captain, having first taken such precautions to insure cor-rect sight setting and clear description of the target or aiming point as the situation permits or requires; thereafter, he gives such additional commands or directions as are necessary to exact compliance with the captain's will. He corrects the sight setting when necessary. He designates an aiming point when the target cannot be seen with the naked eye.

In general, *platoon leaders* observe the target and the effect of their fire and are on the alert for the captain's commands or signals; they observe and regulate the rate of fire. (I. D. R. 252).

720. *The Guides* watch the firing line and check every breach of fire discipline.

721. *The Squad Leaders* transmit commands and signals when neces-sary, observe the conduct of their squads and abate excitement, assist in enforcing fire discipline and participate in the firing.

Every squad leader should place himself just a little in advance of the rest of his squad and by occasionally glancing to the right and left, observe how the men of their squads are doing—whether they are firing at the proper objective, if the sights are apparently properly ad-justed, if they are firing too rapidly, etc. After each shot the squad leader should look toward his platoon leader, and then glance to his right and left to observe his men, and then load and fire again.

722. *The Musicians* assist the captain by observing the enemy, the tar-get, and the fire effect, by transmitting commands or signals, and by watching for signals. (I. D. R. 235).

The Privates will take advantage of cover, exercise care in setting the sights and delivering fire; be on the constant lookout for orders from their leaders; always aim deliberately; observe the enemy carefully, in-creasing the fire when the target is favorable and ceasing firing when the enemy disappears; not neglect a target because it is indistinct; not waste ammunition, but be economical with it; if firing without a leader to retain their presence of mind and direct an efficient fire upon the proper target.

723. Finding the Range. There are various methods of ascertaining the range, viz:

1. By means of mechanical range finders.

2. By means of ranging volleys, whenever the ground near the target is such that the strike of the bullets can be seen from the firing line. (I. D. R. 240).

3. By means of range finders, five or six officers or men selected from the most accurate estimaters in the company, being designated as range finders (I. D. R. 240). This is the method most commonly used.

724. Distribution of Fire. The distribution of fire over the entire target is of the greatest importance; for, a section of the target not covered by fire represents a number of the enemy permitted to fire coolly and effectively. So, remember that all parts of the target are equally important, and care must be taken that the men do not neglect its less visible parts.

The captain allots a part of the target to each platoon, or each platoon leader takes as his target that part which corresponds to his position in the company. Every man is so instructed that he will fire on that part of the target which is directly opposite him.

If the target cannot be seen with the naked eye, platoon leaders select an object in front of or behind it, designate this as the aiming point, and direct a sight setting which will carry the fire into the target. The men aim at the good aiming point or line, but with such an increased or decreased sightsetting, as the case may be, that the bullets will fall on the target instead of on the aiming point.

Distribution of fire is assured by dividing the whole target assigned the company into definite parts or sectors, and allotting these parts or sectors to the various platoons. And, of course, the whole of the target must be kept under fire while the company is advancing. This may be accomplished by one of two methods:

725. Overlapping Method. In this method each sector (target) is covered by more than one fire unit. For example, in a company of four platoons the entire company sector would be divided in two parts, the right part being covered by the first and second platoons and the left part by the third and fourth platoons. When the first platoon ceases fire to advance, the second platoon would replace the lost rifles by firing faster. With three platoons the company sector would be divided into two parts, one being assigned to each flank platoon and the whole company sector to the center platoon. When the first platoon advanced, the center platoon would cover its target, both the center and third platoons increasing their rate of fire. With two platoons, each would cover the whole company sector.

726. Switch Method. The company is divided into a number of parts, one less than the number of platoons in the company. One platoon is designated as the ''switch'', and swings into fire automatically into that sector from which the fire of its assigned unit is withdrawn. For example, with four platoons, and platoon rushes to start from the right, the company sector is divided into three parts assigned to the first, second and third platoons, the fourth being the ''switch''. When number 1 ceases fire to advance, No. 4 fires at No. 1's target; when No. 2 ceases to fire, No. 4 fires at No. 2's target then at No. 3's target, and, finally No. 4 advances.

727. Designation of Target. It is very important that the commanders should be able to describe the objectives to be attacked and the sectors * to be defended, and that individual soldiers should be able to understand and transmit to other soldiers such descriptions. Within the squad, target designation implies ability on the part of the squad leader to understand and transmit to his squad the target designation received from his platoon leader, and also ability on his own part to designate a target intelligently; within the platoon, target designation implies ability on the part of the platoon leader to understand the company commander's designation of the target and to transmit that designation to his platoon in such manner as to insure an equal distribution of its fire within the sector assigned to it; within the company, target designation implies ability on the part of the company commander to designate the targets into which the company sector is divided in such manner that the platoon leaders will have no trouble in understanding him. It also implies ability on the part of the company commander to change the objectives or sectors of his platoons, and his ability to cover the whole target of the company during a forward movement of a part of the company, by the so-called ''switch'' or the ''overlapping'' method, or by any other method which is practicable and accomplishes the desired end. Targets should be designated in a concise, prompt, unmistakable manner, but, as we all know, it is not always an easy matter to describe the location of an object, especially if the object be not conspicuous or readily recognized. This is due to two reasons: First, the unit commander is likely to indulge in vague talk instead of accurate discription, and, second, even if correct terms are used, it is more than likely that all members of the firing line will not be able to grasp the idea, because the commander will be using expressions which, although understood by himself (in some cases perhaps due to the fact that he is looking at the objective), they will not be clear to the men. The secret of prompt, accurate and concise designation of a target lies in the use of simple words and terms with which both the unit commander and the men on the firing line are thoroughly familiar.

Of course, if the target be distinct and clearly defined, it can easily be designated by name, as for example, ''That battery on the hill just in front of us'', ''Cavalry to our right front'', etc.

Generally the designation of a target, if not conspicuous nor readily recognized, will include:

1. A statement of what the target is, or its appearance (shape, color, size, etc.)

2. Where the target is with reference to some easily recognized reference point.

---

* In attack the target is called "objective"; in the defense, "sector."

3. How wide the company sector is.

The following systems of target designation are used at the School of Musketry. Each has its limitations, defects and advantages, under various conditions of ground, etc. A wise selection of one or a combination of two or more, is a material factor in efficiency.

728. *Horizontal Clock Face System.* (Used with visible, distinct targets)

SYSTEM	EXAMPLE
1. Announce direction.	"At one o'clock".
2. Announce range.	"Range 1000".
3. Announce objective.	"A troop of cavalry dismounted".

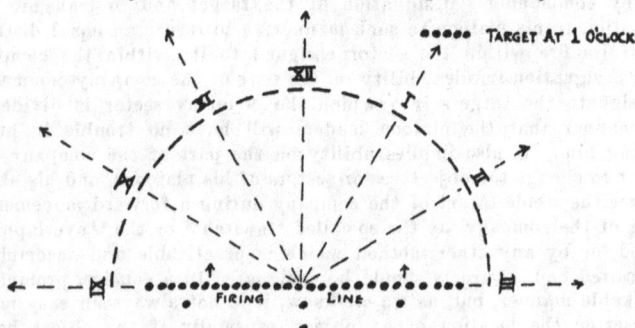

HORIZONTAL CLOCK FACE SYSTEM

PROCEDURE:

1. All look along the line pointing toward one o'clock of a horizontal clock face whose center is at the firing point, and whose 12 o'clock mark is directly perpendicular to the front of the firing line.
2. All look at a point about 1000 yards away on the one o'clock line, and
3. At 1000 yards on the one o'clock line find the objective.

729. *Vertical Clock Face System.* (Used with small or indistinct targets)

SYSTEM	EXAMPLE
1. Announce the general direction of the reference point.	"To our right front" (or "At two o'clock").
2. Designate as a reference point the most prominent object in the zone indicated.	"A stone house with two chimneys".
3. Announce the position of the target with respect to the reference point.	"At three o'clock".

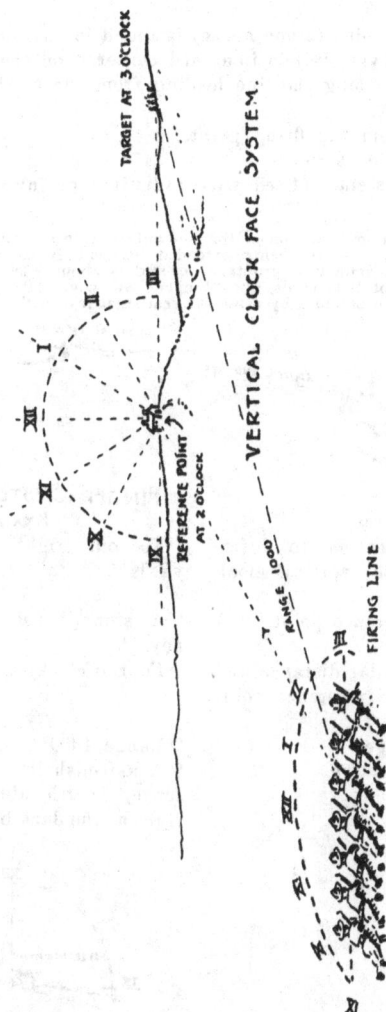

TARGET AT 3 O'CLOCK

VERTICAL CLOCK FACE SYSTEM.

REFERENCE POINT AT 2 O'CLOCK

RANGE 1000

FIRING LINE

4. Announce the range.   "Range 1000".
5. Announce the objective.   "A hostile patrol of four men".
    PROCEDURE:
1. All men look to their right front (or along the two o'clock line).

2.   The reference point (stone house) is found in the indicated direction.

3.   A clock face (vertical) is imagined centered on the reference point, and the men look along the line leading from the clock center through three o'clock, and

4.   1000 yards from the firing point.

5.   Find the hostile patrol.

   730. Finger System. (Used with indistinct or invisible targets and to define sectors).

   (By one "Finger" we mean the amount of frontage that one finger, held vertically, will cover, the arm being extended horizontally to its full length   In the average case this amount of frontage covered is about 1/20 of the range.   For instance, at a range of 1000 yards, one "Finger" will cover fifty yards of the sector   The same result will be obtained by using the rear-sight leaf in the position of aiming)

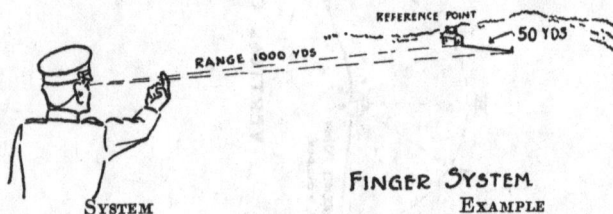

SYSTEM	FINGER SYSTEM EXAMPLE
1. Announce direction to reference point as in the vertical clock face system.	"To our right front, at 1000 yards".
2. Announce reference point.	"A stone house with two chimneys".
3. Announce angular distance and direction from the reference point to the target.	"Four o'clock, three fingers".
4. Announce range.	"Range 1000".
5. Announce objective.	"A skirmish line alongside of the fence, length about two fingers, right at the dark bush".

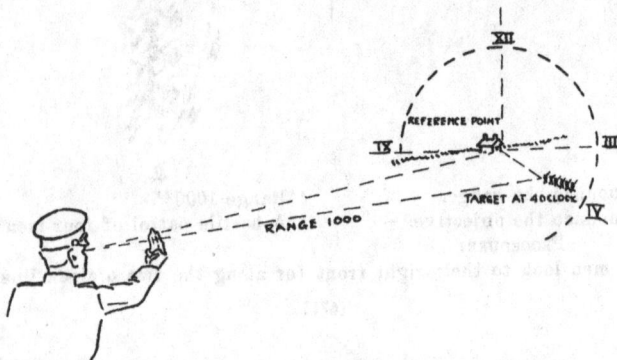

PROCEDURE:

The reference point is found as explained, and the vertical o'clock line upon which the target will be found. The soldiers who do not see the target will extend the arm to its *full extent* palm of the hand upward, finger held vertically with one side of the hand "against" the reference point. The target will be found on the four o'clock line, and touching the third finger, at 1000 yards distance, its right flank at the bush and its left flank about 100 yards farther to the right.

The following case will illustrate more concretely the use of the "Finger" system:

There is a red house about 1½ miles to our front, and to the right of this house and a hundred yards or so to its rear, there is a line of trenches that can be seen with the aid of field glasses, but the trenches are difficult to locate with the unaided eye. There is no prominent landmark in the direction of this line of trenches, or on either flank, except the red house mentioned. The company commander locates the flanks of the line of trenches through his field glasses; he then extends his arm forward horizontally its full length, palm up, raises the fingers of his hand and, sighting on the line of trenches, finds that the trench line has a length of four "finger widths", and that the flank of the line nearest the red house is three "finger widths" from it. He decides to divide the line into two sections of two "fingers" each, and assign one section to each of his two platoons. He then calls his platoon leaders (and range finders, if necessary), and says, for instance: "Center of objective, five inches to the right of that red house, First Platoon, two fingers; Second Platoon, two fingers." The two platoon leaders then estimate the range and give the company commander their estimates independently. The company commander also estimates the range, and, taking the average, then announces the range, say 1200 yards, after which the platoon leaders return to their platoons, and give, for instance, these instructions: "The target is a line of trenches four 'fingers' long, and about 1200 yards away; the center of the target is five 'fingers' to the right of that red house, at about 10 o'clock. We are to fire at the two fingers on the right of the center and the Second Platoon will look after the two fingers on the left of the center." (The leader of the Second Platoon gives similar instructions).

Every man in the platoon figures out the platoon objective and endeavors to fix it with respect to some features of the ground so that he will be able to pick it up promptly after his platoon starts to advance. After fixing well in his mind the platoon objective, he figures out what part of it belongs to his squad, and then selects that portion of the squad objective corresponding to his position in the squad. If during the advance, his particular portion of the target should become hidden from view, he will fire on the nearest portion of the trench line, returning to his own part as soon as it becomes visible.

731. Communication. After the company has been committed to the fire fight, verbal commands cannot be heard, and it is well nigh impossible even to secure attention to signals. It is, therefore, most important that we should train and practice the company as much as possible during time of peace in the rapid and accurate transmission of orders and signals along the firing line.

Matter upon which a commander would need to communicate with his subordinates, in addition to tactical orders, would generally be confined to:

(a) Changes of elevation and deflection.

(b) Changes in the apportionment of the target among the subdivisions.

(c) Changes within the limits of the sector, or objective.

(d) Changes in the rate of fire.

(e) And rarely change of target from one within to one without the limits of the objective or sector.

The following arm signals covering matters upon which a company commander might need to communicate with his subordinates, were gotten up by a board of officers appointed for the purpose at Texas City, Texas:

1. Commence Firing.

Move the arm extended its full length, hand palm down, several times through a horizontal arc in front of the body.

To Fire Slower.

Execute "Commence Firing" slowly.

To Fire Faster.

Execute "Commence Firing" rapidly.

2. Ready. (Are you ready, or I am ready.)

Raise the hand, fingers extended and joined, palm toward the person addressed.

3. What Range Are You Using?

Extend the arm its full length to the front, palm to the right palm to the front, resting on the other hand, fist closed.

4. To Swing the Cone of Fire to the Right or Left.

Extend the arm its full length to the front, palm to the right (left), swinging the arm to the right (left).

The amount of change in deflection may be indicated by exposing a number of fingers on the hand used in the signal equal to the number of "fingers" separating the old and new target.

Where the new objective is plainly visible, point toward it.

5. Fix Bayonets.

For units smaller than a regiment, the following arm signal may be used instead of the bugle signal authorized in paragraph 41, Infantry Drill Regulations:

Simulate the movement "Fix Bayonet."

It may be said that while a company commander could probably generally give his orders directly to the firing line with our present small companies, he should, it is thought, for purposes of practice and instruction, give his orders through his platoon leaders.

732. Fire Discipline. The importance of fire discipline cannot be overestimated. The principal things which fire discipline imply are enumerated in Par. 640.

733. Procedure. The following is given merely as a concrete example of the procedure that might be followed in certain firing exercises—it will not, of course, apply to all cases; it is merely given as a concrete illustration of what might actually be done under certain conditions.

Company Commander. On receiving his instructions from the officer in charge of the exercise, the company commander returns to his company, keeping track of the changing aspect of his target as he does so. Arriving at the center of his company, he is met by his platoon leaders, and range finders, who have assembled in his absence. The company commander says:

"The target is a line of skirmishers, visible in part. It may be seen between us and that long line of green bushes which begins one finger to the right of that red water tower at 11 o'clock and it extends well beyond the bushes both to the right and to the left."

(At this point the range finders begin their estimation and the captain pauses until the senior range finder, or other designated person automatically announces the average estimate of the range, saying for example, "Range 1100").

The captain then resumes, saying:

"The sector assigned to this company is three fingers long and extends from that group one finger to the right of the water tower, to a point four inches to the right of the tower. Each platoon will cover the entire company sector. Range ten-fifty and eleven-fifty. Fire at will at my signal. Posts."

Platoon Leaders. The platoon leaders then hasten to the center of their platoons and "put into execution the commands and directions of the captain, having first taken such precautions to insure a correct sight setting and clear description of the aiming point as the situation permits or requires" (Par. 251 I. D. R.), by saying:

Target: The target is a line of skirmishers about 1100 yards to our front, only parts of which are visible.

Reference point: That long line of bushes about 1300 yards to our left front. The company sector is three fingers long and lies between us and that reference point, extending one-half finger beyond each end of the bushes.

Aiming point: The bottom of the line of bushes.

Range: 1050 and 1150.

As soon as the range is announced each front rank man sets his sight at 1050 and each rear rank man at 1150. Squad leaders assure themselves that sights are set and that the men of their squads understand the aiming point and sector and then raise their hand as a signal that all are ready. Similarly, the platoon leaders raise their hands to show that all of the squads are ready, and when the captain sees that all of his platoons are ready, he drops his hand by his side as a signal to begin firing. At the captain's signal, each platoon leader commands: "Fire At Will".

Firing then begins at a rate of about 3 shots per minute (Par. 14, I. D. R.).

734. *Points To Be Borne in Mind*. Bear in mind the following points in the solution of field firing problems:

1. Combine sights should, as a rule, be used where the estimated range is 1000 yards or more, the two ranges being 50 yards on each side of the estimated range, the even numbers firing at one range, the odd numbers at the other.

2. When aiming points are chosen they should be clearly described. Bushes, bunches of lines of grass, fence posts, etc., should not be designated as aiming points when clear and more definite aiming points are available. The choice of the best of several possible aiming points is of great importance.

3. Have some system of simple signals whereby you may know when all your men are ready to begin firing. Otherwise, you may begin the firing before some of your men have their sights set and before they understand the sector and point of aim. For example, let each squad leader raise his right hand when his squad is ready, and each platoon leader his right hand when his platoon is ready.

4. Platoon leaders must always be sure to designate a definite aiming point. Remember that in the case of an indistinct target, the company commander describes the TARGET to the platoon leaders, and they in turn announce the AIMING POINT. Having seen and located the target, the platoon leader must examine the terrain at, in front of and behind the target, and choose the aiming point for his men. He must then determine the proper sight setting for that particular aiming point. He then announces both aiming point and range.

5. Instead of describing a sector as, for example, extending so many yards (or so many "fingers") north from the reference point, it is

better to describe it as extending from the reference point northward for a definite distance, as "To that tall red house".

The last method is the best, because it leaves no room for guessing on the part of subordinates. So, remember it is always best, when possible, to define the limits of sectors physically, as, extending, for example, from "That house to that windmill", etc.

6. When acting as part of the battalion, always be sure to designate someone (usually one of the musicians) to watch for signals from the battalion commander, and don't fail to repeat back all signals.

7. In advancing by rushes, always allow sufficient time between rushes to recover the loss in fire caused by the cessation of fire. In other words, the next rear unit should not start forward until the one that has just advanced has resumed an effective fire.

8. Remember that in all field firing problems the distribution of hits has big weight. Consequently, it should be definitely understood before hand, that, in the absence of any target designation by the company commander, each platoon leader will look after the sector corresponding to his front, and that each man will fire at the part of the sector corresponding to his front. Should the targets in a given sector disappear, then the platoon leader covering that sector will at once switch his fire to the adjoining sector until the reappearance of the targets in his own sector. For example, let us suppose the company

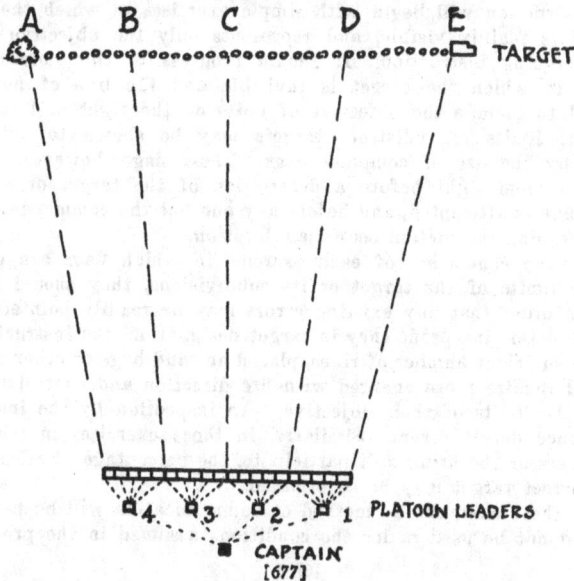

A   B   C   D   E

TARGET

4   3   2   1 PLATOON LEADERS

CAPTAIN
[677]

sector, A-B (the company being on the defense and not advancing) is divided into four parts A-B, B-C, C-D and D-E. Platoon No. 1 would look after everything that appeared in D-E; No. 2, after everything that appeared in C-D; No. 3, everything that appeared in B-C; and No. 4, after everything that appeared in A-B.

Should the target suddenly disappear from D-E, then No. 1 would switch his fire over to C-D, and keep it there until the target reappeared in D-E, and if the targets disappeared from C-D, before reappearing in D-E, then both No. 1, and No. 2, would switch their fire cones to A-C.

735. **Exercises.** The following exercises for the elementary training of individuals and squads were used with success by the troops mobilized on the Texas border:

TARGETS

1. The target will be represented by individual soldiers.

2. With reference to their visibility, the battlefield will present three classes of targets:

    (a) Those which are visible throughout.

    (b) Those which are visible in part.

    (c) Those which are invisible, but whose location might be described.

Targets will be arranged to simulate one of the classes enumerated. Instruction will begin with simple exercises in which the target presented is plainly visible, and represents only the objective of the unit undergoing instruction. It should progress to the more difficult exercises in which the target is invisible and the line of figures is prolonged to include the objective of units on the right or left.

3. The limits of indistinct targets may be shown to unit commanders by the use of company flags. These flags, however, will be withdrawn from sight before a description of the target or estimate of the range is attempted, and before any one but the commander of the unit undergoing instruction sees their location.

4. At the conclusion of each exercise in which flags are used to mark the limits of the target or its subdivisions, they should be displayed, in order that any existing errors may be readily pointed out.

5. To determine proficiency in target designation, the instructor will provide a sufficient number of rifles, placed on sand bags or other suitable rests, and require those charged with fire direction and control to sight them at the limits of their objective. An inspection by the instructor will at once detect errors. Similarly, in those exercises in which all the members of the firing unit participate, the percentage of rifles aimed at the correct target may be determined.

6. In these exercises no method of communication will be permitted that could not be used under the conditions assumed in the problem.

EXERCISE NO. 1—RANGING.

Object: To train the individual to set his sight quickly and accurately for the announced range and windage; and to accustom leaders to the giving of windage data.

Situation: The company is formed in single rank at the ready with the rear sight set at zero and the slide screw normally tight.

Action: The range and windage are announced, sights are set accurately in accordance therewith and as rapidly as may be, each man coming to port arms immediately upon completing the operation.

Time: Time is taken from the last word of the command.

Standard: Sights should be correctly set within 15 seconds.

Note: Of the two elements, time and accuracy, accuracy is the more important.

Par. 411, I. D. R., implies complete use of the rear sight, that is, utilization of the wind gauge, and sight setting to the least reading of the rear sight leaf, i. e., 25 yards. Sight setting therefore in this exercise should include, more often than not, "fractional ranges" and windage data.

EXERCISE NO. 2—RANGING

Object: To familiarize officers and noncommissioned officers in the use of an auxiliary aiming point.

Situation: Two men with the company flags are stationed to mark the enemy's invisible position. This position should be suitably located with reference to a practicable aiming point.

Action: The markers are signalled to display their flags. An officer or noncommissioned officer is called up and the enemy's position is pointed out. The flags are then withdrawn and the officer or noncommissioned officer selects an auxiliary aiming point and gives his commands for firing at that point.

EXERCISE NO. 3—TARGET DESIGNATION

Object: To train the individual soldier to locate a target, from a description solely. To do so quickly and accurately and fire thereon with effect, and to train officers and noncommissioned officers in concise, accurate and clear description of targets.

Situation: The men are so placed as not to be able to see to the target. The instructor places himself so as to see the objective.

Action: The instructor, to one man at a time, describes the objective, and directs him to fire one simulated round. The man immediately moves so as to see the target, locates it, estimates the range and fires one simulated shot.

Standard: For ranges within battle sight, time 20 seconds; beyond battle sight, time 30 seconds. Not more than 15% error in the estimation of the range. Objective correctly located.

735 (contd.)

Note: Arrangements made so that the description of the target is heard by only the man about to fire. After firing the man will not mingle with those waiting to fire.

EXERCISE NO. 4—TARGET DESIGNATION

Object: To train the squad leader in promptly bringing the fire of his squad to bear effectively upon the target presented. To train the individuals of a squad to fire effectively from orders of the squad leader and automatically to obtain effective dispersion.

Situation: The squad is deployed, the squad leader being in the firing line. Position prone. A sighting rest is provided for each rifle.

Action: Upon the appearance of the target the squad leader gives the necessary orders for delivering an effective fire. The men under these orders sight their rifles and. then rise. The instructor then examines the position and sighting of each rifle.

Time: Time is taken from the appearance of the target until the last man has risen.

Target: A squad of men to outline a partially concealed enemy emerges from cover, advances a short distance and lies down.

Standard: 90% of the rifles should be sighted in conformity with the orders of the squad leader and should evenly cover the whole front of the objective. The squad leader's estimate of the range should not be in error over 15%.

Note: The squad leader should not, in general, be allowed to divide the target into sectors but to obtain distribution by training the men to fire at that portion of the objective directly related to the position they occupy in their own line. The exercise should be repeated with the squad leader in rear of the squad and not firing. As to this, it is to be noted that Musketry School experiments prove that in small groups the directed fire of say seven (7) rifles is more effective than the partially undirected fire of eight rifles obtained when the group leader is himself firing.

EXERCISE NO. 5—COMMUNICATION

Object: To teach prompt and accurate transmission of firing data without cessation of fire, and also to teach automatic readjustment of fire distribution.

Situation: A squad deployed in the prone position and with sighting rests, is firing at a designated target.

Action: A squad with sights set at zero is deployed and brought up at the double time into the intervals of the firing line and halted. The firing data is transmitted to them without cessation of fire. At the command **Rise,** given 20 seconds after the command **Halt,** the first squad rises and retires a short distance to the rear. At the same time, the supports cease fire and adjust their rifles in the rests so as to be aimed at the target as they understand it. They then rise and their rifles are examined by the instructor for range and direction.

Standard: 80% of the rifles should be sighted according to the transmitted data and aimed according to the principles of fire distribution.

Target: One target equal to a squad front, which is increased to two squads prior to the arrival of the supports in the firing line.

Note: This exercise should be repeated with the supporting squad reënforcing on a flank. To determine whether the original squad is able to keep its assigned sector during an advance, this exercise should be repeated, the supports being thrown in after a series of short advances by the original squad. Care should be exercised to prevent the transmission of firing data in a manner under which service conditions would be impracticable. (See Exercise No. 6).

EXERCISE NO. 6—COMMUNICATION

Object: To train the squad leader in receiving and transmitting instructions by visual signals alone.

Situation: A squad with its leader in the firing line is deployed in the prone position firing at will.

Action: The instructor, without sound or other cautionary means, signals (visually) to the squad leader at various intervals to,
First: Change elevation.

swing the fire to the right or left.

suspend the firing.

etc., etc.

The squad leader, upon receiving a signal, causes his squad to execute it without verbal command, or exposing himself.

Time: No specified time limit.

Standard: The squad leader should fire with his squad, but after each shot should look towards his platoon leader for any signal, then observe the fire and conduct of his men, then, after glancing again at his platoon leader, fire again. This the squad leader should do without exposing himself. By lying about a head's length ahead of his men he can see his squad front. In transmitting orders he can accomplish it by nudging the men on his right and left and signaling to them with his hand.

Note: This exercise is essential to prepare men for the deafening noise of a heavy action when speech or sound signals are largely futile.

EXERCISE NO. 7—FIRE DISCIPLINE.

Object: To train men to carry out strictly the fire orders given them, and to refrain from starting, repeating or accepting any change therefrom without direct orders from a superior.

735 (contd.)

Situation: A squad deployed in the prone position.

Action: While the squad is firing at an indistinct but specified target, another and clearly visible target appears in the vicinity of the first target but not in the same sector. Upon the appearance of this second target, the instructor sees that the men continue firing at the assigned target. The corporal should check any breach of fire discipline.

Note: Variations of this exercise should be given to test the fire discipline of the men in other phases, such as rate of fire (Par. 147, I. D. R.), etc.

CHAPTER XIX

CAMPING

736. Castrametation. The art of laying out camps is called castrametation.

737. Selection of Site. The following conditions must be considered in the selection of camp sites:

1. Location.

2. Water and wood.

3. Sanitation, and in time of war, defense and safety.

Camps should be on slightly sloping ground, easily drained and subject to sunny exposures.

Closely cropped turf with sandy or gravelly subsoil is best; high banks of rivers are suitable, provided no marshes are near.

In hot summer months, the ground selected should be high, free from underbrush, and shaded with trees if possible.

In cold weather ground sloping to the south, with woods to break the north winds, is desirable.

Old camp grounds and the vicinity of cemeteries are undesirable. Marshy ground and stagnant water are objectionable on account of the damp atmosphere and the annoyance and infection from mosquitoes. Ground near the foot of a hill range generally has a damp subsoil and remains muddy for a long time. Thick forests, dense vegetation, made ground, alluvial soil, punch-bowl depressions, inclosed ravines, and dry beds of streams are unfavorable.

Camp sites should be selected so that troops of one unit need not pass through the camp grounds of another.

As a protection against epidemics, temporary camp sites in the theater of operations should be changed every two or three weeks.

The ground should accommodate the command with as little crowding as possible, be easily drained, and have no stagnant water within 300 yards.

The water supply should be sufficient, pure, and accessible.

There should be good roads to the camp and good interior communication.

Wood, grass, forage, and supplies must be at hand or obtainable.

When camp is established for an indefinite period, drainage should be attended to at once. Each tent should have a shallow trench dug around it and the company and other streets ditched on both sides, all the trenches and ditches connecting with a ditch that carries the water from the camp. All surface drainage from higher ground should be intercepted and turned aside.

737 (contd.)

In front of every camp of a permanent nature, there should be a parade ground for drills and ceremonies, and the sanitary conditions of the camp should be carefully considered.

In camping for the night on a fordable stream that is to be crossed, always cross before going into camp; for a sudden rise, or the appearance of the enemy might prevent the crossing the next morning.

Whenever windstorms are expected, the tent pegs should be secured and additional guy ropes attached to the tents. If the soil be loose or sandy, stones or other hard material should be placed under the tent poles to prevent their working into the soil, thus leaving the tent slack and unsteady. When the soil is so loose that the pegs will not hold at all, fasten the guy ropes to brush, wood or rocks buried in the ground.

Tents may be prevented from blowing down by being made fast at the corners to posts firmly driven into the ground, or by passing ropes over the ridge poles and fastening them to pegs firmly driven into the ground.

While trees add very much to the comfort of a camp, care should exercised not to pitch tents near trees whose branches or trunks might fall.

In a hostile country the capability of defense of a camp site should always be considered.

Form and Dimensions of Camps. The forms of the camp should be such as to facilitate the prompt encampment of troops after a march and their prompt departure when camp is broken. The form of camps will depend upon the tactical situation and the amount and nature of ground available. In certain cases, particularly in one-night halts in the presence of the enemy, camps must of necessity be contracted, while in other cases, where a more extended halt is contemplated and where tactical reasons will permit, better camp sanitation may be secured, and a more comfortable arrangement made by the expansion of camp areas.

The diagram on the opposite page gives the general form, dimensions, and interior arrangements of a camp for a regiment of infantry at war strength.

[684]

CAMP OF A REGIMENT OF INFANTRY, WAR STRENGTH

(19.8 Acres)
280 Yards

345 Yds.

Bath Latrine } Officers

Officers' Latrine

Messes and Offices

Hdqrs Mess

Lt Col.Co.

Field Officer's Line

Battn.Mess

Maj.

Adjt Sup. O

Company Officers' Line

Store Mess

Tents

Pits

1st. Sgt.

A B C D E F G H I K L M

Hdqrs. Co. and Sanitary Detach.

1st. Battalion

2d. Battalion

3d. Battalion

Supply and Machine Gun Cos.

Latrines

Tram Park

Store Tents

Camp Guard

Men's Bath

Animals

Stable B.S. Guard

738. Making Camp. The command should be preceded by the commanding officer or a staff officer, who selects the camp site, and designates, by planting stakes, the lines of tents, the positions of the sinks, guard tent, kitchens, picket line, etc.

After the companies are marched to their proper positions and arms are stacked, the details for guard and to bring wood, water, dig sinks, pitch tents, handle rations, etc., should be made before ranks are broken.

Immediately upon reaching camp and before the men are allowed to go around, patrolling sentinels should be established to prevent men from polluting the camp site or adjoining ground before the sinks are constructed.

Sentinels should be posted over the water supply without delay.

As soon as the tents have been pitched and the sinks dug, the camp should be inspected and all unnecessary sentinels relieved.

The tents should be pitched and the sinks dug simultaneously.

If the weather is at all threatening or if it is intended to camp more than one night, all tents should be ditched.

Should the troops reach camp before the wagons, the companies may be divided into squads and set to work clearing the ground, gathering fire wood, collecting leaves, grass, etc., for beds, etc.

The moment a command reaches camp its officers and men usually want to go here and there under all sorts of pretexts. No one should be allowed to leave camp until all necessary instructions have been given.

Officers should not be allowed to leave camp without permission from the commanding officer, and enlisted men should not be permitted to leave camp without permission of their company commanders.

Sick call should be held as soon as practicable after the tents have been pitched.

Retreat roll call should always be under arms, an officer being with each company and inspecting its arms.

739. Construction of Latrines. The latrines must be dug immediately upon reaching camp—their construction must not be delayed until the camps have been pitched and other duties performed. The exact location of the latrines should be determined by the commanding officer, or by some officer designated by him, the following considerations being observed:

1. They should be so located as not to contaminate the water supply.

2. They should not be placed where they can be flooded by rain water from higher ground, nor should they be so placed that they can pollute the camp by overflow in case of heavy rains.

3. They should be as far from the tents as is compatible with convenience—if too near, they will be a source of annoyance; if too far, some men, especially at night, and particularly if affected with diarrhoea, will defecate before reaching the latrine. Under ordinary circumstances, a distance of about 75 yards is considered sufficient.

Latrines for the men are always located on the opposite side of the camp from the kitchens, generally one for each company unit and one for the officers of a battalion or squadron. They are so placed that the drainage or overflow can not pollute the water supply or camp grounds.

When the camp is for one night only, straddle trenches suffice. In camp of longer duration, and when it is not possible to provide latrine boxes, as for permanent camps, deeper trenches should be dug. These may be used as straddle trenches or a seat improvised. When open trenches are used the excrement must be kept covered at all times with a layer of earth. In more permanent camps the trenches are not over 2 feet wide, 6 feet deep, and 12 feet long, and suitably screened. Seats with lids are provided and covered to the ground to keep flies from reaching the deposits; urinal troughs discharging into the trenches are provided. Each day the latrine boxes are thoroughly cleaned, outside by scrubbing and inside by applying, when necessary, a coat of oil or whitewash. The pit is burned out daily with approximately 1 gallon oil and 15 pounds straw. When filled to within 2 feet of the surface, such latrines are discarded, filled with earth, and their position marked. All latrines and kitchen pits are filled in before the march is resumed. In permanent camps and cantonments, urine tubs may be placed in the company streets at night and emptied after reveille.

All latrines must be filled before marching. The following illustration shows a very simple and excellent latrine seat which can be made and kept in the company permanently for use in camps on the march:

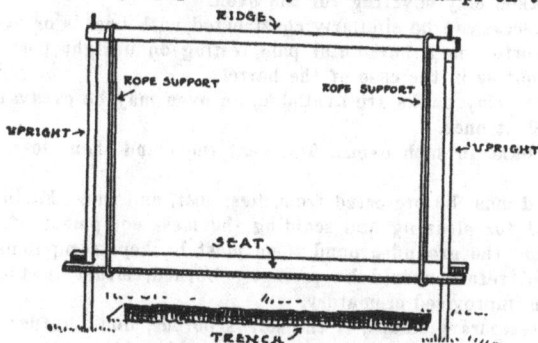

740. Kitchens. Camp kettles can be hung on a support consisting of a green pole lying in the crotches of two upright posts of the same character. A narrow trench for the fire, about 1 foot deep, dug under the pole, not only protects the fire from the wind but saves fuel.

A still greater economy of fuel can be effected by digging a similar trench in the direction of the wind and slightly narrower than the diameter of the kettles. The kettles are then placed on the trench and the space between the kettles filled in with stones, clay, etc., leaving the flue running beneath the kettles. The draft can be improved by building a chimney of stones, clay, etc., at the leeward end of the flue.

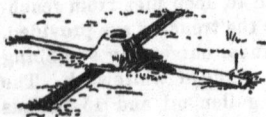

Four such trenches radiating from a common central chimney will give one flue for use whatever may be the direction of the wind.

A slight slope of the flue, from the chimney down, provides for drainage and improves the draft.

The lack of portable ovens can be met by ovens constructed of stone and covered with earth to better retain the heat. If no stone is available, an empty barrel, with one head out, is laid on its side, covered with wet clay to a depth of 6 or more inches and then with a layer of dry earth equally thick. A flue is constructed with the clay above the closed end of the barrel, which is then burned out with a hot fire. This leaves a baked clay covering for the oven.

A recess can be similarly constructed with boards or even brushwood, supported on a horizontal pole resting on upright posts, covered and burnt out as in the case of the barrel.

When clay banks are available, an oven may be excavated therein and used at once.

To bake in such ovens, first heat them and then close flues and ends.

Food must be protected from flies, dust, and sun. Facilities must be provided for cleaning and scalding the mess equipment of the men. Kitchens and the ground around them must be kept scrupulously clean.

Solid refuse should be promptly burned, either in the kitchen fire or in an improvised crematory.

In temporary camps, if the soil is porous, liquid refuse from the kitchens may be strained through gunny sacking into seepage pits dug near the kitchen. Flies must not have access to these pits. Boards or poles, covered with brush or grass and a layer of earth may be used for this purpose. The strainers should also be protected from flies. Pits of this kind, dug in clayey soil, will not operate successfully. All pits should be filled with earth before marching.

(The above regarding kitchens, etc., is from the Infantry Drill Regulations.)

As a precautionary measure against setting the camp on fire, all dry grass, underbrush, etc., in the immediate vicinity of the kitchen should be cut down.

In case of a fire in camp, underbrush, spades, shovels, blankets, etc., are used to beat it out.

Gunny sacks dipped in water are the best fire fighters.

Burning away dried grass and underbrush around exterior of camp is a great protection against fire from outside.

741. Kitchen Pits. Pits of convenient size should be constructed for the liquid refuse from the kitchens. Solid refuse should be burned either in the kitchen fire or at some designated place, depending upon whether the camp is of a temporary or permanent nature. Unless the camp be of a very temporary nature, the pits should be covered with boards or other material in order to exclude the flies.

All pits should be filled in with earth before breaking camp.

742. Incinerators. The incineration pit shown in the following diagram, affords an excellent, simple and economical way of disposing of camp waste and offal, tin cans and dish-water included:

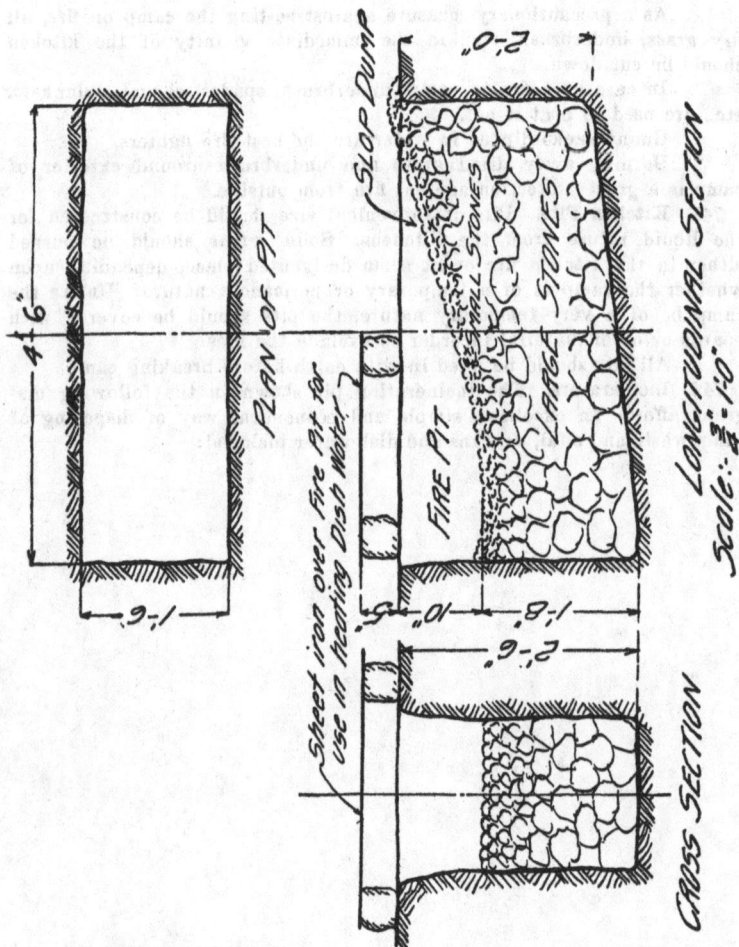

PLAN OF PIT

4'-6"

1'-6"

SLOP DUMP

2'-0"

SMALL STONES

LARGE STONES

LONGITUDINAL SECTION

Scale: 3': 1'0"

FIRE PIT

Sheet iron over Fire Pit for use in Heating Dish Water.

1'-8"

10"

5"

2'-6"

CROSS SECTION

Description:

The pit is about 4½ feet long, 1½ feet wide and 2 feet deep at one end and 2½ at the other. It is partially filled with stones, the larger ones on the bottom and the smaller on the top. At one end of the pit the stones extend a little above the surface, and slope gradually toward the other end until the fire pit is reached ten inches below the surface of the trench. Over the fire pit, about five inches above the ground, is placed a crab or a piece of boiler iron, on which is boiled all the water for washing dishes, etc. The fire pit is only about one-half of the stone surface, as the radiated heat keeps the rest of the stones hot, causing all dish and slop water to evaporate quickly.

Any tin cans that may be thrown into the fire pit are removed after a short exposure to the heat and placed in a trench especially dug for the purpose.

743. The company incinerator shown below was used with great success by some of our troops at Texas City, Texas. The rocks should not be too large. The men should be instructed to drop all liquid on the sides of the incinerator and throw all solid matter on the fire—the liquids will thus be evaporated and the solids burned. Until the men learn how to use the incinerator properly, a noncommissioned officer should be detailed to supervise its use.

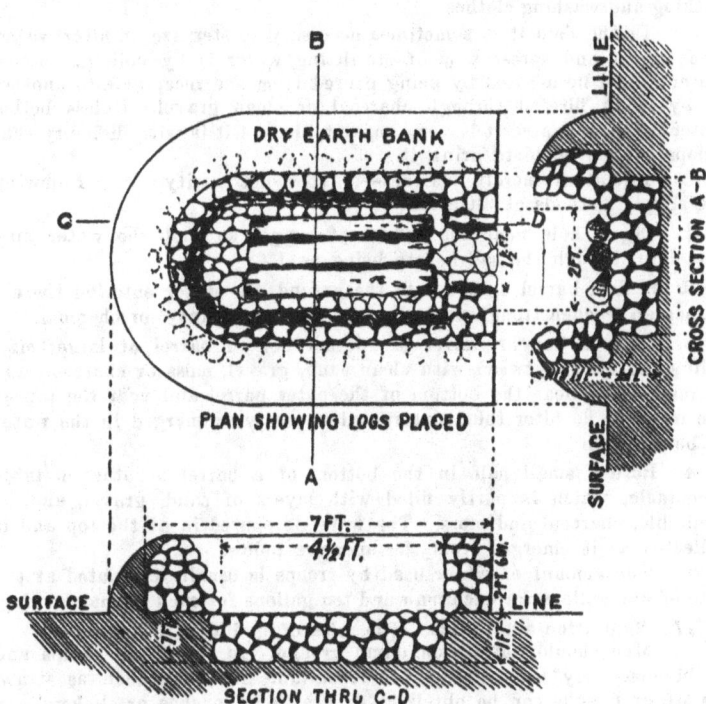

PLAN SHOWING LOGS PLACED

CROSS SECTION A-B

SECTION THRU C-D

744. Bunks. Place a number of small poles about seven feet long close together, the upper ends resting on a cross pole about six inches in diameter and the lower ends resting on the ground; or, the poles may be raised entirely off the ground by being placed on cross poles supported by forked stakes at the corners; on the poles place grass, leaves, etc.

745. Wood. The firewood should be collected, cut and piled near the kitchen. Dry wood is usually found under logs or roots of trees.

If wagons are not heavily loaded it is sometimes a good plan to bring a few sticks of dry wood from the preceding camp, or to pick up good wood en route.

746. Water. Precautionary measures should always be taken to prevent the contamination of the water, and a guard from the first troops reaching camp should be placed over the water supply. Water used for drinking purposes should be gotten from above the camp, and places below this point should be designated for watering the animals, bathing and washing clothes.

In the field it is sometimes necessary to sterilize or filter water. The easiest and surest way of sterilizing water is by boiling. Boiled water should be aerated by being poured from one receptacle to another or by being filtered through charcoal or clean gravel. Unless boiled water be thus aerated it is very unpalatable and it is with difficulty that troops can be made to drink it.

Filtration merely clarifies—it does not purify. The following are simple methods of filtration:

1. Dig a hole near the source of supply so that the water may percolate through the soil before being used.

2. Sink a barrel or box into the ground, the water entering therein through a wooden trough packed with clean sand, gravel or charcoal.

3. Place a box or barrel in another box or barrel of larger size, filling the space between with clean sand, gravel, moss or charcoal, and piercing holes near the bottom of the outer barrel and near the top of the inner. The filter thus constructed is partly submerged in the water to be filtered.

4. Bore a small hole in the bottom of a barrel or other suitable receptacle, which is partly filled with layers of sand, gravel, and, if available, charcoal and moss. The water is poured in at the top and is collected as it emerges from the aperture below.

The amount of water used by troops is usually computed at the rate of five gallons for each man and ten gallons for each animal per day.

747. Sanitation and Police. The rules of sanitation are enforced.

Men should not lie on damp ground. In temporary camps and in bivouac they raise their beds if suitable material, such as straw, leaves, or boughs can be obtained, or use their ponchos or slickers. In cold weather and when fuel is plentiful the ground may be warmed by fires, the men making their beds after raking away the ashes.

When troops are to remain in camp for sometime all underbrush is cleared away and the camp made as comfortable as possible. Watering troughs, shelter in cold weather, and shade in hot, are provided for the animals, if practicable.

In camps of some duration guard and other routine duties follow closely the custom in garrison. The watering, feeding, and grooming of animals take place at regular hours and under the supervision of officers.

The camp is policed daily after breakfast and all refuse matter burned.

Tent walls are raised and the bedding and clothing aired daily, weather permitting.

Arms and personal equipments are kept in the tents of the men. In the cavalry, horse equipments are also usually kept in the tents, but in camps of some duration they may be placed on racks outside and covered with slickers. In the artillery, horse equipments and harness are placed on the poles of the carriages and covered with paulins.

The water supply is carefully guarded. When several commands are encamped along the same stream this matter is regulated by the senior officer.

If the stream is small, the water supply may be increased by building dams. Small springs may be dug out and lined with stone, brick, or empty barrels. Surface drainage is kept off by a curb of clay.

When sterilized water is not provided, or when there is doubt as to the purity of the water, it is boiled 20 minutes, then cooled and aerated.

(The above regarding sanitation and police is from the *Infantry Drill Regulations*.)

CHAPTER XX

INDIVIDUAL COOKING*

748. For such individual cooking as may be necessary for the soldier when thrown upon his own resources, the following bills of fare have been prepared. Where the tin cup and spoon are mentioned, reference is made to those issued with the field mess kit.

Remember that the best fire for cooking is a small, clear one, or better yet, a few brisk coals.

Almost anything that can be cooked at all can be prepared in the mess kit, though the variety is necessarily small and quantities limited on account of few utensils of small capacity.

Company commanders in estimating the amounts that will be required for each meal may assume that one man will consume for one meal about—

1 ounce of sugar.

½ ounce of coffee, 1 ounce chocolate or cocoa, or 1/10 ounce of tea.

4 ounces of dried vegetables.

4 ounces of flour or 4 hardtacks.

8 ounces of fresh vegetables.

4 ounces of sliced bacon or 6 to 8 ounces of fresh meat.

1/5 ounce of salt.

1/50 ounce of pepper.

Bills of fare

	Meats.	Vegetables.	Bread, etc.	Drink.
1	Bacon.......................	Boiled rice............	Flapjack......	Coffee.
2	Meat and vegetable stew.....	....	Flapjack......	Coffee.
3	Broiled steak...............	Fried potatoes and onions.	Hard bread...	Cocoa.
4	Bacon.......................	Stewed tomatoes........	Hoecake......	Coffee.
5	Bacon.......................	Oatmeal...............	Hard bread...	Tea.
6	Bacon.......................	Baked potatoes; rice...	Flapjack......	Chocolate.
7	Fried Steak.................	Boiled potatoes; cold tomatoes.	Hard bread...	Coffee.
	Etc........................	Etc.	Etc.........	Etc.

Or, When Time is More Limited

	Meats.	Vegetables.	Bread, etc.	Drink.
8	Fried bacon................	Fried potatoes...........	Hard bread...	Coffee.
9	Fried bacon................	....	Flapjack...	Coffee
10	Corned beef (cold).........	Tomato stew...........	Hard bread...	Coffee.
11	Fried fish and bacon.......	Baked potatoes........	Hard bread...	Coffee.
12	Meat and vegetable stew....	....	Hoecake....	Tea.
13	Broiled steak..............	Baked potatoes........	Hard bread...	Cocoa.
14	Boiled fish................	Fried potatoes........	Hard bread...	Tea.
	Etc........................	Etc	Etc	Etc.

* From "Manual for Army Cooks," prepared by the Subsistence Department.

SUGGESTIONS FOR HANDLING BILL OF FARE No. 1.

Take two-thirds of a cup of water and bring to a boil. Add 4 spoonfuls of rice and boil until soft, i. e., until it can be mashed by the fingers with but little resistance. This will require about fifteen minutes. Add 2 pinches of salt and, after stirring, pour off the water and empty the rice out on the lid of the mess pan.

Meanwhile, fry 3 slices of bacon until slightly browned, in the mess pan over a brisk fire or hot coals, and lay them on top of the rice, leaving sufficient grease in the pan in which to fry the flap jack.

Take 6 spoonfuls of flour and one-third spoonful of baking powder and mix thoroughly. Add sufficient cold water to make a batter that will drip freely from the spoon. Add a pinch of salt and 2 pinches of sugar and pour the batter into the mess pan, which should contain the grease from the fried bacon. Place over medium hot coals and bake from five to seven minutes; see that it will slip easily in the pan and then, by the quick toss, turn it over and continue the baking from five to seven minutes longer or until, by examination, it is found to be done.

While the batter is frying, wash out the tin cup; two-thirds fill with water and let come to a boil. Add 1 medium heaping spoonful of coffee and stir well, and, if desired, 1 spoonful of sugar and let boil for about five minutes. Let simmer for about ten minutes longer. Settle by a dash of cold water or let stand a few minutes.

A hot meal is now ready to serve. Time about forty minutes.

RECIPES
Drinks
(For one meal for one man.)

Article and amount.	Amount of water.	Add when—	Let boil	Add sugar if desired.	Remarks.
Coffee, 1 heaping spoonful.	Cup 2/3	Water boils	Min. 5	Sp'nful. 1	Stir grains well when adding. Let simmer ten minutes after boiling. Settle with a dash of water or let stand a few minutes. Ready to serve.
Cocoa, 1 heaping spoonful.	2/3	..do..	5	1½	Stir when adding until dissolved. Ready to serve when sufficiently cooled. Do.
Chocolate, 1 cubic inch.	2/3	. do..	5	1½	
Tea, 1/2 level spoonful.	2/3	..do .	0	1	Let stand or "draw" eight minutes. If allowed to stand longer, the tea will get bitter unless separated from the grains.

NOTE.—Coffee made by above recipe is of medium strength and the same as when using 4 ounces to the gallon of water. It is within the limit of the ration if made but twice each day.

Tea.—A little more than medium strength, the same as when using 3/5 ounce to the gallon, and within the ration allowance if made three times per day.

Chocolate and cocoa.—About 1 ounce per man per meal If available, milk should be used in the place of water, and should be kept somewhat below the boiling point. Mix a 1-pound can of evaporated milk with 3½ quarts of water to make 1 gallon of milk of the proper consistency for use in making cocoa or chocolate.

Dried Vegetables

(For one meal for one man)

Article and amount.	Amount of water.	Add when—	Let boil.	Season with pinches of salt.	Add heaping spoonful sugar if desired	Remarks
Rice, 4 heaping spoonfuls.	Cup. 2/3	Water boils.	Hours 1/3	2	1	Should be boiled until grains (while still nicely separated) may be crushed between the fingers with but little resistance. Then drain off the water.
Cornmeal, hominy, fine oatmeal, 4 heaping spoonfuls	1/3	..do..	1/3	2	....	All water should now be taken up by the cornmeal, hominy, or oatmeal, which forms a thick paste.
Dried sweet corn, 4 heaping spoonfuls	1/3	..do..	1/3	2	1	
Lima beans, 4 heaping spoonfuls.	2/3	Water is put on.	2 or 3	1	....	When done the beans should still be whole but soft. Add one small slice of bacon one-half hour before done. Add water as required.
Chili beans and frijolas, 4 heaping spoonfuls.	2/3	..do..	3 or 4	1	....	Above remark applies.
Beans, issue dried green peas, hominy, coarse split peas, 4 heaping spoonfuls.	2/3	..do..	3 or 4	1	....	Not recommended on account of time required for cooking.

NOTE.—By a *heaping spoonful* is meant here all that can readily be taken up.

A *pinch of salt* is the amount that can readily be taken up between the end of the thumb and forefinger.

Meats

Bacon.—Cut side of bacon in half lengthwise. Then cut slices about five to the inch, three of which should generally be sufficient for one man for one meal. Place in a mess pan with about one-half inch of cold water. Let come to a boil and then pour the water off. Fry over a brisk fire, turning the bacon once and quickly browning it. Remove the bacon to lid of mess pan, leaving the grease for frying potatoes, onions, rice flapjacks, etc., according to recipe.

Fresh meat—*To fry*.—To fry, a small amount of grease (1 to 2 spoonfuls) is necessary. Put grease in mess pan and let come to a smoking temperature, then drop in the steak and, if about one-half inch thick, let fry for about one minute before turning—depending upon whether it is desired it shall be rare, medium, or well done. Then turn and fry briskly as before. Salt and pepper to taste.

Applies to beef, veal, pork, mutton, venison, etc.

Fresh meat—*To broil*.—Cut in slices about 1 inch thick, from half as large as the hand to four times that size. Sharpen a stick or branch of convenient length, say from 2 to 4 feet long, and weave the point of the stick through the steak several times so that it may be readily turned over a few brisk coals or on the windward side of a small fire. Allow to brown nicely, turning frequently. Salt and pepper to taste. Meat with considerable fat is preferred, though any meat may be broiled in this manner.

Fresh meat—*To stew*.—Cut into chunks from one-half inch to 1-inch cubes. Fill cup about one-third full of meat and cover with about 1 inch of water. Let boil or simmer about one hour or until tender. Add such fibrous vegetables as carrots, turnips, or cabbage, cut into small chunks, soon after the meat is put on to boil, and potatoes, onions, or other tender vegetables when the meat is about half done. Amount of vegetables to be added, about the same as meat, depending upon supply and taste. Salt and pepper to taste. Applies to all fresh meats and fowls. The proportion of meat and vegetables used varies with their abundance and fixed quantities can not be adhered to. Fresh fish can be handled as above, except that it is cooked much quicker, and potatoes, onions, and canned corn are the only vegetables generally used with it, thus making a chowder. A slice of bacon would greatly improve the flavor. May be conveniently cooked in mess pan or tin cup.

Fresh Vegetables

Potatoes, fried.—Take two medium-sized potatoes or one large one (about one-half pound), peel and cut into slices about one-fourth inch thick and scatter well in the mess pan in which the grease remains after frying the bacon. Add sufficient water to half cover the potatoes, cover with the lid to keep the moisture in, and let come to a boil from fifteen to twenty minutes. Remove the cover and dry as desired. Salt and pepper to taste. During the cooking the bacon already prepared may be kept on the cover, which is most conveniently placed bottom side up over the cooking vegetables.

Onions, fried.—Same as potatoes.

Potatoes, boiled.—Peel two medium-sized potatoes or one large one (about one-half pound), and cut in coarse chunks of about the same size—say 1½-inch cubes. Place in mess pan and three-fourths fill with water. Cover with lid and let boil or simmer for fifteen or twenty minutes. They are done when easily penetrated with a sharp stick. Pour off the water and let dry out for one or two minutes over hot ashes or light coals.

Potatoes, baked.—Take two medium-sized potatoes or one large one cut in half (about one-half pound). Lay in a bed of light coals, cover with same and smother with ashes. Do not disturb for thirty or forty minutes, when they should be done.

Canned Tomatoes.—One 2-pound can is generally sufficient for five men.

Stew. Pour into the mess pan one man's allowance of tomatoes, add about two large hardtacks broken into small pieces, and let come to a boil. Add salt and pepper to taste, or add a pinch of salt and one-fourth spoonful of sugar.

Or, having fried the bacon, pour the tomatoes into the mess pan, the grease remaining, and add, if desired, two broken hardtacks. Set over a brisk fire and let come to a boil.

Or, heat the tomatoes just as they come from the can, adding two pinches of salt and one-half spoonful of sugar if desired.

Or, especially in hot weather, eaten cold with hard bread they are very palatable.

Hot Breads

Flapjack.—Take 6 spoonfuls of flour and one-third spoonful of baking powder and mix thoroughly (or dry mix in a large pan before issue, at the rate of 25 pounds of flour and three half-pound cans of baking powder for 100 men). Add sufficient cold water to make a batter that will drip freely from the spoon, adding a pinch of salt. Pour into the mess pan, which should contain the grease from fried bacon, or a spoonful of butter or fat, and place over medium hot coals sufficient to bake so that in from five to seven minutes the flapjack may be turned over by a quick toss of the pan. Fry from five to seven minutes longer or until, by examination, it is found to be done.

Hoecake.—Hoecake is made exactly the same as a flapjack by substituting *corn meal* for *flour*.

Emergency Ration

Emergency Rations.—Detailed instructions as to the manner of preparing the emergency ration are found on the label with each can. Remember that even a very limited amount of bacon or hard bread,

or both, taken with the emergency ration makes it far more palatable, and greatly extends the period during which it can be consumed with relish. For this reason it would be better to husband the supply of hard bread and bacon to use with the emergency ration when it becomes evident that the latter must be consumed, rather than to retain the emergency ration to the last extremity to be used exclusively for a longer period than two or three days.

www.ingramcontent.com/pod-product-compliance
Lightning Source LLC
Chambersburg PA
CBHW011931190326
41519CB00027B/7481

this screen reigns utter darkness, the buried one's delight. This is capital.

What would happen if, by an artifice, the sideward layer were nowhere thick enough to satisfy the grub? Now, this time, I have the wherewithal to solve the problem, in the shape of a big glass tube, open at both ends, about three feet long and less than an inch wide. I use it to blow the flame of hydrogen in the little chemistry-lessons which I give my children.

I close one end with a cork and fill the tube with fine, dry, sifted sand. On the surface of this long column, suspended perpendicularly in a corner of my study, I install some twenty Sarcophaga-grubs, feeding them with meat. A similar preparation is repeated in a wider jar, with a mouth as broad as one's hand. When they are big enough, the grubs in either apparatus will go down to the depth that suits them. There is no more to be done but to leave them to their own devices.

The worms at last bury themselves and harden into pupæ. This is the moment to consult the two apparatus. The jar gives me the answer which I should have obtained in the open fields. Four inches down, or thereabouts, the worms have found a quiet lodging, protected above by the layer through which they have passed and on every side by the thickness of the vessel's contents. Satisfied with the site, they have stopped there.

It is a very different matter in the tube. The least buried of the pupæ are half a yard down. Others are lower still; most of them even have reached the bottom of the tube and are touching the cork stopper, an insuperable barrier. These last, we can see, would have gone yet deeper if the apparatus had allowed them. Not one of the score of grubs has settled at the customary halting-place; all have travelled farther down the column, until their strength gave way. In their anxious flight, they have dug deeper and ever deeper.

What were they flying from? The light. Above them, the column traversed forms a more than sufficient shelter; but, at the sides, the irksome sensation is still felt through a coat of earth half an inch thick if the descent is made perpendicularly. To escape the disturbing impression, the grub therefore goes deeper and deeper, hoping to obtain lower down the rest which is denied it above. It only ceases to move when worn out with the effort or stopped by an obstacle.

Now, in a soft diffused light, what can be the

radiations capable of acting upon this lover of darkness? They are certainly not the simple luminous rays, for a screen of fine, heaped-up earth, nearly half an inch in thickness, is perfectly opaque. Then, to alarm the grub, to warn it of the over-proximity of the exterior and send it to mad depths in search of isolation, other radiations, known or unknown, must be required, radiations capable of penetrating a screen against which ordinary radiations are powerless. Who knows what vistas the natural philosophy of the maggot might open out to us? For lack of apparatus, I confine myself to suspicions.

To go underground to a yard's depth—and farther if my tube had allowed it—is on the part of the Flesh-fly's grub a vagary provoked by unkind experiment: never would it bury itself so low down, if left to its own wisdom. A hand's-breadth thickness is quite enough, is even a great deal when, after completing the transformation, it has to climb back to the surface, a laborious operation absolutely resembling the task of an entombed well-sinker. It will have to fight against the sand that slips and gradually fills up the small amount of empty space obtained; it will perhaps, without crowbar or pickaxe, have to cut itself a gallery through something tantamount to tufa, that is to say, through earth which a shower has rendered compact. For the descent, the grub has its fangs; for the assent, the Fly has nothing. Only that moment come into existence, she is a weakling, with tissues still devoid of any firmness. How does she manage to get out? We shall know by watching a few pupæ placed at the bottom of a test-tube filled with earth. The method of the Flesh-flies will teach us that of the Greenbottles and the other Flies, all of whom make use of the same means.

Enclosed in her pupa, the nascent Fly begins by bursting the lid of her casket with a hernia which comes between her two eyes and doubles or trebles the size of her head. This cephalic blister throbs: it swells and subsides by turns, owing to the alternate flux and reflux of the blood. It is like the piston of an hydraulic press opening and forcing back the front part of the keg.

The head makes its appearance. The hydrocephalous monster continues the play of her forehead, while herself remaining stationary. Inside the pupa, a delicate work is being performed: the casting of the white nymphal tunic. All through this operation, the hernia

The Grey Flesh-Flies

lengthen and spread. Then, motionless on the
surface of the sand, the Fly matures fully.
Let us set her at liberty. She will go and join
the others on the Snakes in my pans.

CHAPTER XI

THE BUMBLE-BEE FLY

UNDERNEATH the Wasp's brown-paper manor-house, the ground is channelled into a sort of drain for the refuse of the nest. Here are shot the dead or weakly larvæ which a continual inspection roots out from the cells to make room for fresh occupants; here, at the time of the autumn massacre, are flung the backward grubs; here, lastly, lies a good part of the crowd killed by the first touch of winter. During the rack and ruin of November and December, this sewer becomes crammed with animal matter.

Such riches will not remain unemployed. The world's great law which says that nothing edible shall be wasted provides for the consumption of a mere ball of hair disgorged by the Owl. How shall it be with the vast stores of a ruined Wasps' nest! If they have not come yet, the consumers whose task it is to salve this abundant wreckage for nature's markets, they will not tarry in coming and waiting for the manna that will soon descend from

The Bumble-Bee Fly

above. That public granary, lavishly stocked by death, will become a busy factory of fresh life. Who are the guests summoned to the banquet?

If the Wasps flew away, carrying the dead or sickly grubs with them, and dropped them on the ground round about their home, those banqueters would be, first and foremost, the insect-eating birds, the Warblers, all of whom are lovers of small game. In this connection, we will allow ourselves a brief digression. We all know with what jealous intolerance the Nightingales occupy each his own cantonment. Neighbourly intercourse among them is tabooed. The males frequently exchange defiant couplets at a distance; but, should the challenged party draw near, the challenger makes him clear off. Now, not far from my house, in a scanty clump of holly-oaks which would barely give the woodcutter the wherewithal for a dozen faggots, I used, all through the spring, to hear such full-throated warbling of Nightingales that the songs of those virtuosi, all giving voice at once and with no attempt at order, degenerated into a deafening hubbub.

Why did those passionate devotees of solitude come and settle in such large numbers at a spot where custom decrees that there is just room enough for one household only? What reasons have made the recluse become a congregation? I asked the owner of the spinney about the matter.

'It's like that every year,' he said. 'The clump is overrun by Nightingales.'

'And the reason?'

'The reason is that there is a hive close by, behind that wall.'

I looked at the man in amazement, unable to understand what connection there could be between a hive and the thronging Nightingales.

'Why, yes,' he added, 'there are a lot of Nightingales because there are a lot of Bees.'

Another questioning look from my side. I did not yet understand. The explanation came:

'The Bees,' he said, 'throw out their dead grubs. The front of the hive is strewn with them in the mornings; and the Nightingales come and collect them for themselves and their families. They are very fond of them.'

This time I had solved the puzzle. Delicious food, abundant and fresh each day, had brought the songsters together. Contrary to their habit, numbers of Nightingales are living on friendly terms in a cluster of bushes, in order to be near the hive and to have a larger share in the morning distribution of plump dainties.

In the same way, the Nightingale and his gastronomical rivals would haunt the neighbourhood of the Wasps' nests, if the dead grubs were cast out on the surface of the soil; but these delicacies fall inside the burrow and no little bird would dare to enter the murky cave, even if the entrance were not too small to admit it. Other consumers are needed here, small in size and great in daring; the Fly is called for and her maggot, the king of the departed. What the Greenbottles, the Bluebottles and the Flesh-flies do in the open air, at the expense of every kind of corpse, other Flies, narrowing their province, do underground at the Wasps' expense.

Let us turn our attention, in September, to the wrapper of a Wasps' nest. On the outer surface and there alone, this wrapper is strewn with a multitude of big, white, elliptical dots, firmly fixed to the brown paper and measuring about two millimetres and a half long by one and a half wide.[1] Flat below, convex above and of a lustrous white, these dots resemble

[1]About .1 by .06 of an inch.—*Translator's Note.*

very neat drops fallen from a tallow candle.
Lastly, their backs are streaked with faint
transversal lines, an elegant detail perceptible
only with the lens. These curious objects are
scattered all over the surface of the wrapper,
sometimes at a distance from one another,
sometimes gathered into more or less dense
groups. They are the eggs of the Volucella,
or Bumble-bee Fly (*Volucella zonaria*, LIN.).

Also stuck to the brown paper of the outer
wrapper and mixed up with the Volucella's are
a large number of other eggs, chalk-white,
spear-shaped and ridged lengthwise with seven
or eight thin ribs, after the manner of the
seeds of certain Umbelliferæ. The finishing
touch to their delicate beauty is the fine stip-
pling all over the surface. They are smaller
by half than the others. I have seen grubs
come out of them which might easily be the
earliest stage of some pointed maggots which
I have already noticed in the burrows. My
attempts to rear them failed; and I am not
able to say which Fly these eggs belong to.
Enough for us to note the nameless one in
passing. There are plenty of others, which
we must make up our minds to leave un-
labelled, in view of the jumbled crowd of
feasters in the ruined Wasps' nest. We will

The Bumble-Bee Fly

concern ourselves only with the most remarkable, in the front rank of which stands the
Bumble-bee Fly.

She is a gorgeous and powerful Fly; and
her costume, with its brown and yellow bands,
shows a vague resemblance to that of the
Wasps. Our fashionable theorists have
availed themselves of this brown and yellow
to cite the Volucella as a striking instance of
protective mimicry. Obliged, if not on her
own behalf, at least on that of her family, to
introduce herself as a parasite into the Wasp's
home, she resorts, they tell us, to trickery and
craftily dons her victim's livery. Once inside
the Wasps' nest, she is taken for one of the
inhabitants and attends quietly to her business.

The simplicity of the Wasp, duped by a
very clumsy imitation of her garb, and the de-
pravity of the Fly, concealing her identity
under a counterfeit presentment, exceed the
limits of my credulity. The Wasp is not so
silly nor the Volucella so clever as we are assured. If the latter really meant to deceive
the Wasp by her appearance, we must admit
that her disguise is none too successful. Yellow sashes round the abdomen do not make a
Wasp. It would need more than that and,
above all, a slender figure and a nimble

carriage; and the Volucella is thickset and corpulent and sedate in her movements. Never will the Wasp take that unwieldy insect for one of her own kind. The difference is too great.

Poor Volucella, mimesis has not taught you enough. You ought—this is the essential point—to have adopted a Wasp's shape; and that you forgot to do: you remained a fat Fly, easily recognizable. Nevertheless, you penetrate into the terrible cavern; you are able to stay there for a long time, without danger, as the eggs profusely strewn on the wrapper of the Wasps' nest show. How do you set about it?

Let us, first of all, remember that the Bumble-bee Fly does not enter the enclosure in which the combs are heaped: she keeps to the outer surface of the paper rampart and there lays her eggs. Let us, on the other hand, recall the Polistes[1] placed in the company of the Wasps in my vivarium. Here of a surety is one who need not have recourse to mimicry to find acceptance. She belongs to the guild, she is a Wasp herself. Any of us that had not the trained eye of the entomologist would confuse the two species. Well, this stranger, as

[1] A species of Wasp that builds her nest in trees.—
Translator's Note.

long as she does not become too importunate,
is quite readily tolerated by the caged Wasps.
None seeks to pick a quarrel with her. She is
even admitted to the table, the strip of paper
smeared with honey. But she is doomed if
she inadvertently sets foot upon the combs.
Her costume, her shape, her size, which tally
almost exactly with the costume, shape and
size of the Wasp, do not save her from her
fate. She is at once recognized as a stranger
and attacked and slaughtered with the same
vigour as the larvæ óf the Hylotoma Saw-fly
and the Saperda Beetle, neither of which bears
any outward resemblance to the larva of the
Wasps.

Seeing that identity of shape and costume
does not save the Polistes, how will the Volucella fare, with her clumsy imitation? The
Wasp's eye, which is able to discern the dissimilar in the like, will refuse to be caught.
The moment she is recognized, the stranger
is killed on the spot. As to that there is not
the shadow of a doubt.

In the absence of Bumble-bee Flies at the
moment of experimenting, I employ another
Fly, *Milesia fulminans,* who, thanks to her
slim figure and her handsome yellow bands,
presents a much more striking likeness to the

Wasp than does the fat *Volucella zonaria*. Despite this resemblance, if she rashly venture on the combs, she is stabbed and slain. Her yellow sashes, her slender abdomen deceive nobody. The stranger is recognized behind the features of a double.

My experiments under glass, which varied according to the captures which I happened to make, all lead me to this conclusion: as long as there is more propinquity, even around the honey, the other occupants are tolerated fairly well; but, if they touch the cells, they are assaulted and often killed, without distinction of shape or costume. The grubs' dormitory is the sanctum sanctorum which no outsider must enter under pain of death.

With these caged captives I experiment by daylight, whereas the free Wasps work in the absolute darkness of their underground retreat. Where light is absent, colour goes for nothing. Once, therefore, that she has entered the cavern, the Bumble-bee Fly derives no benefit from her yellow bands, which are supposed to be her safeguard. Whether garbed as she is or otherwise, it is easy for her to effect her purpose in the dark, on condition that she avoids the tumultuous interior of the Wasps' nest. So long as she has the prudence

The Bumble-Bee Fly

not to hustle the passers-by, she can dab her eggs, without danger, on the paper wall. No one will know of her presence. The dangerous thing is to cross the threshold of the burrow in broad daylight, before the eyes of those who go in and out. At that moment alone, protective mimicry would be convenient. Now does the entrance of the Volucella into the presence of a few Wasps entail such very great risks? The Wasps' nest in my enclosure, the one which was afterwards to perish in the sun under a bell-glass, gave me the opportunity for prolonged observations, but without any result upon the subject of my immediate concern. The Bumble-bee Fly did not appear. The period for her visits had doubtless passed; for I found plenty of her grubs when the nest was dug up.

Other flies rewarded me for my assiduity. I saw some—at a respectful distance, I need hardly say—entering the burrow. They were insignificant in size and of a dark-grey colour, not unlike that of the House-fly. They had not a patch of yellow about them and certainly had no claim to protective mimicry. Nevertheless, they went in and out as they pleased, calmly, as though they were at home. As long as there was not too great a number at the door, the Wasps left them alone. When there was anything of a crowd, the grey visitors waited near the threshold for a less busy moment. No harm came to them.

Inside the establishment, the same peaceful relations prevail. In this respect I have the evidence of my excavations. In the underground charnel-house, so rich in Fly-grubs, I find no corpses of adult Flies. If the strangers had been slaughtered in passing through the entrance-hall, or lower down, they would fall to the bottom of the burrow anyhow, with the other rubbish. Now in this charnel-house, as I said, there are never any dead Bumble-bee Flies, never a Fly of any sort. The incomers are respected. Having done their business, they go out unscathed.

This tolerance on the part of the Wasps is surprising. And a suspicion comes to one's mind: can it be that the Volucella and the rest are not what the accepted theories of natural history call them, namely, enemies, grub-killers sacking the Wasps' nest? We will look into this by examining them when they are hatched. Nothing is easier, in September and October, than to collect the Volucella's eggs in such numbers as we please. They abound on the outer surface of the Wasps' nest. Moreover, as with the larvæ of the Wasp, it is some time before they are suffocated by the petroleum-fumes; and so most of them are sure to hatch. I take my scissors, cut the most densely-populated bits from the paper wall of the nest and fill a jar with them. This is the warehouse from which I shall daily, for the best part of the next two months, draw my supply of nascent grubs.

The Volucella's egg remains where it is, with its white colour always strongly marked against the brown of the background. The shell wrinkles and collapses; and the fore-end tears open. From it there issues a pretty little white grub, thin in front, swelling slightly in the rear and bristling all over with fleshy protuberances. The creature's papillæ are set on its sides like the teeth of a comb; at the rear, they lengthen and spread into a fan; on the back, they are shorter and arranged in four longitudinal rows. The last section but one carries two short, bright-red breathing-tubes, standing aslant and joined to each other. The fore-part, near the pointed mouth, is of a darker, brownish colour. This is the biting-and motor-apparatus, seen through the skin and consisting of two fangs. Taken all round, the grub is a pretty little thing, with its bristling whiteness, which gives it the appearance of a tiny snow-flake. But this elegance does not last long: grown big and strong, the Bumble-bee Fly's grub becomes soiled with sanies, turns a russety-brown and crawls about in the guise of a hulking porcupine.

What becomes of it when it leaves the egg? This my warehousing-jar tells me, partly. Unable to keep its balance on sloping surfaces, it drops to the bottom of the receptacle, where I find it, daily, as hatched, wandering restlessly. Things must happen likewise at the Wasps'. Incapable of standing on the slant of the paper wall, the new-born grubs slide to the bottom of the underground cavity, which contains, especially at the end of the summer, a heaped-up provender of deceased Wasps and dead larvæ removed from the cells and flung outside the house, all nice and gamy, as proper maggot's food should be.

The Volucella's offspring, themselves maggots, notwithstanding their snowy apparel, find in this charnel-house victuals to their liking, incessantly renewed. Their fall from the high walls might well be not accidental, but rather a means of reaching, quickly and without searching, the good things down at the bottom of the cavern. Perhaps, also, some of the white grubs, thanks to the holes that make the wrapper resemble a spongy cover, manage to slip inside the Wasps' nest. Still, most of the Volucella's grubs, at whatever stage of their development, are in the basement of the burrow, among the carrion remains. The others, those settled in the Wasps' home itself, are comparatively few.

These returns are enough to show us that the grubs of the Bumble-bee Fly do not deserve the bad reputation that has been given them. Satisfied with the spoils of the dead, they do not touch the living; they do not ravage the Wasps' nest: they disinfect it.

Experiment confirms what we have learnt in the actual nests. Over and over again, I bring Wasp-grubs and Volucella-grubs together in small test-tubes, which are easy to observe. The first are well and strong; I have just taken them from their cells. The others are in various stages, from that of the snow-flake born the same day to that of the sturdy porcupine. There is nothing tragic about the encounter. The grubs of the Bumble-bee Fly roam about the test-tube without touching the live tit-bit. The most that they do is to put their mouths for a moment to the morsel; then

265

they take it away again, not caring for the dish.

They want something different: a wounded, a dying grub; a corpse dissolving into sanies. Indeed, if I prick the Wasp-grub with a needle, the scornful ones at once come and sup at the bleeding wound. If I give them a dead grub, brown with putrefaction, the worms rip it open and feast on its humours. Better still: I can feed them quite satisfactorily with Wasps that have turned putrid under their horny rings; I see them greedily suck the juices of decomposing Rosechafer-grubs; I can keep them thriving with chopped-up butcher's meat, which they know how to liquefy by the method of the common maggot. And these unprejudiced ones, who accept anything that comes their way, provided it be dead, refuse it when it is alive. Like the true Flies that they are, frank body-snatchers, they wait, before touching a morsel, for death to do its work.

Inside the Wasps' nest, robust grubs are the rule and weaklings the rare exception, because of the assiduous supervision which eliminates anything that is diseased and like to die. Here, nevertheless, Volucella-grubs are found, on the combs, among the busy Wasps. They are not, it is true, so numerous as in the charnel-house below, but still pretty frequent. Now what do they do in this abode where there are no corpses? Do they attack the healthy? Their continual visits from cell to cell would at first make one think so; but we shall soon be undeceived if we observe their movements closely; and this is possible with my glass-roofed colonies.

I see them fussily crawling on the surface of the combs, curving their necks from side to side and taking stock of the cells. This one does not suit, nor that one either; the bristly creature passes on, still in search, thrusting its pointed fore-part now here, now there. This time, the cell appears to fulfil the requisite conditions. A larva, glowing with health, opens wide its mouth, believing its nurse to be approaching. It fills the hexagonal chamber with its bulging sides.

The gluttonous visitor bends and slides its slender fore-part, a blade of exquisite supple-ness, between the wall and the inhabitant, whose slack rotundity yields to the pressure of this animated wedge. It plunges into the cell, leaving no part of itself outside but its wide hind-quarters, with the red dots of the two breathing-tubes.

It remains in this posture for some time,

occupied with its work at the bottom of the cell. Meanwhile, the Wasps present do not interfere, remain impassive, showing that the grub visited is in no peril. The stranger, in fact, withdraws with a soft, gliding motion. The chubby babe, a sort of indiarubber bag, resumes its original volume without having suffered any harm, as its appetite proves. A nurse offers it a mouthful, which it accepts with every sign of unimpaired vigour. As for the Volucella-grub, it licks its lips after its own fashion, pushing its two fangs in and out; then, without further loss of time, goes and repeats its probing elsewhere.

What it wants down there, at the bottom of the cells, behind the grubs, cannot be decided by direct observation; it must be guessed at. Since the visited larva remains intact, it is not prey that the Volucella-grub is after. Besides, if murder formed part of its plans, why descend to the bottom of the cell, instead of attacking the defenceless recluse straightway? It would be much easier to suck the patient's juices through the actual orifice of the cell. Instead of that, we see a dip, always a dip and never any other tactics.

Then what is there behind the Wasp-grub? Let us try to put it as decently as possible. In spite of its exceeding cleanliness, this grub is not exempt from the physiological ills insepa-rable from the stomach. Like all that eats, it has intestinal waste matter with regard to which its confinement compels it to behave with extreme discretion. Like so many other close-cabined larvæ of Wasps and Bees, it waits until the moment of the transformation to rid itself of its digestive refuse. Then, once and for all, it casts out the unclean accumulation whereof the pupa, that delicate, reborn organ-ism, must not retain the least trace. This is found later, in any empty cell, in the form of a dark-purple plug. But, without waiting for this final purge, this lump, there are, from time to time, slight excretions of fluid, clear as water. We have only to keep a Wasp-grub in a little glass tube to recognize these occasional discharges. Well, I see nothing else to explain the action of the Volucella's grubs when they dip into the cells without wounding the larvæ. They are looking for this liquid, they provoke its emission. It represents to them a dainty which they enjoy over and above the more substantial fare provided by the corpses.

The Bumble-bee Fly, that sanitary inspector of the Vespine city, fulfils a double office: she wipes the Wasp's children and she rids the nest of its dead. For this reason, she is peacefully received, as an auxiliary, when she enters the burrow to lay her eggs; for this reason, her grub is tolerated, nay more, respected, in the very heart of the dwelling, where none might stray with impunity. I remember the brutal reception given to the Saperda- and Hylotoma-grubs when I place them on a comb. Forthwith grabbed, bruised and riddled with stings, the poor wretches perish. It is quite a different matter with the offspring of the Volucella. They come and go as they please, poke about in the cells, elbow the inhabitants and remain unmolested. Let us give some instances of this clemency, which is very strange in the irascible Wasp.

For a couple of hours, I fix my attention on a Volucella-grub established in a cell, side by side with the Wasp-grub, the mistress of the house. The hind-quarters emerge, displaying their papillæ. Sometimes also the fore-part, the head, shows, bending from side to side with sudden, snake-like motions. The Wasps have just filled their crops at the honey-pot; they are dispensing the rations, are very busily at work; and things are taking place in broad daylight, on the table by the window.

As they pass from cell to cell, the nurses

The Bumble-Bee Fly

repeatedly brush against and stride across the Volucella-grub. There is no doubt that they see it. The intruder does not budge, or, if trodden on, curls up, only to reappear the next moment. Some of the Wasps stop, bend their heads over the opening, seem to be making enquiries and then go off, without troubling further about the state of things. One of them does something even more remarkable: she tries to give a mouthful to the lawful occupant of the cell; but the larva, which is being squeezed by its visitor, has no appetite and refuses. Without the least sign of anxiety on behalf of the nursling which she sees in awkward company, the Wasp retires and goes to distribute its ration elsewhere. In vain I prolong my examination: there is no fluster of any kind. The Volucella-grub is treated as a friend, or at least as a visitor that does not matter. There is no attempt to dislodge it, to worry it, to put it to flight. Nor does the grub seem to trouble greatly about those who come and go. Its tranquillity tells us that it feels at home.

Here is some further evidence: the grub has plunged, head downwards, into an empty cell, which is too small to contain the whole of it. Its hind-quarters stick out, very visibly. For long hours, it remains motionless in this position. At every moment, Wasps pass and re-pass close by. Three of them, at one time together, at another separately, come and nibble at the edges of the cell; they break off particles which they reduce to paste for a new piece of work. The passers-by, intent upon their business, may not perceive the intruder; but these three certainly do. During their work of demolition, they touch the grub with their legs, their antennæ, their palpi; and yet none of them minds it. The fat grub, so easily recognized by its queer figure, is left alone; and this in broad daylight, where everybody can see it. What must it be when the profound darkness of the burrows protects the visitor with its mysteries!

I have been experimenting all along with big Volucella-grubs, coloured with the dirty red which comes with age. What effect will pure white produce? I sprinkle on the surface of the combs some larvæ that have lately left the egg. The tiny, snow-white grubs make for the nearest cells, go down into them, come out again and hunt elsewhere. The Wasps peaceably let them go their way, as heedless of the little white invaders as of the big red ones. Sometimes, when it enters an occupied

cell, the little creature is seized by the owner,
the Wasp-grub, which nabs it and turns and
returns it between its mandibles. Is this a defensive bite? No, the Wasp-grub has merely
blundered, taking its visitor for a proffered
mouthful. There is no great harm done.
Thanks to its suppleness, the little grub
emerges from the grip intact and continues its
investigations.

It might occur to us to attribute this tolerance to some lack of penetration in the Wasps'
vision. What follows will undeceive us: I
place separately, in empty cells, a grub of
Saperda scalaria and a Volucella-grub, both
of them white and selected so as not to fill the
cell entirely. Their presence is revealed only
by the paleness of the hind-part which serves
as a plug to the opening. A superficial examination would leave the nature of the recluse undecided. The Wasps make no mistake: they extirpate the Saperda-grub, kill it,
fling it on the dust-heap; they leave the Volucella-grub in peace.

The two strangers are quite well recognized
in the secrecy of the cells: one is the intruder
that must be turned out; the other is the regular visitor that must be respected. Sight helps,
for things take place in the daylight, under

glass; but the Wasps have other means of information in the dimness of the burrow. When I produce darkness by covering the apparatus with a screen, the murder of the trespassers is accomplished just the same. For so say the police-regulations of the Wasps' nest: any stranger discovered must be slain and thrown on the midden.

To thwart this vigilance, the real enemies need to be masters of the art of deceptive immobility and cunning disguise. But there is no deception about the Volucella-grub. It comes and goes, openly, wheresoever it will; it looks round amongst the Wasps for cells to suit it. What has it to make itself thus respected? Strength? Certainly not. It is a harmless creature, which the Wasp could rip open with a blow of her shears, while a touch of the sting would mean lightning death. It is a familiar guest, to whom no denizen of a Wasps' nest bears any ill-will. Why? Because it renders good service: so far from working mischief, it does the scavenging for its hosts. Were it an enemy or merely an intruder, it would be exterminated; as a deserving assistant, it is respected.

Then what need is there for the Volucella to disguise herself as a Wasp? Any Fly,

The Bumble-Bee Fly

whether clad in drab or motley, is admitted to the burrow directly she makes herself useful to the community. The mimicry of the Bumble-bee Fly, which was said to be one of the most conclusive cases, is, after all, a mere childish notion. Patient observation, continually face to face with facts, will have none of it and leaves it to the arm-chair naturalists, who are too prone to look at the animal world through the illusive mists of theory.

CHAPTER XII

MATHEMATICAL MEMORIES: NEWTON'S BINOMIAL THEOREM.

THE Spider's web is a glorious mathematical problem. I should enjoy working it out in all its details, were I not afraid of wearying the reader's attention. Perhaps I have even gone too far in the little that I have said,[1] in which case I owe him some compensation:

'Would you like me,' I will ask him, 'would you like me to tell you how I acquired sufficient algebra to master the logarithmic systems and how I became a surveyor of Spiders' webs? Would you? It will give us a rest from natural history.'

I seem to catch a sign of acquiescence. The story of my village-school, visited by the chicks and the porkers, has been received with some indulgence; why should not my harsh school of solitude possess its interest as well?

[1] Cf. *The Life of the Spider:* chaps. ix to xii and appendix.—*Translator's Note.*

Newton's Binomial Theorem

Let us try to describe it. And who knows? Perhaps, in doing so, I shall revive the courage of some other poor derelict hungering after knowledge.

I was denied the privilege of learning with a master. I should be wrong to complain. Solitary study has its advantages: it does not cast you in the official mould; it leaves you all your originality. Wild fruit, when it ripens, has a different taste from hot-house produce: it leaves on a discriminating palate a bitter-sweet flavour whose virtue is all the greater for the contrast. Yes, if it were in my power, I would start afresh, face to face with my only counsellor, the book itself, not always a very lucid one; I would gladly resume my lonely watches, my struggles with the darkness whence, at last, a glimmer appears as I continue to explore it; I should retraverse the irksome stages of yore, stimulated by the one desire that has never failed me, the desire of learning and of afterwards bestowing my mite of knowledge on others.

When I left the normal school, my stock of mathematics was of the scantiest. How to extract a square root, how to calculate and prove the surface of a sphere: these represented to me the culminating points of the subject.

The Life of the Fly

Those terrible logarithms, when I happened to open a table of them, made my head swim, with their columns of figures; actual fright, not unmixed with respect, overwhelmed me on the very threshold of that arithmetical cave. Of algebra I had no knowledge whatever. I had heard the name; and the syllables represented to my poor brain the whole whirling legion of the abstruse.

Besides, I felt no inclination to decipher the alarming hieroglyphics. They made one of those indigestible dishes which we confidently extol without touching them. I greatly preferred a fine line of Virgil, whom I was now beginning to understand; and I should have been surprised indeed had any one told me that, for long years to come, I should be an enthusiastic student of the formidable science. Good fortune procured me my first lesson in algebra, a lesson given and not received, of course.

A young man of about my own age came to me and asked me to teach him algebra. He was preparing for his examination as a civil engineer; and he came to me because, ingenuous youth that he was, he took me for a well of learning. The guileless applicant was very far out in his reckoning.

Newton's Binomial Theorem

His request gave me a shock of surprise, which was forthwith repressed on reflection:

'*I* give algebra-lessons?' said I to myself. 'It would be madness: I don't know anything about the subject!'

And I left it at that for a moment or two, thinking hard, drawn now this way, now that with indecision:

'Shall I accept? Shall I refuse?' continued the inner voice.

Pooh, let's accept! An heroic method of learning to swim is to leap boldly into the sea. Let us hurl ourselves head first into the algebraical gulf; and perhaps the imminent danger of drowning will call forth efforts capable of bringing me to land. I know nothing of what he wants. It makes no difference: let's go ahead and plunge into the mystery. I shall learn by teaching.

It was a fine courage that drove me full tilt into a province which I had not yet thought of entering. My twenty-year-old confidence was an incomparable lever.

'Very well,' I replied. 'Come the day after to-morrow, at five, and we'll begin.'

This twenty-four hours' delay concealed a plan. It secured me the respite of a day, the

blessed Thursday, which would give me time
to collect my forces.

Thursday comes. The sky is grey and
cold. In this horrid weather, a grate well
filled with coke has its charms. Let's warm
ourselves and think.

Well, my boy, you've landed yourself in a
nice predicament! How will you manage to-
morrow? With a book, plodding all through
the night, if necessary, you might scrape up
something resembling a lesson, just enough
to fill the dread hour more or less. Then you
could see about the next: sufficient for the
day is the evil thereof. But you haven't the
book. And it's no use running out to the
bookshop. Algebraical treatises are not current wares. You'll have to send for one,
which will take a fortnight at least. And I've
promised for to-morrow, for to-morrow certain! Another argument and one that admits
of no reply: funds are low; my last pecuniary
resources lie in the corner of a drawer. I
count the money: it amounts to twelve sous,
which is not enough.

Must I cry off? Rather not! One resource suggests itself: a highly improper one,
I admit, not far removed indeed from larceny. O quiet paths of algebra, you are my

excuse for this venial sin! Let me confess the temporary embezzlement.

Life at my college is more or less cloistered. In return for a modest payment, most of us masters are lodged in the building; and we take our meals at the principal's table. The science-master, who is the big gun of the staff and lives in the town, has nevertheless, like ourselves, his own two cells, in addition to a balcony, or leads, where the chemical preparations give forth their suffocating gases in the open air. For this reason, he finds it more convenient to hold his class here during the greater part of the year. The boys come to these rooms in winter, in front of a grate stuffed full of coke, like mine, and there find a blackboard, a pneumatic trough, a mantelpiece covered with glass receivers, panoplies of bent tubes on the walls, and, lastly, a certain cupboard in which I remember seeing a row of books, the oracles consulted by the master in the course of his lessons.

'Among those books,' said I to myself, 'there is sure to be one on alegebra. To ask the owner for the loan of it does not appeal to me. My amiable colleague would receive me superciliously and laugh at my ambitious aims. I am sure he would refuse my request.'

The Life of the Fly

The future was to show that my distrust
was justified. Narrow-mindedness and petty
jealousy prevail everywhere alike.

I decide to help myself to this book, which
I should never get by asking. This is the
half-holiday. The science-master will not put
in an appearance to-day; and the key of my
room is practically the same as his. I go,
with eyes and ears on the alert. My key does
not quite fit; it sticks a little, then goes in;
and an extra effort makes it turn in the lock.
The door opens. I inspect the cupboard and
find that it does contain an algebra-book, one
of the big, fat books which men used to write
in those days, a book nearly half a foot thick.
My legs give way beneath me. You poor
specimen of a housebreaker, suppose you
were caught at it! However, all goes well.
Quick, let's lock the door again and go back
to our own quarters with the pilfered volume.

And now we are together, O mysterious
tome, whose Arab name breathes a strange
mustiness of occult lore and claims kindred
with the sciences of almagest and alchemy.
What will you show me? Let us turn the
leaves at random. Before fixing one's eyes on
a definite point in the landscape, it is well to
take a summary view of the whole. Page

Newton's Binomial Theorem

follows swiftly upon page, telling me nothing.
A chapter catches my attention in the middle
of the volume; it is headed, *Newton's Binomial Theorem.*

The title allures me. What can a binomial
theorem be, especially one whose author is
Newton, the great English mathematician
who weighed the worlds? What has the
mechanism of the sky to do with this? Let
us read and seek for enlightenment. With
my elbows on the table and my thumbs behind my ears, I concentrate all my attention.

I am seized with astonishment, for I understand! There are a certain number of letters,
general symbols which are grouped in all
manner of ways, taking their places here,
there and elsewhere by turns; there are, as
the text tells me, arrangements, permutations
and combinations. Pen in hand, I arrange,
permute and combine. It is a very diverting
exercise, upon my word, a game in which the
test of the written result confirms the anticipa-
tions of logic and supplements the short-
comings of one's thinking-apparatus.

'It will be plain sailing,' said I to myself,
'if algebra is no more difficult than this.'

I was to recover from the illusion later,
when the binomial theorem, that light, crisp

biscuit, was followed by heavier and less digestible fare. But, for the moment, I had no foretaste of the future difficulties, of the pitfall in which one becomes more and more entangled, the longer one persists in struggling. What a delightful afternoon that was, before my grate, amid my permutations and combinations! By the evening, I had nearly mastered my subject. When the bell rang, at seven, to summon us to the common meal at the principal's table, I went downstairs puffed up with the joys of the newly-initiated neophyte. I was escorted on my way by *a*, *b* and *c*, intertwined in cunning garlands.

Next day, my pupil is there. Blackboard and chalk, everything is ready. Not quite so ready is the master. I bravely broach my binomial theorem. My hearer becomes interested in the combinations of letters. Not for a moment does he suspect that I am putting the cart before the horse and beginning where we ought to have finished. I relieve the dryness of my explanations with a few little problems, so many halts at which the mind takes breath awhile and gathers strength for fresh flights.

We try together. Discreetly, so as to leave him the merit of the discovery, I shed a little

Newton's Binomial Theorem

light on the path. The solution is found. My pupil triumphs; so do I, but silently, in my inner consciousness, which says:

'You understand, because you succeed in making another understand.'

The hour passed quickly and very pleasantly for both of us. My young man was contented when he left me; and I no less so, for I perceived a new and original way of learning things.

The ingenious and easy arrangement of the binomial gave me time to tackle my algebra-book from the proper commencement. In three or four days, I had rubbed up my weapons. There was nothing to be said about addition and subtraction: they were so simple as to force themselves upon one at first sight. Multiplication spoilt things. There was a certain rule of signs which declared that minus multiplied by minus made plus. How I toiled over that wretched paradox! It would seem that the book did not explain this subject clearly, or rather employed too abstract a method. I read, reread and meditated in vain: the obscure text retained all its obscurity. That is the drawback of books in general: they tell you what is printed in them and nothing more. If you fail to understand, they never advise you, never suggest an attempt along another road which might lead you to the light. The merest word would sometimes be enough to put you on the right track; and that word the books, hide-bound in a regulation phraseology, never give you.

How greatly preferable is the oral lesson! It goes forward, goes back, starts afresh, walks around the obstacle and varies the methods of attack until, at long last, light is shed upon the darkness. This incomparable beacon of the master's word was what I lacked; and I went under, without hope of succour, in that treacherous pool of the rule of signs.

My pupil was bound to suffer the effects. After an attempt at an explanation in which I made the most of the few gleams that reached me I asked him:

'Do you understand?'

It was a futile question, but useful for gaining time. Myself not understanding, I was convinced beforehand that he did not understand either.

'No,' he replied, accusing himself, perhaps, in his simple mind, of possessing a brain incapable of taking in those transcendental verities.

'Let us try another method.'

And I start again this way and that way and yet another way. My pupil's eyes serve as my thermometer and tell me of the progress of my efforts. A blink of satisfaction announces my success. I have struck home, I have found the joint in the armour. The product of minus multiplied by minus delivers its mysteries to us.

And thus we continued our studies: he, the passive receiver, taking in the ideas acquired without effort; I, the fierce pioneer, blasting my rock, the book, with the aid of much sitting up at night, to extract the diamond, truth. Another and no less arduous task fell to my share: I had to cut and polish the recondite gem, to strip it of its ruggedness and present it to my companion's intelligence under a less forbidding aspect. This diamond-cutter's work, which admitted a little light into the precious stone, was the favourite occupation of my leisure; and I owe a great deal to it.

The ultimate result was that my pupil passed his examination. As for the book borrowed by stealth, I restored it to the shelves and replaced it by another, which, this time, belonged to me.

At my normal school, I had learnt a little elementary geometry under a master. From the first few lessons onwards, I rather enjoyed the subject. I divined in it a guide for one's reasoning faculties through the thickets of the imagination; I caught a glimpse of a search after truth that did not involve too much stumbling on the way, because each step forward rests solidly upon the step already taken; I suspected geometry to be what it pre-eminently is: a school of intellectual fencing.

The truth demonstrated and its application matter little to me; what rouses my enthusiasm is the process that sets the truth before us. We start from a brilliantly-lighted spot and gradually get deeper and deeper in the darkness, which, in its turn, becomes self-illuminated by kindling new lights for a higher ascent. This progressive march of the known toward the unknown, this conscientious lantern lighting what follows by the rays of what comes before: that was my real business.

Geometry was to teach me the logical progression of thought; it was to tell me how

Newton's Binomial Theorem

the difficulties are broken up into sections
which, elucidated consecutively, together form
a lever capable of moving the block that resists any direct efforts; lastly, it showed me
how order is engendered, order, the base of
clarity. If it has ever fallen to my lot to
write a page or two which the reader has run
over without excessive fatigue, I owe it, in
great part, to geometry, that wonderful
teacher of the art of directing one's thought.
True, it does not bestow imagination, a delicate flower blossoming none knows how and
unable to thrive on every soil; but it arranges
what is confused, thins out the dense, calms
the tumultuous, filters the muddy and gives
lucidity, a superior product to all the tropes
of rhetoric.

Yes, as a toiler with the pen, I owe much
to it. Wherefore my thoughts readily turn
back to those bright hours of my novitiate,
when, retiring to a corner of the garden in
recreation-time, with a bit of paper on my
knees and a stump of pencil in my fingers, I
used to practise deducing this or that property correctly from an assemblage of straight
lines. The others amused themselves all
around me; I found my delight in the frus-
trum of a pyramid. Perhaps I should have

done better to strengthen the muscles of my thighs by jumping and leaping, to increase the suppleness of my loins with gymnastic contortions. I have known some contortionists who have prospered beyond the thinker.

See me then entering the lists as an instructor of youth, fairly well acquainted with the elements of geometry. In case of need, I could handle the land-surveyor's stake and chain. There my views ended. To cube the trunk of a tree, to gauge a cask, to measure the distance of an inaccessible point appeared to me the highest pitch to which geometrical knowledge could hope to soar. Were there loftier flights? I did not even suspect it, when an unexpected glimpse showed me the puny dimensions of the little corner which I had cleared in the measureless domain.

At that time, the college in which, two years before, I had made my first appearance as a teacher, had just halved the size of its classes and largely increased its staff. The newcomers all lived in the building, like myself, and we had our meals in common at the principal's table. We formed a hive where, in our leisure time, some of us, in our respective cells, worked up the honey of algebra

and geometry, history and physics, Greek and
Latin most of all, sometimes with a view to
the class above, sometimes and oftener with
a view to acquiring a degree. The university
titles lacked variety. All my colleagues were
bachelors of letters, but nothing more. They
must, if possible, arm themselves a little better to make their way in the world. We all
worked hard and steadily. I was the youngest
of the industrious community and no less
eager than the rest to increase my modest
equipment.

Visits between the different rooms were
frequent. We would come to consult one
another about a difficulty, or simply to pass
the time of day. I had as a neighbour, in the
next cell to mine, a retired quartermaster
who, weary of barrack life, had taken refuge
in education. When in charge of the books
of his company he had become more or less
familiar with figures; and it became his ambition to take a mathematical degree. His
cerebrum appears to have hardened while he
was with his regiment. According to my dear
colleagues, those amiable retailers of the mis-
fortunes of others, he had already twice been
plucked. Stubbornly, he returned to his

books and exercises, refusing to be daunted
by two reverses.

It was not that he was allured by the beauties of mathematics, far from it; but the step
to which he aspired favoured his plans. He
hoped to have his own boarders and dispense
butter and vegetables to lucrative purpose.
The lover of study for its own sake and the
persistent trapper hunting a diploma as he
would something to put in his mouth were not
made to understand or to see much of each
other. Chance, however, brought us together.

I had often surprised our friend sitting in
the evening, by the light of a candle, with his
elbows on the table and his head between his
hands, meditating at great length in front of
a big exercise-book crammed with cabalistic
signs. From time to time, when an idea came
to him, he would take his pen and hastily put
down a line of writing wherein letters, large
and small, were grouped without any grammatical sense. The letters x and y often recurred, intermingled with figures. Every row
ended with the sign of equality and a nought.
Next came more reflection, with closed eyes,
and a fresh row of letters arranged in a different order and likewise followed by a

nought. Page after page was filled in this queer fashion, each line winding up with o.

'What are you doing with all those rows of figures amounting to zero?' I asked him one day.

The mathematician gave me a leery look, picked up in barracks. A sarcastic droop in the corner of his eye showed how he pitied my ignorance. My colleague of the many noughts did not, however, take an unfair advantage of his superiority. He told me that he was working at analytical geometry.

The phrase had a strange effect upon me. I ruminated silently to this purpose: there was a higher geometry, which you learnt more particularly with combinations of letters in which x and y played a prominent part. When my next-door neighbour reflected so long, clutching his forehead between his hands, he was trying to discover the hidden meaning of his own hieroglyphics; he saw the ghostly translation of his sums dancing in space. What did he perceive? How would the alphabetical signs, arranged first in one and then in another manner, give an image of the actual things, an image visible to the eyes of the mind alone? It beat me.

The Life of the Fly

'I shall have to learn analytical geometry some day,' I said. 'Will you help me?'

'I'm quite willing,' he replied, with a smile in which I read his lack of confidence in my determination.

No matter; we struck a bargain that same evening. We would together break up the stubble of algebra and analytical geometry, the foundation of the mathematical degree; we would make common stock: he would bring long hours of calculation, I my youthful ardour. We would begin as soon as I had finished with my arts degree, which was my main preoccupation for the moment.

In those far-off days it was the rule to make a little serious literary study take precedence of science. You were expected to be familiar with the great minds of antiquity, to converse with Horace and Virgil, Theocritus and Plato, before touching the poisons of chemistry or the levers of mechanics. The niceties of thought could only be the gainers by these preparations. Life's exigencies, ever harsher as progress afflicts us with its increasing needs, have changed all that. A fig for correct language! Business before all!

This modern hurry would have suited my impatience. I confess that I fumed against

Newton's Binomial Theorem

the regulation which forced Latin and Greek
upon me before allowing me to open up relations with the sine and cosine. To-day, wiser,
ripened by age and experience, I am of a
different opinion. I very much regret that
my modest literary studies were not more
carefully conducted and further prolonged.
To fill up this enormous blank a little, I re-
spectfully returned, somewhat late in life, to
those good old books which are usually sold
second-hand with their leaves hardly cut.
Venerable pages, annotated in pencil during
the long evenings of my youth, I have found
you again and you are more than ever my
friends. You have taught me that an obliga-
tion rests upon whoso wields the pen: he
must have something to say that is capable
of interesting us. When the subject comes
within the scope of natural science, the interest
is nearly always assured; the difficulty, the
great difficulty, is to prune it of its thorns and
to present it under a prepossessing aspect.
Truth, they say, rises naked from a well.
Agreed; but admit that she is all the better
for being decently clothed. She craves, if not
the gaudy furbelows borrowed from rhetoric's
wardrobe, at least a vine-leaf. The geometers
alone have the right to refuse her that modest

garment; in theorems, plainness suffices. The others, especially the naturalist, are in duty bound to drape a gauze tunic more or less elegantly around her waist.

Suppose I say:

'Baptiste, give me my slippers.'

I am expressing myself in plain language, a little poor in variants. I know exactly what I am saying and my speech is understood.

Others—and they are numerous—contend that this rudimentary method is the best in all things. They talk science to their readers as they might talk slippers to Baptiste. Kaffir syntax does not shock them. Do not speak to them of the value of a well-selected term, set down in its right place, still less of a lilting construction, sounding rather well. Childish nonsense they call all that; the fiddling of a short-sighted mind!

Perhaps they are right: the Baptiste idiom is a great economiser of time and trouble. This advantage does not tempt me; it seems to me that an idea stands out better if expressed in lucid language, with sober imagery. A suitable phrase, placed in its correct position and saying without fuss the things we want to say, necessitates a choice, an often laborious choice. There are drab words, the commonplaces of colloquial speech; and there are, so to speak, coloured words, which may be compared with the brush-strokes strewing patches of light over the grey background of a painting. How are we to find those picturesque words, those striking features which arrest the attention? How are we to group them into a language heedful of syntax and not displeasing to the ear?

I was taught nothing of this art. For that matter, is it ever taught in the schools? I greatly doubt it. If the fire that runs through our veins, if inspiration do not come to our aid, we shall flutter the pages of the thesaurus in vain: the word for which we seek will refuse to come. Then to what masters shall we have recourse to quicken and develop the humble germ that is latent within us? To books.

As a boy, I was always an ardent reader; but the niceties of a well-balanced style hardly interested me: I did not understand them. A good deal later, when close upon fifteen, I began vaguely to see that words have a physiognomy of their own. Some pleased me better than others by the distinctness of their meaning and the resonance of their rhythm; they produced a clearer image

The Life of the Fly

in my mind; after their fashion, they gave me
a picture of the object described. Coloured
by its adjective and vivified by its verb, the
name became a living reality: what it said I
saw. And thus, gradually, was the magic of
words revealed to me, when the chances of
my undirected reading placed a few easy
standard pages in my way.

CHAPTER XIII

MATHEMATICAL MEMORIES: MY LITTLE
TABLE

IT is time to start our analytical geometry.
He can come now, my partner, the mathematician: I think I shall understand what he
says. I have already run through my book
and noticed that our subject, whose beautiful
precision makes work a recreation, bristles
with no very serious difficulties.

We begin in my room, in front of a blackboard. After a few evenings, prolonged
into the peaceful watches of the night, I become aware, to my great surprise, that my
teacher, the past master in those hieroglyphics, is really, more often than not, my
pupil. He does not see the combinations of
the abscissæ and ordinates very clearly. I
make bold to take the chalk in hand myself,
to seize the rudder of our algebraical boat. I
comment on the book, interpret it in my own
fashion, expound the text, sound the reefs
until daylight comes and leads us to the haven
of the solution. Besides, the logic is so irre-

sistible, it is all such easy going and so lucid that often one seems to be remembering rather than learning.

And so we proceed, with our positions reversed. I dig into the hard rock, crumble it, loosen it until I make room for thought to penetrate. My comrade—I can now allow myself to speak of him on equal terms—my comrade listens, suggests objections, raises difficulties which we try to solve in unison. The two combined levers, inserted in the fissure, end by shaking and overturning the rocky mass.

I no longer see in the corner of the quartermaster's eye the leery droop that greeted me at the start. Cordial frankness now reigns, the infectious high spirits imparted by success. Little by little, dawn breaks, very misty as yet, but laden with promises. We are both greatly amazed; and my share in the satisfaction is a double one, for he sees twice over who makes others see. Thus do we pass half the night, in delightful hours. We cease when sleep begins to weigh too heavily on our eyelids.

When my comrade returns to his room, does he sleep, careless for the moment of the shifting scene which we have conjured up?

My Little Table

He confesses to me that he sleeps soundly. This advantage I do not possess. It is not in my power to pass the sponge over my poor brain even as I pass it over the blackboard. The network of ideas remains and forms as it were a moving cobweb in which repose wriggles and tosses, incapable of finding a stable equilibrium. When sleep does come at last, it is often but a state of somnolence which, far from suspending the activity of the mind, actually maintains and quickens it more than waking would. During this torpor, in which night has not yet closed upon the brain, I sometimes solve mathematical difficulties with which I struggled unsuccessfully the day before. A brilliant beacon, of which I am hardly conscious, flares in my brain. Then I jump out of bed, light my lamp again and hasten to jot down my solutions, the recollection of which I should have lost on awakening. Like lightning-flashes, those gleams vanish as suddenly as they appear.

Whence do they come? Probably from a habit which I acquired very early in life: to have food always there for my mind, to pour the never-failing oil constantly into the lamp of thought. Would you succeed in the things of the mind? The infallible method is to be always thinking of them. This method I practised more sedulously than my comrade; and hence, no doubt, arose the interchange of positions, the disciple turned into the master. It was not, however, an overwhelming infatuation, a painful obsession; it was rather a recreation, almost a poetic feast. As our great lyric writer put it in the preface to his volume, *Les Rayons et les ombres:*[1]

'Mathematics play their part in art as well as in science. There is algebra in astronomy: astronomy is akin to poetry; there is algebra in music: music is akin to poetry.'

Is this poetic exaggeration? Surely not: Victor Hugo spoke truly. Algebra, the poem of order, has magnificent flights. I look upon its formulæ, its strophes as superb, without feeling at all astonished when others do not agree. My colleague's satirical look came back when I was imprudent enough to confide my extrageometrical raptures to his ears:

'Nonsense,' said he, 'pure stuff and nonsense! Let's get on with our tangents.'

The quartermaster was right: the strict severity of our approaching examination allowed of no such dreamer's outbursts. Was I, on my side, very wrong? To warm chill

[1] Published in 1840.—*Translator's Note.*

My Little Table

calculation by the fire of the ideal, to lift one's thought above mere formulæ, to brighten the caverns of the abstract with a spark of life: was this not to ease the effort of penetrating the unknown? Where my comrade plodded on, scorning my viaticum, I performed a journey of pleasure. If I had to lean on the rude staff of algebra, I had for my guide that voice within me, urging me to lofty flights. Study became a joy.

It became still more interesting when, after the angularities of a combination of straight lines, I learnt to portray the graces of a curve. How many properties were there of which the compass knew nothing, how many cunning laws lay contained in embryo within an equation, the mysterious nut which must be artistically cracked to extract the rich kernel, the theorem! Take this or that term, place the $+$ sign before it and forthwith you have the ellipse, the trajectory of the planets, with its two friendly foci, transmitting pairs of vectors whose sum is constant; substitute the $-$ sign and you have the hyperbola with the antagonistic foci, the desperate curve that dives into space with infinite tentacles, approaching nearer and nearer to straight lines, the asymptotes, but never succeeding in meeting them. Suppress that term and you have the parabola, which vainly seeks in infinity its lost second focus; you have the trajectory of the bombshell; you have the path of certain comets which come one day to visit our sun and then flee to depths whence they never return. Is it not wonderful thus to formulate the orbit of the worlds? I thought so then and I think so still.

After fifteen months of this exercise, we went up together for our examination at Montpellier; and both of us received our degrees as bachelors of mathematical science. My companion was a wreck: I, on the other hand, had refreshed myself with analytical geometry.

Utterly worn out by his course of conic sections, my chum declares that he has had enough. In vain I hold out the glittering prospect of a new degree, that of licentiate of mathematical science, which would lead us to the splendours of the higher mathematics and initiate us into the mechanics of the heavens: I cannot prevail upon him, cannot make him share my audacity. He calls it a mad scheme, which will exhaust us and come to nothing. Without the advice of an experienced pilot, with no other compass than a book, which is not always very clear, because of its laconic adherence to set terms, our poor bark is bound to be wrecked on the first reef. One might as well put out to sea in a nutshell and defy the billows of the vasty deep. He does not use these actual words, but his gloomy estimate of the extreme difficulties to be encountered is enough to explain his refusal. I am quite free to go and break my neck in far countries; he is more prudent and will not follow me.

I suspect another reason, which the deserter does not confess. He has obtained the title needed for his plans. What does he care for the rest? Is it worth while to sit up late at night and wear one's self out in toil for the mere pleasure of learning? He must be a madman who, without the lure of profit, lends an ear to the blandishments of knowledge. Let us retreat into our shell, close our lid to the importunities of the light and lead the life of a mussel. There lies the secret of happiness.

This philosophy is not mine. My curiosity sees in a stage accomplished no more than the preparation for a new stage towards the retreating unknown. My partner, therefore, leaves me. Henceforth, I am alone, alone and wretched. There is no one left with whom I can sit up and thresh the subject out in exhilarating discussion. There is no one near me to understand me, no one who can even passively oppose his ideas to mine and take part in the conflict whence the light will spring, even as a spark is born of the concussion of two flints. When a difficulty arises, steep as a cliff, there is no friendly shoulder to support me in my attempt to climb it. Alone, I have to cling to the roughness of the jagged rock, to fall, often, and pick myself up, covered with bruises, and renew the assault; alone, I must give my shout of triumph, without the least echo of encouragement, when, reaching the summit and broken in the effort, I am at last allowed to see a little way beyond.

My mathematical campaign will cost me much stubborn thought: I am aware of this after the first few lines of my book. I am entering upon the domain of the abstract, rough ground that can only be cleared by the insistent plough of reflection. The blackboard, excellent for the curves of analytical geometry studied in my friend's company, is now neglected. I prefer the exercise-book, a quire of paper bound in a cover. With this

confidant, which allows one to remain seated
and rests the muscles of the legs, I can com-
mune nightly under my lamp-shade, until a
late hour, and keep going the forge of thought
wherein the intractable problem is softened
and hammered into shape.

My study-table, the size of a pocket-hand-
kerchief, occupied on the right by the ink-
stand—a penny bottle—and on the left by the
open exercise-book, gives me just the room
which I need to wield the pen. I love that
little piece of furniture, one of the first ac-
quisitions of my early married life. It is
easily moved where you wish: in front of
the window, when the sky is cloudy; into the
discreet light of a corner, when the sun is troublesome. In winter, it allows you to come
close to the hearth, where a log is blazing.

Poor little walnut board, I have been faith-
ful to you for half a century and more. Ink-
stained, cut and scarred with the pen-knife,
you lend your support to-day to my prose as
you once did to my equations. This variation
in employment leaves you indifferent; your
patient back extends the same welcome to the
formulæ of algebra and the formulæ of
thought. I cannot boast this placidity; I find
that the change has not increased my peace

of mind; hunting for ideas troubles the brain even more than hunting for the roots of an equation.

You would never recognize me, little friend, if you could give a glance at my grey mane. Where is the cheerful face of former days, bright with enthusiasm and hope? I have aged, I have aged. And you, what a falling off, since you came to me from the dealer's, gleaming and polished and smelling so good with your bees-wax! Like your master, you have wrinkles, often my work, I admit; for how many times, in my impatience, have I not dug my pen into you, when, after its dip in the muddy inkpot, the nib refused to write decently!

One of your corners is broken off; the boards are beginning to come loose. Inside you, I hear, from time to time, the plane of the Death-watch, who despoils old furniture. From year to year, new galleries are excavated, endangering your solidity. The old ones show on the outside in the shape of tiny round holes. A stranger has seized upon the latter, excellent quarters, obtained without trouble. I see the impudent intruder run nimbly under my elbow and penetrate forthwith into the tunnel abandoned by the Death-watch,

My Little Table

She is after game, this slender huntress, clad in black, busy collecting Wood-lice for her grubs. A whole nation is devouring you, you old table; I am writing on a swarm of insects! No support could be more appropriate to my entomological notes.

What will become of you when your master is gone? Will you be knocked down for a franc, when the family come to apportion my poor spoils? Will you be turned into a stand for the pitcher beside the kitchen-sink? Will you be the plank on which the cabbages are shredded? Or will my children, on the contrary, agree and say:

'Let us preserve the relic. It was where he toiled so hard to teach himself and make himself capable of teaching others; it was where he so long consumed his strength to find food for us when we were little. Let us keep the sacred plank.'

I dare not believe in such a future for you. You will pass into strange hands, O my old friend; you will become a bedside-table, laden with bowl after bowl of linseed-tea, until, decrepit, rickety and broken down, you are chopped up to feed the flames for a brief moment under the simmering saucepan. You will vanish in smoke to join my labours in

that other smoke, oblivion, the ultimate resting-place of our vain agitations.

But let us return, little table, to our young
days; those of your shining varnish and of
my fond illusions. It is Sunday, the day of
rest, that is to say, of continuous work, uninterrupted by my duties in the school. I greatly
prefer Thursday, which is not a general holiday and more propitious to studious calm.
Such as it is, for all its distractions, the Lord's
day gives me a certain leisure. Let us make
the most of it. There are fifty-two Sundays
in the year, making a total that is almost
equivalent to the long vacation.

It so happens that I have a glorious question to wrestle with to-day; that of Kepler's[1]
three laws, which, when explored by the
calculus, are to show me the fundamental
mechanism of the heavenly bodies. One of
them says:

'The area swept out in a given time by the
radius vector of the path of a planet is pro-
portional to the time taken.'

From this I have to deduce that the force
which confines the planet to its orbit is di-

[1] Johan Kepler (1571-1630) announced the first two of
his three laws of planetary motion in 1609 and discovered
the third in 1618.—*Translator's Note.*

rected towards the sun. Gently entreated by
the differential and integral calculus, already
the formula is beginning to voice itself. My
concentration redoubles, my mind is set upon
seizing the radiant dawn of truth.

Suddenly, in the distance, br-r-r-rum!
Br-r-r-rum! Br-r-r-rum! The noise comes
nearer, grows louder. Woe upon me! And
plague take the Pagoda!

Let me explain. I live in a suburb, at the
beginning of the Pernes Road, far from the
tumult of the town.[1] Twenty yards in front
of my house, some pleasure-gardens have been
opened, bearing a sign-board inscribed, 'The
Pagoda.' Here, on Sunday afternoons, the
lads and lasses from the neighbouring farms
come to disport themselves in country-dances.
To attract custom and push the sale of re-
freshments, the proprietor of the ball ends
the Sunday hop with a tombola. Two hours
beforehand, he has the prizes carried along
the public roads, preceded by fifes and drums.
From a beribboned pole, borne by a stalwart
fellow in a red sash, dangle a plated goblet, a
handkerchief of Lyons silk, a pair of candle-
sticks and some packets of cigars. Who

[1] The town of Carpentras, where Fabre was a master
at the college. Pernes is about a mile from Carpentras.—*Translator's Note.*

would not enter the pleasure-gardens, with such a bait?

'Br-r-r-rum! Br-r-r-rum! Br-r-r-rum!' goes the procession.

It comes just under my window, wheels to the right and marches into the establishment, a huge wooden booth, hung with evergreens. And now, if you dislike noise, flee, flee as far as you can. Until nightfall, the ophicleides will bellow, the fifes tootle and the cornets bray. How would you deduce the steps of Kepler's laws to the accompaniment of that nigger orchestra! It is enough to drive one mad. Let us be off with all speed.

A mile away, I know a flinty waste beloved of the Wheatear and the Locust. Here reigns perfect calm; moreover, there are some clumps of evergreen oak which will lend me their scanty shade. I take my book, a few sheets of paper and a pencil and fly to this solitude. What beauteous silence, what exquisite quiet! But the sun is overwhelming, under the meagre cover of the bushes. Cheerily, my lad! Have at your Kepler's laws in the company of the blue-winged Locusts. You will return home with your problems solved, but with a blistered skin. An overdose of sun in the neck shall be the outcome of grasping the law of the areas. One thing makes up for another.

During the rest of the week, I have my Thursdays and the evenings, which I employ in study until I drop with sleep. All told I have no lack of time, despite the drudgery of my college ties. The great thing is not to be discouraged by the unavoidable difficulties encountered at the outset. I, lose my way easily in that dense forest overgrown with creepers that have to be cut away with the axe to obtain a clearing. A fortunate turn or two; and I once more know where I am. I lose my way again. The stubborn axe makes its opening without always letting in sufficient light.

The book is just a book, that is to say, a set text, saying not a word more than it is obliged to, exceedingly learned, I admit, but, alas, often obscure! The author, it seems, wrote it for himself. He understood; therefore others must. Poor beginners, left to yourselves, you manage as best you can! For you, there shall be no retracing of steps in order to tackle the difficulty in another way; no circuit easing the arduous road and preparing the passage; no supplementary aperture to admit a glimmer of daylight. Incom-

parably inferior to the spoken word, which begins again with fresh methods of attack and
is ready to vary the paths that lead to the
open, the book says what it says and nothing
more. Having finished its demonstration,
whether you understand or no, the oracle is
inexorably dumb. You reread the text and
ponder it obstinately; you pass and repass
your shuttle through the woof of figures.
Useless efforts all: the darkness continues.
What would be needed to supply the illumin-
ating ray? Often enough, a trifle, a mere
word; and that word the book will not speak.

Happy is he who is guided by a master's
teaching! His progress does not know the
misery of those wearisome break-downs.
What was I to do before the disheartening
wall that every now and then rose up and
barred my road? I followed d'Alembert's[1]
precept in his advice to young mathematical
students:

'Have faith and go ahead,' said the great
geometrician.

Faith I had; and I went on pluckily. And
it was well for me that I did, for I often
found behind the wall the enlightenment

[1] Jean Baptiste le Rond d'Alembert (1717-1783), editor
of the *Encyclopédie* and perpetual secretary of the French
Academy.—*Translator's Note.*

My Little Table

which I was seeking in front of it. Giving up the bad patch as hopeless, I would go on and, after I had left it behind, discover the dynamite capable of blasting it. 'Twas a tiny grain at first, an insignificant ball rolling and increasing as it went. From one slope to the other of the theorems, it grew to a heavy mass; and the mass became a mighty projectile which, flung backwards and retracing its course, split the darkness and spread it into one vast sheet of light.

D'Alembert's precept is good and very good, provided you do not abuse it. Too much precipitation in turning over the intractable page might expose you to many a disappointment. You must have fought the difficulty tooth and nail before abandoning it. This rough skirmishing leads to intellectual vigour.

Twelve months of meditation in the company of my little table at last won me my degree as a licentiate of mathematical science; and I was now qualified to perform, half a century later, the eminently lucrative functions of an inspector of Spiders' webs![1]

[1]Cf. *The Life of the Spider:* chaps. ix and x.—*Translator's Note.*

CHAPTER XIV

THE BLUEBOTTLE: THE LAYING

TO PURGE the earth of death's impurities and cause deceased animal matter to be once more numbered among the treasures of life there are hosts of sausage-queens, including, in our part of the world, the Bluebottle (*Calliphora vomitaria*, LIN.) and the Chequered Flesh-fly (*Sarcophaga carnaria*, LIN.). Every one knows the first, the big, dark-blue Fly who, after effecting her designs in the ill-watched meat-safe, settles on our window-panes and keeps up a solemn buzzing, anxious to be off in the sun and ripen a fresh emission of germs. How does she lay her eggs, the origin of the loathsome maggot that battens poisonously on our provisions, whether of game or butcher's meat? What are her stratagems and how can we foil them? This is what I propose to investigate.

The Bluebottle frequents our homes during autumn and a part of winter, until the cold becomes severe; but her appearance in the fields dates back much earlier. On the first

The Bluebottle: The Laying

fine day in February, we shall see her warming herself, chillily, against the sunny walls. In April, I notice her in considerable numbers on the laurestinus. It is here that she seems to pair, while sipping the sugary exudations of the small white flowers. The whole of the summer season is spent out of doors, in brief flights from one refreshment-bar to the next. When autumn comes, with its game, she makes her way into our houses and remains until the hard frosts.

This suits my stay-at-home habits and especially my legs, which are bending under the weight of years. I need not run after the subjects of my present study; they call on me. Besides, I have vigilant assistants. The household knows of my plans. Every one brings me, in a little screw of paper, the noisy visitor just captured against the panes.

Thus do I fill my vivarium, which consists of a large, bell-shaped cage of wire-gauze, standing in an earthenware pan full of sand. A mug containing honey is the dining-room of the establishment. Here the captives come to recruit themselves in their hours of leisure. To occupy their maternal cares, I employ small birds—Chaffinches, Linnets, Sparrows

—brought down, in the enclosure, by my son's gun.

I have just served up a Linnet shot two days ago. I next place in the cage a Bluebottle, one only, to avoid confusion. Her fat belly proclaims the advent of a laying-time. An hour later, when the excitement of being put in prison is allayed, my captive is in labour. With eager, jerky steps, she explores the morsel of game, goes from the head to the tail, returns from the tail to the head, repeats the action several times and at last settles near an eye, a dimmed eye sunk into its socket.

The ovipositor bends at a right angle and dives into the junction of the beak, straight down to the root. Then the eggs are emitted for nearly half an hour. The layer, utterly absorbed in her serious business, remains stationary and impassive and is easily observed through my lens. A movement on my part would doubtless scare her; but my restful presence gives her no anxiety. I am nothing to her.

The discharge does not go on continuously until the ovaries are exhausted; it is intermittent and performed in so many packets. Several times over, the Fly leaves the bird's beak and comes to take a rest upon the wire-gauze.

where she brushes her hind-legs one against
the other. In particular, before using it again,
she cleans, smoothes and polishes her laying-
tool, the probe that places the eggs. Then,
feeling her womb still teeming, she returns to
the same spot at the joint of the beak. The
delivery is resumed, to cease presently and
then begin anew. A couple of hours are thus
spent in alternate standing near the eye and
resting on the wire-gauze.

At last, it is over. The Fly does not go
back to the bird, a proof that her ovaries are
exhausted. The next day, she is dead. The
eggs are dabbed in a continuous layer, at the
entrance to the throat, at the root of the
tongue, on the membrane of the palate. Their
number appears considerable; the whole inside
of the gullet is white with them. I fix a little
wooden prop between the two mandibles of
the beak, to keep them open and enable me to
see what happens.

I learn in this way that the hatching takes
place in a couple of days. As soon as they are
born, the young vermin, a swarming mass,
leave the place where they are and disappear
down the throat. To enquire further into the
work is useless for the moment. We shall

learn more about it later, under conditions that make examination easier.

The beak of the bird invaded was closed at the start, as far as the natural contact of the mandibles allowed. There remained a narrow slit at the base, sufficient at most to admit the passage of a horse-hair. It was through this that the laying was performed. Lengthening her ovipositor like a telescope, the mother inserted the point of her implement, a point slightly hardened with a horny armour. The fineness of the probe equals the fineness of the aperture. But, if the beak were entirely closed, where would the eggs be laid then?

With a tied thread, I keep the two mandibles in absolute contact; and I place a second Bluebottle in the presence of the Linnet, which the colonists have already entered by the beak. This time, the laying takes place on one of the eyes, between the lid and the eye-ball. At the hatching, which again occurs a couple of days later, the grubs make their way into the fleshy depths of the socket. The eyes and the beak, therefore, form the two chief entrances into feathered game.

There are others; and these are the wounds. I cover the Linnet's head with a paper hood which will prevent invasion through the beak

The Bluebottle: The Laying

and eyes. I serve it, under the wire-gauze bell, to a third egg-layer. The bird has been struck by a shot in the breast, but the sore is not bleeding: no outer stain marks the injured spot. Moreover, I am careful to arrange the feathers, to smooth them with a hair-pencil, so that the bird looks quite smart and has every appearance of being untouched.

The Fly is soon there. She inspects the Linnet from end to end; with her front tarsi she fumbles at the breast and belly. It is a sort of auscultation by sense of touch. The insect becomes aware of what is under the feathers by the manner in which these react. If scent comes to her assistance, it can only be very slightly, for the game is not yet high. The wound is soon found. No drop of blood is near it, for it is closed by a plug of down rammed into it by the shot. The Fly takes up her position without separating the feathers or uncovering the wound. She remains here for two hours without stirring, motionless, with her abdomen concealed beneath the plumage. My eager curiosity does not distract her from her business for a moment.

When she has finished, I take her place. There is nothing either on the skin or at the mouth of the wound. I have to withdraw the downy plug and dig to some depth before discovering the eggs. The ovipositor has therefore lengthened its extensible tube and pushed beyond the feather stopper driven in by the lead. The eggs are in one packet; they number about three hundred.

When the beak and eyes are rendered inaccessible, when the body, moreover, has no wounds, the laying still takes place, but, this time, in a hesitating and niggardly fashion. I pluck the bird completely, the better to watch what happens; also, I cover the head with a paper hood to close the usual means of access. For a long time, with jerky steps, the mother explores the body in every direction; she takes her stand by preference on the head, which she sounds by tapping on it with her front tarsi. She knows that the openings which she needs are there, under the paper; but she also knows how frail are her grubs, how powerless to pierce their way through the strange obstacle which stops her as well and interferes with the work of her ovipositor. The cowl inspires her with profound distrust. Despite the tempting bait of the veiled head, not an egg is laid on the wrapper, slight though it may be.

Weary of vain attempts to compass this obstacle, the Fly at last decides in favour of

The Bluebottle: The Laying

other points, but not on the breast, belly or back, where the hide would seem too tough and the light too intrusive. She needs dark hiding-places, corners where the skin is very delicate. The spots chosen are the cavity of the axilla, corresponding with our armpit, and the crease where the thigh joins the belly. Eggs are laid in both places, but not many, showing that the groin and the axilla are adopted only reluctantly and for lack of a better spot.

With an unplucked bird, also hooded, the same experiment failed: the feathers prevent the Fly from slipping into those deep places. Let us add, in conclusion, that, on a skinned bird, or simply on a piece of butcher's meat, the laying is effected on any part whatever, provided that it be dark. The gloomiest corners are the favourite ones.

It follows from all this that, to lay the eggs, the Bluebottle picks out either naked wounds or else the mucuous membranes of the mouth or eyes, which are not protected by a skin of any thickness. She also needs darkness. We shall see the reasons for her preference later on.

The perfect efficiency of the paper bag, which prevents the inroads of the worms

through the eye-sockets or the beak, suggests a
similar experiment with the whole bird. It
is a matter of wrapping the body in a sort of
artificial skin which will be as discouraging to
the Fly as the natural skin. Linnets, some
with deep wounds, others almost intact, are
placed one by one in paper envelopes similar
to those in which the nursery-gardener keeps
his seeds, envelopes just folded, without being
stuck. The paper is quite ordinary and of
average thickness. Torn pieces of newspaper
serve the purpose.

These sheaths with the corpses inside them
are freely exposed to the air, on the table in
my study, where they are visited, according to
the time of day, in dense shade and in bright
sunlight. Attracted by the effluvia from the
dead meat, the Bluebottles haunt my laboratory, the windows of which are always open.
I see them daily alighting on the envelopes
and very busily exploring them, apprised of
the contents by the gamy smell. Their inces-
sant coming and going is a sign of intense cu-
pidity; and yet none of them decides to lay on
the bags. They do not even attempt to slide
their ovipositor through the slits of the folds.
The favourable season passes and not an egg
is laid on the tempting wrappers. All the

The Bluebottle: The Laying

mothers abstain, judging the slender obstacle of the paper to be more than the vermin will be able to overcome.

This caution on the Fly's part does not at all surprise me: motherhood everywhere has gleams of great perspicacity. What does astonish me is the following result. The parcels containing the Linnets are left for a whole year uncovered on the table; they remain there for a second year and a third. I inspect the contents from time to time. The little birds are intact, with unrumpled feathers, free from smell, dry and light, like mummies. They have become not decomposed, but mummified.

I expected to see them putrefying, running into sanies, like corpses left to rot in the open air. On the contrary, the birds have dried and hardened, without undergoing any change. What did they want for their putrefaction? Simply the intervention of the Fly. The maggot, therefore, is the primary cause of dissolution after death; it is, above all, the putrefactive chemist.

A conclusion not devoid of value may be drawn from my paper game-bags. In our markets, especially in those of the South, the game is hung unprotected from the hooks on the stalls. Larks strung up by the dozen with

a wire through their nostrils, Thrushes, Plo-
vers, Teal, Partridges, Snipe, in short, all the
glories of the spit which the autumn migration
brings us, remain for days and weeks at the
mercy of the Flies. The buyer allows himself
to be tempted by a goodly exterior; he makes
his purchase and, back at home, just when the
bird is being prepared for roasting, he discovers that the promised dainty is alive with
worms. O horror! There is nothing for it
but to throw the loathsome, verminous thing
away.

The Bluebottle is the culprit here. Everybody knows it; and nobody thinks of seriously
shaking off her tyranny: not the retailer, nor
the wholesale dealer, nor the killer of the
game. What is wanted to keep the maggots
out? Hardly anything: to slip each bird into
a paper sheath. If this precaution were taken
at the start, before the Flies arrive, any game
would be safe and could be left indefinitely to
attain the degree of ripeness required by the
epicure's palate.

Stuffed with olives and myrtleberries, the
Corsican Blackbirds are exquisite eating. We
sometimes receive them at Orange, layers of
them, packed in baskets through which the air
circulates freely and each contained in a paper

wrapper. They are in a state of perfect preservation, complying with the most exacting demands of the kitchen. I congratulate the nameless shipper who conceived the bright idea of clothing his Blackbirds in paper. Will his example find imitators? I doubt it.

There is, of course, a serious objection to this method of preservation. In its paper shrould, the article is invisible; it is not enticing; it does not inform the passer-by of its nature and qualities. There is one resource left which would leave the bird uncovered: simply to case the head in a paper cap. The head being the part most threatened, because of the mucus membrane of the throat and eyes, it would be sufficient, as a rule, to protect the head, in order to keep off the Flies and to thwart their attempts.

Let us continue to study the Bluebottle, while varying our means of information. A tin, about four inches deep, contains a piece of butcher's meat. The lid is not put in quite straight and leaves a narrow slit at one point of its circumference, allowing, at most, of the passage of a fine needle. When the bait begins to give off a gamy scent, the mothers come. Singly or in numbers. They are attracted by the odour which, transmitted through a thin crevice, hardly reaches my nostrils.

They explore the metal receptacle for some time, seeking an entrance. Finding naught that enables them to reach the coveted morsel, they decide to lay their eggs on the tin, just beside the aperture. Sometimes, when the width of the passage allows of it, they insert the ovipositor into the tin and lay the eggs inside, on the very edges of the slit. Whether outside or in, the eggs are dabbed down in a fairly regular and absolutely white layer. I as it were shovel them up with a little paper scoop. I thus obtain all the germs that I require for my experiments, eggs bearing no trace of the stains which would be inevitable if I had to collect them on tainted meat.

We have seen the Bluebottle refusing to lay her eggs on the paper bag, notwithstanding the carrion fumes of the Linnet enclosed; yet now, without hesitation, she lays them on a sheet of metal. Can the nature of the floor make any difference to her? I replace the tin lid by a paper cover stretched and pasted over the orifice. With the point of my knife, I make a narrow slit in this new lid. That is quite enough: the parent accepts the paper.

The Bluebottle: The Laying

What determined her, therefore, is not simply the smell, which can easily be perceived even through the uncut paper, but, above all, the crevice, which will provide an entrance for the vermin, hatched outside, near the narrow passage. The maggots' mother has her own logic, her prudent foresight. She knows how feeble her wee grubs will be, how powerless to cut their way through an obstacle of any resistance; and so, despite the temptation of the smell, she refrains from laying so long as she finds no entrance through which the new-born worms can slip unaided.

I wanted to know whether the colour, the shininess, the degree of hardness and other qualities of the obstacle would influence the decision of a mother obliged to lay her eggs under exceptional conditions. With this object in view, I employed small jars, each baited with a bit of butcher's meat. The respective lids were made of different-coloured paper, of oil-skin, or of some of that tin-foil, with its gold or coppery sheen, which is used for sealing liqueur-bottles. On not one of these covers did the mothers stop, with any desire to deposit their eggs; but, from the moment that the knife had made the narrow slit, all the lids were, sooner or later, visited and all of them, sooner or later, received the white shower somewhere near the gash. The look of the obstacle, therefore, does not count; dull or brilliant, drab or coloured: these are details of no importance; the thing that matters is that there should be a passage to allow the grubs to enter.

Though hatched outside, at a distance from the coveted morsel, the new-born worms are well able to find their refectory. As they release themselves from the egg, without hesitation, so accurate is their scent, they slip beneath the edge of the ill-joined lid, or through the passage cut by the knife. Behold them entering upon their promised land, their reeking paradise.

Eager to arrive, do they drop from the top of the wall? Not they! Slowly creeping, they make their way down the side of the jar; they use their fore-part, ever in quest of information, as a crutch and grapnel in one. They reach the meat and at once instal themselves upon it.

Let us continue our investigation, varying the conditions. A large test-tube, measuring nine inches high, is baited at the bottom with a lump of butcher's meat. It is closed with wire-gauze, whose meshes, two millimetres[1]

[1] ·078 inch.—*Translator's Note.*

The Bluebottle: The Laying

wide, do not permit of the Fly's passage.
The Bluebottle comes to my apparatus,
guided by scent rather than sight. She hastens
to the test-tube whose contents are veiled under an opaque cover with the same alacrity as
to the open tube. The invisible attracts her
quite as much as the visible.

She stays a while on the lattice of the
mouth, inspects it attentively; but, whether because circumstances have failed to serve me, or
because the wire network inspires her with distrust, I never saw her dab her eggs upon it for
certain. As her evidence was doubtful, I had
recourse to the Flesh-fly (*Sarcophaga carnaria*).

This Fly is less finikin in her preparations,
she has more faith in the strength of her
worms, which are born ready-formed and vigorous, and easily shows me what I wish to see.
She explores the trellis-work, chooses a mesh
through which she inserts the tip of her abdomen and, undisturbed by my presence,
emits, one after the other, a certain number of
grubs, about ten or so. True, her visits will
be repeated, increasing the family at a rate of
which I am ignorant.

The new-born worms, thanks to a slight
viscidity, cling for a moment to the wire-

gauze; they swarm, wriggle, release themselves and leap into the chasm. It is a nine-inch drop at least. When this is done, the mother makes off, knowing for a certainty that her offspring will shift for themselves. If they fall on the meat, well and good; if they fall elsewhere, they can reach the morsel by crawling.

This confidence in the unknown factor of the precipice, with no indication but that of smell, deserves fuller investigation. From what height will the Flesh-fly dare to let her children drop? I top the test-tube with another tube, the width of the neck of a claret-bottle. The mouth is closed either with wire-gauze, or with a paper cover with a slight cut in it. Altogether, the apparatus measures twenty-five inches in height. No matter: the fall is not serious for the lithe backs of the young grubs; and, in a few days, the test-tube is filled with larvæ, in which it is easy to recognize the Flesh-fly's family by the fringed coronet that opens and shuts at the maggot's stern like the petals of a little flower. I did not see the mother operating: I was not there at the time; but there is no doubt possible of her coming nor of the great dive taken by the

The Bluebottle: The Laying

family: the contents of the test-tube furnish me with a duly authenticated certificate.

I admire the leap and, to obtain one better still, I replace the tube by another, so that the apparatus now stands forty-six inches high. The column is erected at a spot frequented by Flies, in a dim light. Its mouth, closed with a wire-gauze cover, reaches the level of various other appliances, test-tubes and jars, which are already stocked or awaiting their colony of vermin. When the position is well-known to the Flies, I remove the other tubes and leave the column, lest the visitors should turn aside to easier ground.

From time to time, the Bluebottle and the Flesh-fly perch on the trellis-work, make a short investigation and then decamp. Throughout the summer season, for three whole months, the apparatus remains where it is, without the least result: never a worm. What is the reason? Does the stench of the meat not spread, coming from that depth? Certainly it spreads: it is unmistakable to my dulled nostrils and still more so to the nostrils of my children, whom I call to bear witness. Then why does the Flesh-fly, who but now was dropping her grubs from a goodly height, refuse to let them fall from the top of a column

The Life of the Fly

twice as high? Does she fear lest her worms should be bruised by an excessive drop? There is nothing about her to point to anxiety aroused by the length of the shaft. I never see her explore the tube or take its size. She stands on the trellised orifice; and there the matter ends. Can she be apprised of the depth of the chasm by the comparative faintness of the offensive odours that arise from it? Can the sense of smell measure the distance and judge whether it be acceptable or not? Perhaps.

The fact remains that, despite the attraction of the scent, the Flesh-fly does not expose her worms to disproportionate falls. Can she know beforehand that, when the chrysalids break, her winged family, knocking with a sudden flight against the sides of a tall chimney, will be unable to get out? This foresight would be in agreement with the rules which order maternal instinct according to future needs.

But when the fall does not exceed a certain depth, the budding worms of the Flesh-fly are dropped without a qualm, as all our experiments show. This principle has a practical application which is not without its value in matters of domestic economy. It is as well that

The Bluebottle: The Laying

the wonders of entomology should sometimes give us a hint of commonplace utility.

The usual meat-safe is a sort of large cage with a top and bottom of wood and four wire-gauze sides. Hooks fixed into the top are used whereby to hang pieces which we wish to protect from the Flies. Often, so as to employ the space to the best advantage, these pieces are simply laid on the floor on the cage. With these arrangements, are we sure of warding off the Fly and her vermin?

Not at all. We may protect ourselves against the Bluebottle, who is not much inclined to lay her eggs at a distance from the meat; but there is still the Flesh-fly, who is more venturesome and goes more briskly to work and who will slip the grubs through a hole in the meshes and drop them inside the safe. Agile as they are and well able to crawl, the worms will easily reach anything on the floor; the only things secure from their attacks will be the pieces hanging from the ceiling. It is not in the nature of maggots to explore the heights, especially if this implies climbing down a string in addition.

People also use wire-gauze dish-covers. The trellised dome protects the contents even less than does the meat-safe. The Flesh-fly takes no heed of it. She can drop her worms through the meshes on the covered joint.

Then what are we to do? Nothing could be simpler. We need only wrap the birds which we wish to preserve—Thrushes, Partridges, Snipe and so on—in separate paper envelopes; and the same with our beef and mutton. This defensive armour alone, while leaving ample room for the air to circulate, makes any invasion by the worms impossible, even without a cover or a meat-safe: not that paper possesses any special preservative virtues, but solely because it forms an impenetrable barrier. The Bluebottle carefully refrains from laying her eggs upon it and the Fleshfly from bringing forth her offspring, both of them knowing that their new-born young are incapable of piercing the obstacle.

Paper is equally successful in our strife against the Moths, those plagues of our furs and clothes. To keep away these wholesale ravages, people generally use camphor, naphthaline, tobacco, bunches of lavender and other strong-scented remedies. Without wishing to malign those preservatives, we are bound to admit that the means employed are none too effective. The smell does very little to prevent the havoc of the Moths.

The Bluebottle: The Laying

I would therefore counsel our housewives, instead of all this chemist's stuff, to use newspapers of a suitable shape and size. Take whatever you wish to protect—your furs, your flannel or your clothes—and pack each article carefully in a newspaper, joining the edges with a double fold, well-pinned. If this joining is properly done, the Moth will never get inside. Since my advice has been taken and this method employed in my household, the old damage has never been repeated.

To return to the Fly. A piece of meat is hidden in a jar under a layer of fine, dry sand, a finger's-breadth thick. The jar has a wide mouth and is left quite open. Let whoso come that will, attracted by the smell. The Bluebottles are not long in inspecting what I have prepared for them: they enter the jar, go out and come back again, enquiring into the invisible thing revealed by its fragrance. A diligent watch enables me to see them fussing about, exploring the sandy expanse, tapping it with their feet, sounding it with their proboscis. I leave the visitors undisturbed for a fortnight or three weeks. None of them lays any eggs.

This is a repetition of what the paper bag, with its dead bird, showed me. The Flies refuse to lay on the sand, apparently for the same reasons. The paper was considered an obstacle which the frail vermin would not be able to overcome. With sand, the case is worse. Its grittiness would hurt the new-born weaklings, its dryness would absorb the moisture indispensable to their movements. Later, when preparing for the metamorphosis, when their strength has come to them, the grubs will dig the earth quite well and be able to descend; but, at the start, that would be very dangerous for them. Knowing these difficulties, the mothers, however greatly tempted by the smell, abstain from breeding. As a matter of fact, after long waiting, fearing lest some packets of eggs may have escaped my attention, I inspect the contents of the jar from top to bottom. Meat and sand contain neither larvæ nor pupæ: the whole is absolutely deserted.

The layer of sand being only a finger's-breadth thick, this experiment requires certain precautions. The meat may expand a little, in going bad, and protrude in one or two places. However small the fleshy eyots that show above the surface, the Flies come to them and breed. Sometimes also the juices oozing from the putrid meat soak a small extent of the

The Bluebottle: The Laying

sandy floor. That is enough for the maggot's first establishment. These causes of failure are avoided with a layer of sand about an inch thick. Then the Bluebottle, the Flesh-fly and other Flies whose grubs batten on dead bodies are kept at a proper distance.

In the hope of awakening us to a proper sense of our insignificance, pulpit orators sometimes make an unfair use of the grave and its worms. Let us put no faith in their doleful rhetoric. The chemistry of man's final dissolution is eloquent enough of our emptiness: there is no need to add imaginary horrors. The worm of the sepulchre is an invention of cantankerous minds, incapable of seeing things as they are. Covered by but a few inches of earth, the dead can sleep their quiet sleep: no Fly will ever come to take advantage of them.

At the surface of the soil, exposed to the air, the hideous invasion is possible; ay, it is the invariable rule. For the melting down and remoulding of matter, man is no better, corpse for corpse, than the lowest of the brutes. Then the Fly exercises her rights and deals with us as she does with any ordinary animal refuse. Nature treats us with magnificent indifference in her great regenerating-factory: placed in her crucibles, animals and

339

men, beggars and kings are one and all alike.
There you have true equality, the only equality
in this world of ours: equality in the presence
of the maggot.

CHAPTER XV

THE BLUEBOTTLE: THE GRUB

THE larvæ of the Bluebottle hatch within two days in the warm weather. Whether inside my apparatus, in direct contact with the piece of meat, or outside, on the edge of a slit that enables them to enter, they set to work at once. They do not eat, in the strict sense of the word, that is to say, they do not tear their food, do not chew it by means of implements of mastication. Their mouth-parts do not lend themselves to this sort of work. These mouth-parts are two horny spikes, sliding one upon the other, with curved ends that do not face, thus excluding the possibility of any function such as seizing and grinding.

The two guttural grapnels serve for walking much rather than for feeding. The worm plants them alternately in the road traversed and, by contracting its crupper, advances just that distance. It carries in its tubular throat the equivalent of our iron-tipped sticks which give support and assist progress.

Thanks to this machinery of the mouth, the maggot not only moves over the surface, but also easily penetrates the meat: I see it disappear as though it were dipping into butter. It cuts its way, levying, as it goes, a preliminary toll, but only of liquid mouthfuls. Not the smallest solid particle is detached and swallowed. That is not the maggot's diet. It wants a broth, a soup, a sort of fluid extract of beef which it prepares itself. As digestion, after all, merely means liquefaction, we may say, without being guilty of paradox, that the grub of the Bluebottle digests its food before swallowing it.

With the object of relieving gastric troubles, our manufacturing chemists scrape the stomachs of the Pig and Sheep and thus obtain pepsin, a digestive agent which possesses the property of liquefying albuminous matters and lean meat in particular. Why cannot they rasp the stomach of the maggot! They would obtain a product of the highest quality, for the carnivorous worm also owns its pepsin, pepsin of a singularly active kind, as the following experiments will show us.

I divide the white of a hard-boiled egg into tiny cubes and place them in a little test-tube. On the top of the contents, I sprinkle the eggs of the Bluebottle, eggs free from the least stain, taken from those laid on the outside of

tins baited with meat and not absolutely shut.
A similar test-tube is filled with white of egg,
but receives no germs. Both are closed with a
plug of cotton-wool and left in a dark corner.

In a few days, the tube swarming with new-
born vermin contains a liquid as fluid and
transparent as water. Not a drop would remain in the tube if I turned it upside down.
All the white of egg has disappeared, lique-
fied. As for the worms, which are already a
fair size, they seem very ill at ease. Deprived
of a support whence to attain the outer air,
most of them dive into the broth of their own
making, where they perish by drowning.
Others, endowed with greater vigour, crawl
up the glass to the plug and manage to make
their way through the wadding. Their pointed
front, armed with grappling-irons, is the nail
that penetrates the fibrous mass.

In the other test-tube, standing beside the
first and subjected to the same atmospheric influences, nothing striking has occurred. The
hard-boiled white of egg has retained its deadwhite colour and its firmness. I find it as I left
it. The utmost that I observe is a few traces
of must. The result of this first experiment
is patent : the Bluebottle's grub is the medium
that converts coagulated albumen into a liquid.

The Life of the Fly

The value of chemist's pepsin is estimated by the quantity of hard-boiled white of egg which a gramme of that agent can liquefy. The mixture has to be exposed in an oven to a temperature of 140° F. and also to be frequently shaken. My preparation, in which the Bluebottle's eggs are hatched, is neither shaken nor subjected to the heat of an oven; everything happens in quietness and under the thermometrical conditions of the surrounding air; nevertheless, in a few days, the coagulated albumen, treated by the vermin, runs like water.

The reagent that causes this liquefaction escapes my endeavours to detect it. The worms must disgorge it in infinitesimal doses, while the spikes in their throats, which are in continual movement, emerge a little way from the mouth, re-enter and reappear. Those piston-thrusts, those quasi-kisses, are accompanied by the emission of the solvent: at least, that is how I picture it. The maggot spits on its food, places on it the wherewithal to make it into broth. To appraise the quantity of the matter expectorated is beyond my powers: I observe the result, but do not perceive the leavening agent.

The Bluebottle: The Grub

Well, this result is really astounding, when
we consider the scantiness of the means. No
Pig's or Sheep's pepsin can rival that of the
worm. I have a bottle of pepsin that comes
from the School of Chemistry at Montpellier.
I lavishly powder some pieces of hard-boiled
white of egg with the potent drug, just as I
did with the eggs of the Bluebottle. The oven
is not brought into play, neither is distilled
water added, nor hydrochloric acid: two
auxiliaries which are recommended. The experiment is conducted in exactly the same way
as that of the tubes with the vermin. The
result is entirely different from what I expected. The white of egg does not liquefy.
It simply becomes moist on the surface; and
even this moisture may come from the pepsin,
which is highly absorbent. Yes, I was right:
if the thing were feasible, it would be an advantage for the chemists to collect their digestive drug from the stomach of the maggot.
The worm, in this case, beats the Pig and the
Sheep.

The same method is followed for the remaining experiments. I put the Bluebottle's
eggs to hatch on a piece of meat and leave the
worms to do their work as they please. The
lean tissues, whether of mutton, beef or pork,

no matter which, are not turned into liquid; they become a pea-soup of a clarety brown. The liver, the lung, the spleen are attacked to better purpose, without, however, getting beyond the state of a semi-fluid jam, which easily mixes with water and even appears to dissolve in it. The brains do not liquefy either: they simply melt into a thin gruel.

On the other hand, fatty substances, such as beef-suet, lard and butter, do not undergo any appreciable change. Moreover, the worms soon dwindle away, incapable of growing. This sort of food does not suit them. Why? Apparently because it cannot be liquefied by the reagent disgorged by the worms. In the same way, ordinary pepsin does not attack fatty substances; it takes pancreatin to reduce them to an emulsion. This curious analogy of properties, positive for albuminous, negative for fatty matter, proclaims the similarity and perhaps the identity of the dissolvent discharged by the grubs and the pepsin of the higher animals.

Here is another proof: the usual pepsin does not dissolve the epidermis, which is a material of a horny nature. That of the maggots does not dissolve it either. I can easily rear Bluebottle-grubs on dead Crickets whose

The Bluebottle: The Grub

bellies I have first opened, but I do not succeed
if the morsel be left intact: the worms are unable to perforate the succulent paunch; they
are stopped by the cuticle, on which their reagent refuses to act. Or else I give them
Frogs' hind-legs, stripped of their skin. The
flesh turns to broth and disappears to the bone.
If I do not peel the legs, they remain intact
in the midst of the vermin. Their thin skin
is sufficient to protect them.

This failure to act upon the epidermis explains why the Bluebottle at work on the animal declines to lay her eggs on the first part
that comes handy. She needs the delicate
membrane of the nostrils, eyes or throat, or
else some wound in which the flesh is laid bare.
No other place suits her, however excellent
for flavour and darkness. At most, finding
nothing better when my stratagems interfere,
she persuades herself to dab a few eggs under
the axilla of a plucked bird or in the groin,
two points at which the skin is thinner than
elsewhere.

With her maternal foresight, the Bluebottle
knows to perfection the choice surfaces, the
only ones liable to soften and run under the
influence of the reagent dribbled by the new-
born grubs. The chemistry of the future is

The Life of the Fly

familiar to her, though she does not use it for her own feeding; motherhood, that great inspirer of instinct, teaches her all about it.

Scrupulous though she be in choosing exactly where to lay her eggs, the Bluebottle does not trouble about the quality of the provisions intended for her family's consumption. Any dead body suits her purpose. Redi,[1] the Italian scientist who first exploded the old, foolish notion of worms begotten of corruption, fed the vermin in his laboratory with meat of very different kinds. In order to make his tests the more conclusive, he exaggerated the largess of the dining-hall. The diet was varied with tiger- and lion-flesh, bear and leopard, fox and wolf, mutton and beef, horse-flesh, donkey-flesh and many others, supplied by the rich menagerie of Florence. This wastefulness was unnecessary: wolf and mutton are all the same to an unprejudiced stomach.

A distant disciple of the maggot's biographer, I look at the problem in a light which Redi never dreamt of. Any flesh of one of the higher animals suits the Fly's family. Will it

[1]Francesco Redi (1626-1698), the Italian naturalist and poet, author of *Esperienze intorno alla generazione degli insetti.—Translator's Note.*

be the same if the food supplied be of a lower organism and consist of fish, for instance, of Frog, Mollusc, insect, Centipede? Will the worms accept these viands and, above all, can they manage to liquefy them, which is the first and foremost condition?

I serve a piece of raw Whiting. The flesh is white, delicate, partly translucent, easy for our stomachs to digest and no less suited to the grub's dissolvent. It turns into an opalescent fluid, which runs like water. In fact, it liquefies in much the same way as hard-boiled white of egg. The worms at first wax fat, as long as the conditions allow of some solid eyots remaining; then, when foothold fails, threatened with drowning in the too-fluid broth, they creep up the side of the glass, anxious and restless to be off. They climb to the cotton-wool stopper of the test-tube and try to bolt through the wadding. Endowed with stubborn perseverance, nearly all of them decamp in spite of the obstacle. The test-tube with the white of egg showed me a similar exodus. Although the fare suits them, as their growth witnesses, the worms cease feeding and make a point of escaping when death by drowning is imminent.

With other fish, such as Skate and Sardines,

with the flesh of Frogs and Tree-frogs, the meat simply dissolves into a porridge. Hashes of Slug, Scolopendra or Praying Mantis furnish the same result.

In all these preparations, the dissolving-agent of the worms is as much in evidence as when butcher's meat is employed. Moreover, the grubs seem satisfied with the queer dish which my curiosity prescribes for them; they thrive amidst the victuals and undergo their transformation into pupæ.

The conclusion, therefore, is much more general than Redi imagined. Any meat, no matter whether of a higher or lower order, suits the Bluebottle for the settlement of her family. The carcasses of furred and feathered animals are the favourite victuals, probably because of their richness, which allows of plentiful layings; but, should the occasion demand it, the others are also accepted, without inconvenience. Any carrion that has lived the life of an animal comes within the domain of these scavengers.

What is their number to one mother? I have already spoken of a deposit of three hundred, counted egg by egg. A quite fortuitous circumstance enabled me to go much farther. In the first week of January 1905, we experi-

The Bluebottle: The Grub

enced a sudden short cold snap of a severity very exceptional in my part of the country. The thermometer fell to twelve degrees below zero.[1] While a fierce north wind was raging and beginning to redden the leaves of the olive-trees, came one and brought me a Barn- or Screech-owl, which he had found on the ground, exposed to the air, not far from my house. My reputation as a lover of animals made the donor believe that I should be pleased with his gift.

I was, as a matter of fact, but for reasons whereof the finder certainly never dreamt. The Owl was untouched, with trim feathers and not the least wound that showed. Perhaps he had died of cold. What made me gratefully accept the present was exactly that which would have inclined any one but myself to refuse it. The Owl's eyes, glazed in death, were hidden under a thick mass of eggs, which I recognized as a Bluebottle's. Similar masses occupied the vicinity of the nostrils. If I wanted maggots, here, of a certainty, was a richer crop than I had ever beheld.

I place the corpse on the sand of a pan, with a wire-gauze cover, and leave events to take their course. The laboratory in which I

[1] Centigrade; *i.e.,* 10° Fahrenheit.—*Translator's Note,*

351

instal my bird is none other than my study. It is as cold in there, or nearly, as outside, so much so that the water in the aquarium in which I used to rear Caddis-worms has frozen into a solid block of ice. Under these conditions of temperature, the Owl's eyes keep their white veil of germs unchanged. Nothing stirs, nothing swarms. Weary of waiting, I pay no more attention to the carcass; I leave the future to decide whether the cold has exterminated the Fly's family or not.

Before the end of March, the packets of eggs have disappeared, I know not how long. The bird, for that matter, seems to be intact. On the ventral surface, which is turned to the air, the feathers keep their smooth arrangement and their fresh colouring. I lift the thing. It is light, very dry and gives a hard sound, like an old shoe tanned by the summer sun in the fields. There is no smell. The dryness has vanquished the stench, which, in any case, was never offensive during that time of frost. On the other hand, the back, which touched the sand, is a loathsome wreck, partly deprived of its feathers. The quills of the tail are bare-barrelled; a few whitened bones show, deprived of their muscles. The skin has turned into a dark leather, pierced with

round holes like those of a sieve. It is all
hideously ugly, but most instructive.

The wretched Owl, with his shattered back-
bone, teaches us, first of all, that a temperature
of twelve degrees of frost does not endanger
the existence of the Bluebottle's germs. The
worms were born without accident, despite the
rude blast; they feasted copiously on extract of
meat; then, growing big and fat, they descended into the earth by piercing round holes
in the bird's skin. Their pupæ must now be
in the sand of the pan.

They are, in point of fact, and in such numbers that I have to resort to sifting in order to
collect them. If I used the forceps, I should
never have done sorting so great a quantity.
The sand passes through the meshes of the
sieve, the pupæ remain above. To count them
would wear out my patience. I measure them
by the bushel, that is to say, with a thimble of
which I know the holding-capacity in pupæ.
The result of my calculation is not far short
of nine hundred.

Does this family proceed from one mother?
I am quite ready to admit it, so unlikely is it
that the Bluebottle, who is so rare inside our
houses during the severe cold of winter, should
be frequent enough outside to form into

groups and to do business in common while an
icy blast is raging. A belated specimen, the
plaything of the north wind, and one alone
must have deposited the burden of her ovaries
on the Owl's eyes. This laying of nine hundred eggs, an incomplete laying perhaps, bears
witness to the mighty part played by the Fly
as a liquidator of corpses.

Before throwing away the Screech-owl
treated by the worms, let us overcome our
repugnance and give a glance inside the bird.
We see a tortuous cavity, fenced in by nameless ruins. Muscles and bowels have disappeared, converted into broth and gradually
consumed by the teeming throng. In every
part, what was wet has become dry, what was
solid muddy. In vain my forceps ransacks
every nook and corner: it does not hit upon a
single pupa. All the worms have emigrated,
all, without exception. From first to last, they
have forsaken the refuge of the corpse, so
soft to their delicate skins; they have left the
velvet for the hard ground. Is dryness necessary to them at this stage? They had it in the
carcass, which was thoroughly drained. Would
they protect themselves against the cold and
rain? No shelter could suit them better than
the thick quilt of the feathers, which has re-

mained wholly undamaged on the belly, the breast and every part that was not in touch with the ground. It looks as though they had fled from comfort to seek a less kindly dwelling-place. When the hour of transformation came, all left the Owl, that most excellent lodging; all dived into the sand.

The exodus from the mortuary tabernacle was made through the round holes wherewith the skin is pierced. Those holes are the worms' work: of that there is no doubt; and yet we have lately seen the mothers refuse as a bed for their eggs any part whereat the flesh is protected by a skin of some thickness. The reason is the failure of the pepsin to act on epidermic substances. In the absence of liquefaction at such points, the nourishing gruel is unprocurable. On the other hand, the tiny worms are not able—or, at least, do not know how—to dig through the integument with their pair of guttural harpoons, to rend it and reach the liquefiable flesh. The new-born lack strength and, above all, purpose. But, as the time comes for descending into the earth, the worms, now powerful and suddenly versed in the necessary art, well know how to eat away patiently and clear themselves a passage. With the hooks of their spikes they dig, scratch and tear. Instinct has flashes of inspiration. What the animal did not know how to do at the start it learns without apprenticeship when the time comes to practise this or that industry. The maggot ripe for burial perforates a membranous obstacle which the grub intent upon its broth would not even have attempted to attack with either its pepsin or its grapnels.

Why does the worm quit the carcass, that capital shelter? Why does it go and take up its abode in the ground? As the leading disinfector of dead things, it works at the most important matter, the suppression of the infection; but it leaves a plentiful residuum, which does not yield to the reagents of its analytical chemistry. These remains have to disappear in their turn. After the Fly, anatomists come hastening, who take up the dry relic, nibble skin, tendons and ligaments and scrape the bones clean.

The greatest expert in this work is the Dermestes Beetle, an enthusiastic gnawer of animal remains. Sooner or later, he will come to the joint already exploited by the Fly. Now what would happen if the pupæ were there? The answer is obvious. The Dermestes, who loves hard food, would dig his teeth into the

horny little kegs and demolish them at a bite.
Even though he did not touch the contents, a
live thing which he probably dislikes, he would
at least test the flavour of that lifeless substance, the container. The future Fly would
be lost, because her casing would be pierced.
Even so, in the store-rooms of our silk-mills,
a certain Dermestes (*Dermestes vulpinus,*
FABR.) digs into the cocoons to attack the
horny covering of the chrysalis.

The maggot foresees the danger and makes
itself scarce before the other arrives. In what
sort of memory does it house so much wisdom,
indigent, headless creature that it is, for it is
only by extension that we can give the name of
head to the animal's pointed fore-part? How
did it learn that, to safeguard the pupa, it
must desert the carcass and that, to safeguard
the Fly, it must not bury itself too far down?

To emerge from underground after the perfect insect is hatched, the Bluebottle's device
consists in disjointing her head into two movable halves, which, each distended with its
great red eye, by turns separate and reunite.
In the intervening space, a large, glassy hernia
rises and disappears, disappears and rises.
When the two move asunder, with one eye
forced back to the right, the other to the left,

it is as though the insect were splitting its brain-pan in order to expel the contents. Then the hernia rises, blunt at the end and swollen into a great knob. Next, the forehead closes and the hernia retreats, leaving visible only a kind of shapeless muzzle. In short, a frontal pouch, with deep pulsations momentarily renewed, becomes the instrument of deliverance, the pestle wherewith the newly-hatched Bluebottle bruises the sand and causes it to crumble. Gradually the legs push the rubbish back and the insect advances so much toward the surface.

A hard task, this exhumation by dint of the blows of a cleft and palpitating head. Moreover, the exhausting effort has to be made at the moment of greatest weakness, when the insect leaves that protecting casket, its pupa. It emerges from it pale, flabby and unsightly, sorrily clad in the wings which, folded lengthwise and made shorter by their scalloped edge, only just cover the top of the back. Wildly bristling with hairs and coloured ashen-grey, it is a piteous sight. The large set of wings, suitable for flight, will spread later. For the moment, it would only be in the way amid the obstacles to be passed through. Later also will come the fautless dress wherein the iridescent indigo-blue stands out against the severity of the black.

The frontal hernia that crumbles the sand with its impact has a tendency to make play for some time after the emergence from the ground. Take hold with the forceps of one of the hind-legs of a newly-released Fly. Forthwith, the implement of the head begins to work, swelling and subsiding as energetically as a moment ago, when it had to make a hole in the sand. The insect, hampered in its movements as when it was underground, struggles as best it can against the only obstacle that it knows. With its heaving knob, it pounds the air even as but now it pounded the earthy barrier. In all unpleasant circumstances, its one resource is to cleave its head and produce its cranial hernia, which moves out and in, in and out. For nearly two hours, interspersed with halts due to fatigue, the little machine keeps throbbing in my forceps.

In the meantime, however, the desperate one is hardening her skin; she spreads wide the sail of her wings and dons her deep mourning of black and darkest blue. Then her eyes, warped sideways, come together and resume their normal position. The cleft forehead closes; the delivering blister goes in, never to show itself again. But there is one precaution to be taken first. With its front tarsi, the insect carefully brushes the bump about to disappear from view, lest grit should lodge in the cranium when the two halves of the head are joined for good.

The maggot is aware of the trials that await it when, as a Fly, it will have to come up from under ground; it knows beforehand how difficult the ascent will be with the feeble instrument at its disposal, so difficult, in fact, as to become fatal should the journey be at all prolonged. It foresees the dangers ahead of it and averts them as well as it can. Gifted with two iron-shod sticks in its throat, it can easily descend to such depths as it pleases. The need for greater quiet and a less trying temperature calls for the deepest possible home: the lower down it is, the better for the welfare of the worm and the pupa, on condition that descent be practicable. It is, perfectly; and yet, though free to obey its inspiration, the grub refrains. I rear it in a deep pan, full of fine, dry sand, easy to excavate. The interment never goes very far. About a hand's-breadth is all that the most progressive digger ventures upon. Most of the interred remain nearer still to the surface. Here, un-

The Bluebottle: The Grub

der a thin layer of sand, the grub's skin hardens and becomes a coffin, a casket, wherein the transformation-sleep is slept. A few weeks later, the buried one awakes, transfigured but weak, having naught wherewith to unearth herself but the throbbing hernia of her open forehead.

What the maggot denies itself it is open to me to realize, should I care to know the depth whence the Fly is able to mount. I place fifteen Bluebottle-pupæ, obtained in winter, at the bottom of a wide tube closed at one end. Above the pupæ is a perpendicular column of fine, dry sand, the height of which varies in different tubes. April comes and the hatching begins.

A tube with six centimetres[1] of sand, the shallowest of the columns under experiment, yields the best result. Of the fifteen subjects interred in the pupa stage, fourteen easily reach the surface when they become Flies. Only one of them perishes, one who has not even attempted the ascent. With twelve centimetres[2] of sand, four emerge. With twenty centimetres,[3] two, no more. The other Flies,

[1] 2.34 inches.—*Translator's Note.*
[2] 4.68 inches.--*Translator's Note.*
[3] 7.8 inches.—*Translator's Note.*

jaded with their exertions, have died at a higher or lower stage of the road. Lastly, with yet another tube wherein the column of sand measured sixty centimetres,[1] I obtained the liberation of only a single Fly. The plucky creature must have had a hard struggle to mount from so great a depth, for the other fourteen did not even manage to burst the lid of their caskets.

I presume that the looseness of the sand and the consequent pressure in every direction, similar to that exercised by fluids, have a certain bearing on the difficulties of the exhumation. Two more tubes are prepared, but this time supplied with fresh mould, lightly heaped up, which has not the incoherence of sand, with the attendant drawback of pressure. Six centimetres of mould give me eight Flies for fifteen pupæ buried; twenty centimetres give me only one. There is less success than with the sandy column. My device has diminished the pressure, but, at the same time, increased the passive resistance. The sand falls of itself under the impact of the frontal rammer; the unyielding mould demands the cutting of a gallery. In fact, I perceive, on the road followed, a shaft which continues indefinitely

[1] 23.4 inches.—*Translator's Note.*

The Bluebottle: The Grub

such as it is. The Fly has bored it with the
temporary blister that throbs between her
eyes.

In every medium, therefore, whether sand,
mould or any earthy combination, great are
the sufferings that attend the exhumation of
the Fly. And so the maggot shuns the depths
which a desire for additional security might
seem to recommend. The worm has its own
prudence: foreseeing the dangers ahead, it refrains from making great descents that might
promote the welfare of the moment. It
neglects the present for the sake of the future.

CHAPTER XVI

A PARASITE OF THE MAGGOT

THE dangers of the exhumation are not the only ones; the Bluebottle must be acquainted with others. Life, when all is said, is a knacker's yard wherein the devourer of to-day becomes the devoured of to-morrow; and the robber of the dead cannot fail to be robbed of her own life when the time comes. I know that she has one exterminator in the person of the tiny Saprinus Beetle, a fisher of fat sausages on the edge of the pools formed by liquescent corpses. Here swarm in common the grubs of the Greenbottle, the Flesh-fly and the Bluebottle. The Saprinus draws them to him from the bank and gobbles them indiscriminately. They represent to him morsels of equal value.

This banquet can be observed only in the open country, under the rays of a hot sun. Saprini and Greenbottles never enter our houses; the Flesh-fly visits us but discreetly, does not feel at home with us; the only one who comes fussing along is the Bluebottle,

A Parasite of the Maggot

who thus escapes the tribute due to the consumer of plump sausages. But, in the fields, where she readily lays her eggs upon any carcass that she finds, she, as well as the others, sees her vermin swept away by the gluttonous Saprinus.

In addition, graver disasters decimate her family, if, as I do not doubt, we can apply to the Bluebottle what I have seen happen in the case of her rival, the Flesh-fly. So far, I have had no opportunity of actually perceiving with the first what I have to tell of the second; still, I do not hesitate to repeat about the one what observation has taught me about the other, for the larval analogies between the two Flies are very close.

Here are the facts. I have gathered a number of pupæ of the Flesh-fly in one of my vermin-jars. Wishing to examine the pupa's hinder-end, which is hollowed into a cup and scalloped into a coronet, I stave in one of the little barrels and force open the last segments with the point of my pocket-knife. The horny keg does not contain what I expected to find: it is full of tiny grubs packed one atop the other with the same economy of space as anchovies in a bottle. Save for the skin, which has hardened into a brown shell, the substance of the maggot has disappeared, changed into a restless swarm.

There are thirty-five occupants. I replace them in their casket. The rest of my harvest, wherein, no doubt, are other pupæ similarly stocked, is arranged in tubes that will easily show me what happens. The thing to discover is what genus of parasites the grubs enclosed belong to. But it is not difficult, without waiting for the hatching of the adults, to recognize their nature merely by their mode of life. They form part of the family of Chalcididæ, who are microscopic ravagers of living entrails.

Not long ago, in winter, I took from the chrysalis of a Great Peacock Moth four hundred and forty-nine parasites belonging to the same group. The whole substance of the future Moth had disappeared, all but the nymphal wrapper, which was intact and formed a handsome Russia-leather wallet. The worm-grubs were here heaped up and squeezed together to the point of sticking to one another. The hair-pencil extracts them in bundles and cannot separate them without some difficulty. The holding-capacity is strained to the utmost; the substance of the vanished Moth would not fill it better. That

A Parasite of the Maggot

which died has been replaced by a living mass of equal dimensions, but subdivided. The price of this colony's existence is the conversion of the chrysalis into a sort of milk-food of doubtful constitution. The enormous udder has been drained outright.

You shudder when you think of that budding flesh nibbled bit by bit by four or five hundred gormandizers; the horrified imagination refuses to picture the anguish suffered by the tortured wretch. But is there really any pain? We have leave to doubt it. Pain is a patent of nobility; it is more pronounced in proportion as the sufferer belongs to a higher order. In the lower ranks of animal life, it must be greatly reduced, perhaps even *nil,* especially when life, in the throes of evolution, has not yet acquired a stable equilibrium. The white of an egg is living matter, but endures the prick of a needle without a quiver. Would it not be the same with the chrysalis of the Great Peacock, dissected cell by cell by hundreds of infinitesimal anatomists? Would it not be the same with the pupa of the Flesh-fly? These are organisms put back into the crucible, reverting to the egg-state for a second birth. There is reason to believe, therefore, that their destruction crumb by crumb is merciful.

Towards the end of August, the parasite of the Flesh-fly's grubs makes her appearance out of doors in the adult form. She is a Chalcidid, as I expected. She issues from the barrel through one or two little round holes which the prisoners have pierced with a patient tooth. I count some thirty to each pupa. There would not be enough room in the abode if the family were larger.

The imp is a slim and elegant creature, but oh, how small! She measures hardly two millimetres.[1] Her garb is bronzed-black, with pale legs and a heart-shaped, pointed, slightly pedunculate abdomen, with never a trace of a probe for inoculating the eggs. The head is transversal, the width exceeding the length.

The male is only half the size of the female; he is also very much less numerous. Perhaps pairing is here, as we see elsewhere, a secondary matter from which it is possible to abstain, in part, without injuring the prospects of the race. Nevertheless, in the tube wherein I have housed the swarm, the few males lost among the crowd ardently woo the passing fair. There is much to be done out-

[1] .078 inch.—*Translator's Note.*

A Parasite of the Maggot

side, as long as the Flesh-fly's season lasts; things are urgent; and each pigmy hurries as fast as she can to take up her part as an exterminator.

How is the parasite's inroad into the Flesh-fly's pupæ effected? Truth is always veiled in a certain mystery. The good fortune that secured me the ravaged pupæ taught me nothing concerning the tactics of the ravager. I have never seen the Chalcidid explore the contents of my appliances; my attention was engaged elsewhere and nothing is so difficult to see as a thing not yet suspected. But, though direct observation be lacking, logic will tell us approximately what we want to know.

It is evident, to begin with, that the invasion cannot have been made through the sturdy amour of the pupæ. This is too hard to be penetrated by the means at the pigmy's disposal. Naught but the delicate skin of the maggots lends itself to the introduction of the germs. An egg-laying mother, therefore, appears, inspects the surface of the pool of sanies swarming with grubs, selects the one that suits her and perches on it; then, with the tip of her pointed abdomen, whence emerges, for an instant, a short probe kept hidden until then, she operates on the patient, perforating his paunch with a dexterous wound into which the germs are inserted. Probably, a number of pricks are administered, as the presence of thirty parasites seems to demand.

Anyway, the maggot's skin is pierced at either one point or many; and this happens while the grub is swimming in the pools formed by the putrid flesh. Having said this, we are faced with a question of serious interest. To set it forth necessitates a digression which seems to have nothing to do with the subject in hand and is nevertheless connected with it in the closest fashion. Without certain preliminaries, the remainder would be unintelligible. So now for the preliminaries.

I was in those days busy with the poison of the Languedocian Scorpion[1] and its action upon insects. To direct the sting toward this or the other part of the victim and moreover to regulate its emission would be absolutely impossible and also very dangerous, as long as the Scorpions were allowed to act as they pleased. I wished to be able myself to choose the part to be wounded; I likewise wished to vary the dose of poison at will. How to set

[1] Cf. *The Life and Love of the Insect*: chaps. xv and xvi.—*Translator's Note*.

about it? The Scorpion has no jarlike receptacle in which the venom is accumulated and stored, like that possessed, for instance, by the Wasp and the Bee. The last segment of the tail, gourd-shaped and surmounted by the sting, contains only a powerful mass of muscles along which lie the delicate vessels that secrete the poison.

In default of a poison-jar which I would have placed on one side and drawn upon at my convenience, I detach the last segment, forming the base of the sting. I obtain it from a dead and already withered Scorpion. A watch-glass serves as a basin. Here, I tear and crush the piece in a few drops of water and leave it to steep for four-and-twenty hours. The result is the liquid which I propose to use for the inoculation. If any poison remained in my animal's caudal gourd, there must be at least some traces of it in the infusion in the watch-glass.

My hypodermic syringe is of the simplest. It consists of a little glass tube, tapering sharply at one end. By drawing in my breath, I fill it with the liquid to be tested; I expel the contents by blowing. Its point is almost as fine as a hair and enables me to regulate the dose to the degree which I want. A cubic millimetre[1] is the usual charge. The injection has to be made at parts that are generally covered with horn. So as not to break the point of my fragile instrument, I prepare the way with a needle, with which I prick the victim at the spot required. I insert the tip of the loaded injector in the hole thus made and I blow. The thing is done in a moment, very neatly and in an orthodox fashion, favourable to delicate experiments. I am delighted with my modest apparatus.

I am equally delighted with the results. The Scorpion himself, when wounding with his sting, in which the poison is not diluted as mine is in the watch-glass, would not produce effects like those of my pricks. Here is something more brutal, producing more convulsion in the sufferer. The virus of my contriving excels the Scorpion's.

The test is several times repeated, always with the same mixture, which, drying up by spontaneous evaporation, then made to serve again by the addition of a few drops of water, once more drained and once more moistened, does duty for an indefinite length of time. Instead of abating, the virulence increases. Moreover, the corpses of the insects operated

[1] .175 minim.—*Translator's Note.*

upon undergo a curious change, unknown in my earlier observations. Then the suspicion comes to me that the actual poison of the Scorpion does not enter into the matter at all. What I obtain with the end joint of the tail, with the gland at the base of the sting, I ought to obtain with any other part of the animal.

I crush in a few drops of water a joint of the tail taken from the front portion, far from the poison-glands. After soaking it for twenty-four hours, I obtain a liquid whose effects are absolutely the same as those before, when I used the joint that bears the sting. I try again with the Scorpion's claws, the contents of which consist solely of muscle. The results are just the same. The whole of the animal's body, therefore, no matter which fragment be submitted to the steeping-process, yields the virus that so greatly pricks my curiosity.

Every part of the Spanish Fly,[1] inside and out, is saturated with the blistering element; but there is nothing like this in the Scorpion, who localizes his venom in his caudal gland and has none of it elsewhere. The cause of the effects which I observe is therefore con-

[1] A Beetle known also as the Cantharis, or Blistering Beetle.—*Translator's Note.*

nected with general properties which I ought to find in any insect, even the most harmless.

I consult *Oryctes nasicornis,* the peaceable Rhinoceros Beetle, on this subject. To get at the exact nature of the materials, instead of pulverizing the whole insect in a mortar, I use merely the muscular tissue obtained by scraping the inside of the dried Oryctes' corselet. Or else I extract the dry contents of the hind-legs. I do the same with the desiccated corpses of the Cockchafer, the Capricorn, or Cerambyx Beetle, and the Cetonia, or Rosechafer. Each of my gleanings, with a little water added, is left to soften for a couple of days in a watch-glass and yields to the liquid whatever can be extracted from it by crushing and dissolving.

This time, we take a great step forward. All my preparations, without distinction, are horribly virulent. Let the reader judge. I select as my first patient the Sacred Beetle, *Scarabæus sacer,* who thanks to his size and sturdiness, lends himself admirably to an experiment of this kind. I operate upon a dozen, in the corselet, on the breast, on the belly and, by preference, on one of the hind-legs, far removed from the impressionable nervous cen-

A Parasite of the Maggot

tres. No matter what part my injector attacks, the effect produced is the same, or nearly. The insect falls as though struck by lightning. It lies on its back and wriggles its legs, especially the hind-legs. If I set it on its feet again, I behold a sort of St. Vitus' dance. *Scarabæus* lowers his head, arches his back, draws himself up on his twitching legs. He marks time with his feet on the ground, moves forward a little, moves as much backward, leans to the right, leans to the left, in wild disorder, incapable of keeping his balance or making progress. And this happens with sudden jerks and jolts, with a vigour no whit inferior to that of the animal in perfect health. It is a displacement of all the works, a storm that uproots the mutual relations of the muscles.

Seldom have I witnessed such sufferings, in my career as a cross-examiner of animals and, therefore, as a torturer. I should feel a scruple, did I not foresee that the grain of sand shifted to-day may one day help us by taking its place in the edifice of knowledge. Life is everywhere the same, in the Dung-beetle's body as in man's. To consult it in the insect means consulting it in ourselves, means moving towards vistas which we cannot afford to neglect. That hope justifies my cruel studies, which, though apparently so puerile, are in reality worthy of serious consideration.

Of my dozen sufferers, some rapidly succumb, others linger for a few hours. They are all dead by to-morrow. I leave the corpses on the table, exposed to the air. Instead of drying and stiffening, like the asphyxiated insects intended for our collections, my patients, on the contrary, turn soft and slacken in the joints, notwithstanding the dryness of the surrounding air; they become disjointed and separate into loose pieces, which are easily removed.

The results are the same with the Capricorn, the Cockchafer, the Procrustes,[1] the Carabus.[2] In all of them there is a sudden break-up, followed by speedy death, a slackening of the joints and swift putrefaction. In a non-horny victim, the quick chemical changes of the tissues are even more striking. A Cetonia-grub, which resists the Scorpion's sting, even though repeatedly administered, dies in a very short time if I inject a tiny drop of my terrible fluid into any part of its body. Moreover, it turns very brown and, in

[1] A large Ground-beetle.—*Translator's Note.*
[2] The True Ground-beetle, including the Gold Beetle.
—*Translator's Note.*

376

a couple of days, becomes a mass of black putrescence.

The Great Peacock, that large Moth who
recks little of the Scorpion's poison, is no more
able to resist my inoculations than the Sacred
Beetle and the others. I prick two in the
belly, a male and a female. At first, they
seem to bear the operation without distress.
They grip the trellis-work of the cage and
hang without moving, as though indifferent.
But soon the disease has them in its grip.
What we see is not the tumultuous ending of
the Sacred Beetle; it is the calm advent of
death. With wings slackly quivering, softly
they die and drop from the wires. Next day,
both corpses are remarkably lax; the segments
of the abdomen separate and gape at the least
touch. Remove the hairs and you shall see
that the skin, which was white, has turned
brown and is changing to black. Corruption
is quickly doing its work.

This would be a good opportunity to speak
of bacteria and cultures. I shall do nothing
of the sort. On the hazy borderland of the
visible and the invisible, the microscope inspires
me with suspicion. It so easily replaces the
eye of reality by the eye of imagination; it is
so ready to oblige the theorists with just what

they want to see. Besides, supposing the microbe to be found, if that were possible, the question would be changed, not solved. For the problem of the collapse of the structure through the fact of a prick there would be substituted another no less obscure: how does the said microbe bring about that collapse? In what way does it go to work? Where lies its power?

Then what explanation shall I give of the facts which I have just set forth? Why, none, absolutely none, seeing that I do not know of any. As I am unable to do better, I will confine myself to a pair of comparisons or images, which may serve as a brief resting-place for the mind on the dark billows of the unknown.

All of us, as children, have amused ourselves with the game of "card friars." A number of cards, as many as possible, are bent lengthwise into a semicylinder. They are placed on a table, one behind the other, in a winding row, the spaces in which are suitably disposed. The performance pleases the eye by its curved lines and its regular arrangement. It possesses order, which is a condition of all animated matter. You give a little tap to the first card. It falls and overturns the second, which, in the same way, topsyturvies the third;

and so on, right to the end of the row. In less than no time, the capsizing wave spreads and the handsome edifice is shattered. Order is succeeded by disorder, I might almost say, by death. What was needed thus to upset the procession of friars? A very, very slight first push, out of all proportion to the toppled mass.

Again, take a glass balloon containing a solution of alum supersaturated by heat. It is closed, during the process of boiling, with a cork and is then allowed to cool. The contents remain fluid and limpid for an indefinite period. Mobility is here represented by a fant semblance of life. Remove the cork and drop in a solid particle of alum, however infinitesimal. Suddenly, the liquid thickens into a solid lump and gives off heat. What has happened? This: crystallization has set in at the first contact of the particle of alum, the centre of attraction; next, it has spread bit by bit, each solidified particle producing the solidification of those around. The impulse comes from an atom; the mass impelled is boundless. The very small has revolutionized the immense.

Of course, in the comparison between these two instances and the effects of my injections, the reader must see no more than a figure of speech, which, without explaining anything, tries to throw a glimmer of light upon it. The long procession of card friars is knocked down by the mere touch of the little finger to the first; the voluminous solution of alum suddenly turns solid under the influence of an invisible particle. In the same way, the victims of my operations succumb, thrown into convulsions by a tiny drop of insignificant size and harmless appearance.

Then what is there in that terrible liquid? First of all, there is water, inactive in itself and simply a vehicle of the active agent. If a proof were needed of its innocuousness, here is one: I inject into the thigh of any one of the Sacred Beetle's six legs a drop of pure water larger than that of the fatal inoculations. As soon as he is released, he makes off and trots about as nimbly as usual. He is quite firm on his legs. When put back to his pellet,[1] he rolls it with the same zeal as before the experiment. My injection of water makes no difference to him.

What else is there in the mixture in my

---

[1] The Dung-beetles roll cattle-droppings into pills for their own consumption and that of their grubs. See *Insect Life:* chaps. i and ii; and *The Life and Love of the Insect:* chaps. i to iv.—*Translator's Note.*

A Parasite of the Maggot

watch-glasses? There is the disintegrated
matter of the corpse, especially shreds of dried
muscles. Do these substances yield certain
soluble elements to water? Or are they simply reduced to a fine dust in the crushing? I
will not decide this question, nor is it really of
importance. The fact remains that the poison
proceeds from those substances and from them
alone. Animal matter, therefore, which has
ceased to live is an agent of destruction within
the organism. The dead cell kills the living
cell; in the delicate statics of life, it is the grain
of sand which, refusing its support, entails the
collapse of the whole edifice.

In this connection, we may recall those
dreadful dissecting-room accidents. Through
awkwardness, a student of anatomy pricks
himself with his scalpel in the course of his
work; or else, by inadvertence, he has an insignificant scratch on his hand. A cut which
one would hardly notice, produced by the
point of a pocket-knife, a scratch of no account, from a thorn or otherwise, now becomes
a mortal wound, if powerful antispetics do not
speedily remedy the ill. The scalpel is soiled
by its contact with the flesh of the corpse; so
are the hands. That is quite enough. The
virus of corruption is introduced; and, if not

treated in time, the wound proves fatal. The dead has killed the living. This also reminds us of the so-called Carbuncle-flies, the lancet of whose mouth-parts, contaminated with the sanies of corpses, produces such terrible accidents.

My dealings as against insects are, when all is said, nothing but dissecting-room wounds and Carbuncle-flies' stings. In addition to the gangrene that soon impairs and blackens the tissues, I obtain convulsions similar to those produced by the Scorpion's sting. In its convulsive effects, the venomous fluid emitted by the sting bears a close resemblance to the muscular infusions with which I fill my injector. We are entitled, therefore, to ask ourselves if poisons, generally speaking, are not themselves a produce of demolition, a casting of the organism perpetually renewed, waste matter, in short, which, instead of being gradually expelled, is stored for purposes of attack and defence. The animal, in that case, would arm itself with its own refuse in the same way as it sometimes builds itself a home with its intestinal recrement. Nothing is wasted; life's detritus is used for self-defence.

All things considered, my preparations are meat-extracts. If I replace the flesh of the

A Parasite of the Maggot

insect by that of another animal, the ox, for
instance, shall I obtain the same results? Logic
says yes; and logic is right. I dilute with a
few drops of water a little Liebig's extract,
that precious stand-by of the kitchen. I
operate with this fluid on six Cetoniæ or Rose-
chafers, four in the grub stage, two in the
adult stage. At first, the patients move about
as usual. Next day, the two Cetoniæ are
dead. The larvæ resist longer and do not die
until the second day. All show the same re-
laxed muscles, the same blackened flesh, signs
of putrefaction. It is probable, therefore,
that, if injected into our own veins, the same
fluid would likewise prove fatal. What is excellent in the digestive tubes would be appal-
ling in the arteries. What is food in one case
is poison in the other.

A Liebig's extract of a different kind, the
broth in which the liquefier puddles, is of a
virulence equal, if not superior, to that of my
products. All those operated upon, Capri-
corns, Sacred Beetles, Ground-beetles, die in
convulsions. This brings us back, after a long
way round, to our starting-point, the maggot
of the Flesh-fly. Can the worm, constantly
floundering in the sanies of a carcass, be itself

in danger of inoculation by that whereon it
grows fat? I dare not rely upon experiments
conducted by myself: my clumsy implements
and my shaky hand make me fear that, with
subjects so small and delicate, I might inflict
deep wounds which of themselves would bring
about death.

Fortunately, I have a collaborator of incomparable skill in the parasitic Chalcidid.
Let us apply to her. To introduce her germs,
she has perforated the maggot's paunch, has
even done so several times over. The holes
are extremely small, but the poison all around
is excessively subtle and has thus been able, in
certain cases, to penetrate. Now what has
happened? The pupæ, all from the same apparatus, are numerous. They can be divided
into three not very unequal classes, according
to the results supplied. Some give me the
adult Flesh-fly, others the parasite. The rest,
nearly a third, give me nothing, neither this
year nor next.

In the first two cases, things have taken
their normal course: the grub has developed
into a Fly, or else the parasite has devoured
the grub. In the third case, an accident has
occurred. I open the barren pupæ. They
are coated inside with a dark glaze, the re-

mains of the dead maggot converted into
black rottenness. The grub, therefore, has
undergone inoculation by the virus through the
fine openings effected by the Chalcidid. The
skin has had time to harden into a shell; but it
was too late, the tissues being already infected.

There you see it: in its broth of putrefaction, the worm is exposed to grave dangers.
Now there is a need for maggots in this world,
for maggots many and voracious, to purge the
soil as quickly as possible of death's impurities.
Linnæus tells us that *'Tres muscæ consumunt
cadaver equi æque cito ac leo.'*[1] There is no
exaggeration about the statement. Yes, of a
certainty, the offspring of the Flesh-fly and the
Bluebottle are expeditious workers. They
swarm in a heap, always seeking, always snuf-
fling with their pointed mouths. In those tumultuous crowds, mutual scratches would be
inevitable if the worms, like the other flesh-eat-
ers, possessed mandibles, jaws, clippers
adapted for cutting, tearing and chopping;
and those scratches, poisoned by the dreadful
gruel lapping them, would all be fatal.

How are the worms protected in their horrible work-yard? They do not eat: they drink

---

[1] "Three Flies consume the carcass of a Horse as
quickly as a Lion could do it."

their fill; by means of a pepsin which they disgorge, they first turn their foodstuffs into soup;
they practise a strange and exceptional art of
feeding, wherein those dangerous carving-
implements, the scalpels with their dissecting-
room perils, are superfluous. Here ends, for
the present, the little that I know or suspect
of the maggot, the sanitary inspector in the
service of the public health.

CHAPTER XVII

RECOLLECTIONS OF CHILDHOOD

ALMOST as much as insects and birds—
the former so dear to the child, who loves
to rear his Cockchafers and Rose-beetles on a
bed of hawthorn in a box pierced with holes;
the latter an irresistible temptation, with their
nests and their eggs and their little ones opening tiny yellow beaks—the mushroom early
won my heart with its varied shapes and
colours. I can still see myself as an innocent
small boy sporting my first braces and beginning to know my way through the cabalistic
mazes of my reading-book, I see myself in
ecstasy before the first bird's-nest found and
the first mushroom gathered. Let us relate
these grave events. Old age loves to medi-
tate the past.

O happy days when curiosity awakens and
frees us from the limbo of unconsciousness,
your distant memory makes me live my best
years over again. Disturbed at its siesta by
some wayfarer, the Partridge's young brood
hastily disperses. Each pretty little ball of

down scurries off and disappears in the brushwood; but, when quiet is restored, at the first
summoning note they all return under the
mother's wing. Even so, recalled by memory, do my recollections of childhood return,
those other fledglings which have lost so many
of their feathers on the brambles of life.
Some, which have hardly come out of the
bushes, have aching heads and tottering steps;
some are missing, stifled in some dark corner
of the thicket; some remain in their full freshness. Now of those which have escaped the
clutches of time the liveliest are the first-born.
For them the soft wax of childish memory has
been converted into enduring bronze.

On that day, wealthy and leisured, with an
apple for my lunch and all my time to myself,
I decided to visit the brow of the neighbouring
hill, hitherto looked upon as the boundary of
the world. Right at the top is a row of trees
which, turning their backs to the wind, bend
and toss about as though to uproot themselves
and take to flight. How often, from the little
window in my home, have I not seen them
bowing their heads in stormy weather; how
often have I not watched them writhing like
madmen amid the snow-dust which the north
wind's besom raises and smooths along the

hill-side! What are they doing up there, those desolate trees? I am interested in their supple backs, to-day still and upright against the blue of the sky, to-morrow shaken when the clouds pass overhead. I am gladdened by their calmness; I am distressed by their terrified gestures. They are my friends. I have them before my eyes at every hour of the day. In the morning, the sun rises behind their transparent screen and ascends in its glory. Where does it come from? I am going to climb up there and perhaps I shall find out.

I mount the slope. It is a lean grass-sward close-cropped by the sheep. It has no bushes, fertile in rents and tears, for which I should have to answer on returning home, nor any rocks, the scaling of which involves like dangers; nothing but large, flat stones, scattered here and there. I have only to go straight on, over smooth ground. But the sward is as steep as a sloping roof. It is long, ever so long; and my legs are very short. From time to time, I look up. My friends, the trees on the hill-top, seem to be no nearer. Cheerly, sonnie! Scramble away!

What is this at my feet? A lovely bird has flown from its hiding-place under the eaves of a big stone. Bless us, here's a nest made of hair and fine straw! It's the first I have ever found, the first of the joys which the birds are to bring me. And in this nest are six eggs, laid prettily side by side; and those eggs are a magnificent blue, as though steeped in a dye of celestial azure. Overpowered with happiness, I lie down on the grass and stare.

Meanwhile, the mother, with a little clap of her gullet—'Tack! Tack!'—flies anxiously from stone to stone, not far from the intruder. My age knows no pity, is still too barbarous to understand maternal anguish. A plan is running in my head, a plan worthy of a little beast of prey. I will come back in a fortnight and collect the nestlings before they can fly away. In the meantime, I will just take one of those pretty blue eggs, only one, as a trophy. Lest it should be crushed, I place the fragile thing on a little moss in the scoop of my hand. Let him cast a stone at me that has not, in his childhood, known the rapture of finding his first nest.

My delicate burden, which would be ruined by a false step, makes me give up the remainder of the climb. Some other day I shall see the trees on the hill-top over which the sun rises. I go down the slope again. At the bottom, I meet the parish-priest's curate read-

ing his breviary as he takes his walk. He sees
me coming solemnly along, like a relic-bearer;
he catches sight of my hand hiding something
behind my back:

'What have you there, my boy?' he asks.

All abashed, I open my hand and show my
blue egg on its bed of moss.

'Ah!' says his reverence. 'A Saxicola's
egg! Where did you get it?'

'Up there, father, under a stone.'

Question follows question; and my pecca-
dillo stands confessed. By chance I found a
nest which I was not looking for. There were
six eggs in it. I took one of them—here it is
—and I am waiting for the rest to hatch. I
shall go back for the others when the young
birds have their quill-feathers.

'You mustn't do that, my little friend,' replies the priest. 'You mustn't rob the mother
of her brood; you must respect the innocent
little ones; you must let God's birds grow up
and fly from the nest. They are the joy of the
fields and they clear the earth of its vermin.
Be a good boy, now, and don't touch the
nest.'

I promise and the curate continues his walk.
I come home with two good seeds cast on the
fallows of my childish brain. An author-

itative word has taught me that spoiling birds'-nests is a bad action. I did not quite understand how the bird comes to our aid by destroying vermin, the scourge of the crops; but I felt, at the bottom of my heart, that it is wrong to afflict the mothers.

'Saxicola,' the priest had said, on seeing my find.

'Hullo!' said I to myself. 'Animals have names, just like ourselves. Who named them? What are all my different acquaintances in the woods and meadows called? What does Saxicola mean?'

Years passed and Latin taught me that Saxicola means an inhabitant of the rocks. My bird, in fact, was flying from one rocky point to the other while I lay in ecstasy before its eggs; its house, its nest, had the rim of a large stone for a roof. Further knowledge gleaned from books taught me that the lover of stony hill-sides is also called the *Motteux*, or Clodhopper,[1] because, in the ploughing-season, she flies from clod to clod, inspecting the furrows rich in unearthed grubworms. Lastly, I came upon the Provençal expression *Cul-blanc*,

[1] I do not know that the Saxicola is actually called a Clodhopper in English. Her English names are Stonechat, Wheat-ear, Whin-chat, Fallow-chat, Fallow-finch and White-tail, which last corresponds with the *Cul-blanc* of the Provençal dialect.—*Translator's Note.*

which is also a picturesque term, suggesting the patch on the bird's rump which spreads out like a white Butterfly flitting over the fields.

Thus did the vocabulary come into being that would one day allow me to greet by their real names the thousand actors on the stage of the fields, the thousand little flowers that smile at us from the wayside. The word which the curate had spoken without attaching the least importance to it revealed a world to me, the world of plants and animals designated by their real names. To the future must belong the task of deciphering some pages of the immense lexicon; for to-day I will content myself with remembering the Saxicola, or Stone-chat.

On the west, my village crumbles into an avalanche of garden-patches, in which plums and apples ripen. Low bulging walls, blackened with the stains of lichens and mosses, support the terraces. The brook runs at the foot of the slope. It can be cleared almost everywhere at a bound. In the wider parts, flat stones standing out of the water serve as a foot-bridge. There is no such thing as a whirlpool, the terror of mothers when the children are away; it is nowhere more than knee-deep. Dear little brook, so tranquil, cool and clear, I have seen majestic rivers since, I have seen the boundless sea; but nothing in my memories equals your modest falls. About you clings all the hallowed pleasure of my first impressions.

A miller has bethought him of putting the brook, which used to flow so gaily through the fields, to work. Half-way up the slope, a water-course, economizing the gradient, diverts part of the water and conducts it into a large reservoir, which supplies the mill-wheels with motor power. This basin stands beside a frequented path and is walled off at the end.

One day, hoisting myself on a play-fellow's shoulders, I looked over the melancholy wall, all bearded with ferns. I saw bottomless stagnant waters, covered with slimy green. In the gaps in the sticky carpet, a sort of dumpy, black-and-yellow reptile was lazily swimming. To-day, I should call it a Salamander; at that time, it appeared to me the offspring of the Serpent and the Dragon, of whom we were told such blood-curdling tales when we sat up at night. Hoo! I've seen enough: let's get down again, quick!

The brook runs below. Alders and ash, bending forward on either bank, mingle their branches and form a verdant arch. At their feet, behind a porch of great twisted roots, are watery caverns prolonged by gloomy corridors. On the threshold of these fastnesses shimmers a glint of sunshine, cut into ovals by the leafy sieve above.

This is the haunt of the red-necktied Minnows. Come along very gently, lie flat on the ground and look. What pretty little fish they are, with their scarlet throats! Clustering side by side, with their heads turned against the stream, they puff their cheeks out and in, rinsing their mouths incessantly. To keep their stationary position in the running water, they need naught but a slight quiver of their tail and of the fin on their back. A leaf falls from the tree. Whoosh! The whole troop has disappeared.

On the other side of the brook is a spinney of beeches, with smooth, straight trunks, like pillars. In their majestic, shady branches sit chattering Crows, drawing from their wings old feathers replaced by new. The ground is padded with moss. At one's first step on the downy carpet, the eye is caught by a mushroom, not yet full-spread and looking like an egg dropped there by some vagrant Hen. It is the first that I have picked, the first that have I turned round and round in my fingers, enquiring into its structure with that vague curiosity which is the first awakening of observation.

Soon, I find others, differing in size, shape and colour. It is a real treat for my prentice eyes. Some are fashioned like bells, like extinguishers, like cups; some are drawn out into spindles, hollowed into funnels, rounded into hemispheres. I come upon some that are broken and are weeping milky tears; I step on some that, instantly, become tinged with blue; I see some big ones that are crumbling into rot and swarming with worms. Others, shaped like pears, are dry and open at the top with a round hole, a sort of chimney whence a whiff of smoke escapes when I prod their under side with my finger. These are the most curious. I fill my pockets with them to make them smoke at my leisure, until I exhaust the contents, which are at last reduced to a kind of tinder.

What fun I had in that delightful spinney! I returned to it many a time after my first find; and here, in the company of the Crows, I received my first lessons in mushroom-lore. My harvests, I need hardly say, were not admitted to the house. The mushroom, or the *bouturel,* as we called it, had a bad reputation for poisoning people. That was enough to

make mother banish it from the family-table.
I could scarcely understand how the *bouturel*,
so attractive in appearance, came to be so
wicked; however, I accepted the experience of
my elders; and no disaster ever ensued from
my rash friendship with the poisoner.

As my visits to the beech-clump were repeated, I managed to divide my finds into
three categories. In the first, which was the
most numerous, the mushroom was furnished
underneath with little radiating leaves. In the
second, the lower surface was lined with a
thick pad pricked with hardly visible holes.
In the third, it bristled with tiny spots similar
to the papillæ on a Cat's tongue. The need of
some order to assist the memory made me invent a classification for myself.

Very much later there fell into my hands
certain small books from which I learnt that
my three categories were well-known; they
even had Latin names, which fact was far
from displeasing to me. Ennobled by Latin
which provided me with my first exercises and
translations, glorified by the ancient language
which the rector used in saying his mass, the
mushroom rose in my esteem. To deserve so
learned an appellation, it must possess a genuine importance.

The Life of the Fly

The same books told me the name of the one that had amused me so much with its smoking chimney. It is called the puff-ball in English, but its French name is the *vesse-de-loup*. I disliked the expression, which to my mind smacked of bad company. Next to it was a more decent denomination: *Lycoperdon;* but this was only so in appearance, for Greek roots sooner or later taught me that *Lycoperdon*[1] means *vesse-de-loup* and nothing else. The history of plants abounds in terms which it is not always desirable to translate. Bequeathed to us by earlier ages less reticent than ours, botany has often retained the brutal frankness of words that set propriety at defiance.

How far-off are those blessed times when my childish curiosity sought solitary exercise in making itself acquainted with the mushroom! *'Eheu! Fugaces labuntur anni!'* said Horace. Ah, yes, the years glide fleeting by, especially when they are nearing their end! They were the merry brook that dallies among the willows on imperceptible slopes; to-day, they are the torrent swirling a thou-

---

[1] It was so called by Joseph Pitton de Tournefort (1656-1708), the French botanist, with the object of improving upon the old Latin name, *Crepitus lupi*, by making it less generally intelligible.—*Translator's Note.*

Recollections of Childhood

sand straws along, as it rushes towards the abyss. Fleeting though they be, let us make the most of them. At nightfall, the wood-cutter hastens to bind his last fagots. Even so, in my declining days, I, a humble woodcutter in the forest of science, make haste to put my bundle of sticks in order. What will remain of my researches on the subject of instinct? Not much, apparently; at most, one or two windows opened on a world that has not yet been explored with all the attention which it deserves.

A worse destiny awaits the mushrooms, which were my botanical joys from my earliest youth. I have never ceased to keep up my acquaintance with them. To this day, for the mere pleasure of renewing it, I go, with a halt-ing step, to visit them on fine autumn after-noons. I still love to see the fat heads of the boleti, the tops of the agarics and the coral-red tufts of the clavaria emerge above the carpet pink with heather.

At Sérignan, my last stage, they have lav-ished their seductions upon me, so plentiful are they on the neighbouring hills, wooded with holm-oak, arbutus and rosemary. During these latter years, their wealth inspired me with an insane plan: that of collecting in effigy what I was unable to keep in its natural state in an herbarium. I began to paint life-size pictures of all the species in my neighbourhood, from the largest to the smallest. I know nothing of the art of painting in water-colours. No matter: what I have never seen practised I will invent, managing badly at first, then a little better, at last well. The paint-brush will make a change from the strain of my daily output of prose.

I end by possessing some hundreds of sheets representing the mushrooms of the neighbourhood in their natural size and colours. My collection has a certain value. If it lacks artistic finish, at least it boasts the merit of accuracy. It brings me visitors on Sundays, country-people, who stare at it in all simplicity, astounded that such fine pictures should be done by hand, without a copy and without compasses. They at once recognize the mushroom represented; they tell me its popular name, thus proving the fidelity of my brush.

Well, what will become of this great pile of drawings, the object of so much work? No doubt, my family will keep the relic for a time; but, sooner or later, taking up too much space, shifted from cupboard to cupboard, from attic to attic, gnawed by the rats, foxed,

Recollections of Childhood

dirtied and stained, it will fall into the hands of some little grand-nephew, who will cut it into squares to make paper caps. It is the universal rule. What our illusions have most fondly cherished comes to a pitiful end under the claws of ruthless reality.

CHAPTER XVIII

INSECTS AND MUSHROOMS

IT were out of place to recall my long relations with the boletus and the agaric if the
insect did not here enter into a question of
grave interest. Several mushrooms are edible,
some even enjoy a great reputation; others are
formidable poisons. Short of botanical studies
that are not within everybody's reach, how are
we to distinguish the harmless from the venomous? There is a widespread belief which
says that any mushroom which insects, or,
more frequently, their larvæ, their grubs, accept can be accepted without fear; any mushroom which they refuse must be refused.
What is wholesome food for them cannot fail
to be the same for us; what is poisonous to
them is bound to be equally baneful to ourselves. This is how people argue, with apparent logic, but without reflecting upon the
very different capabilities of stomachs in the
matter of diet. After all, may there not be
some justification for the belief? That is what
I purpose examining.

Insects and Mushrooms

The insect, especially in the larval stage, is the principal devourer of the mushroom. We must distinguish between two groups of consumers. The first really eat, that is to say, they break their food into little bits, chew it and reduce it to a mouthful which is swallowed just as it is; the second drink, after first turning their food into a broth, like the Blue-bottles. The first are the less numerous. Confining myself to the results of my observations in the neighbourhood, I count, all told, in the group of chewers, four Beetles and a Moth-caterpillar. To these may be added the Mollusc, as represented by a Slug, or, more specifically, an Arion, of medium size, brown and adorned with a red edge to his mantle. A modest corporation, when all is said, but active and enterprising, especially the Moth.

At the head of the mushroom-loving Beetles, I will place a Staphylinid (*Oxyporus rufus*, LIN.), prettily garbed in red, blue and black. Together with his larva, which walks with the aid of a crutch at its back, he haunts the fungus of the poplar (*Pholiota ægerita*, FRIES). He specializes in an exclusive diet. I often come across him, both in spring and autumn, and never any elsewhere than on this mushroom. For that matter, he had made a wise choice, the epicure! This popular fungus is one of our best mushrooms, despite its colour of a doubtful white, its skin which is often wrinkled and its gills soiled with rusty brown at the spores. We must not judge people by appearances, nor mushrooms either. This one, magnificent in shape and colour, is poisonous; that other, so poor to look at, is excellent.

Here are two more specialist Beetles, both of small size. One is the Triplax (*Triplax russica*, LIN.), who has an orange head and corselet and black wing-cases. His grub tackles the hispid polyporus (*Polyporus hispidus*, BULL.), a coarse and substantial dish, bristling at its top with stiff hairs and clinging by its side to the old trunks of mulberry-trees, sometimes also of walnut- and elm-trees. The other is the cinnamon-coloured Anisotoma (*Anisotoma cinnamomea*, PANZ.). His larva lives exclusively in truffles.

The most interesting of the mushroom-eating Beetles is the Bolboceras (*Bolboceras gallicus*, MUL.). I have described elsewhere[1] his manner of living, his little song that sounds like the chirping of a bird, his perpendicular

[1] In an essay not yet published in English.—*Translator's Note.*

wells sunk in search of an underground mushroom (*Hydnocystis orenaria,* TUL.), which
constitutes his regular nourishment. He is
also an ardent lover of truffles. I have taken
from between his legs, at the bottom of his
manor-house, a real truffle the size of a hazel-
nut (*Tuber Requienii,* TUL.). I tried to rear
him in order to make the acquaintance of his
grub; I housed him in a large earthen pan
filled with fresh sand and enclosed in a bell-
cover. Possessing neither hydnocistes nor
truffles, I served him up sundry mushrooms of
a rather firm consistency, like those of his
choice. He refused them all, helvellæ and cla-
variæ, chanterelles and pezizæ alike.

With a rhizopogon, a sort of little fungoid
potato, which is frequent in pine-woods at a
moderate depth and sometimes even on the
surface, I achieved complete success. I had
strewn a handful of them on the sand of my
breeding-pan. At nightfall, I often surprised
the Bolboceras issuing from his well, exploring the stretch of sand, choosing a piece not
too big for his strength and gently rolling it
towards his abode. He would go in again,
leaving the rhizopogon, which was too large to
take inside, on the threshold, where it served

the purpose of a door. Next day, I found the piece gnawed, but only on the under side.

The Bolboceras does not like eating in public, in the open air; he needs the discreet retirement of his crypt. When he fails to find his food by burrowing under ground, he comes up to look for it on the surface. Meeting with a morsel to his taste, he takes it home when its size permits; if not, he leaves it on the threshold of his burrow and gnaws at it from below, without reappearing outside. Up to the present, hydnocistes, truffles and rhizopoga are the only food that I have known him to eat. These three instances tell us at any rate that the Bolboceras is not a specialist like the Oxyporus and the Triplax; he is able to vary his diet; perhaps he feeds on all the underground mushrooms indiscriminately.

The Moth enlarges her domain yet further. Her Caterpillar is a grub five or six millimetres long,[1] white, with a black shiny head. Colonies of it abound in most mushrooms. It attacks by preference the top of the stem, for epicurean reasons that escape me; thence it spreads throughout the cap. It is the habitual boarder of the boleti, agarics, lactarii and russulæ. Apart from certain species and certain

[1] About one-fifth of an inch.—*Translator's Note.*

406

groups, everything suits it. This puny grub, which will spin itself an infinitesimal cocoon of white silk under the piece attacked and will later become an insignificant Moth, is the primordial ravager.

Let us next mention the Arion, that voracious Mollusc who also tackles most mushrooms of some size. He digs himself spacious niches inside them and there sits blissfully eating. Few in numbers, compared with the other devourers, he usually sets up house alone. He has, by way of a set of jaws, a powerful plane which creates great breaches in the object of his depredations. It is he whose havoc is most apparent.

Now all these gnawers can be recognized by their leavings, such as crumbs and worm-holes. They dig clean passages, they slash and crumble without a slimy trail, they are the pinkers. The others, the liquefiers, are the chemists; they dissolve their food by means of reagents. All are the grubs of Flies and belong to the commonalty of the Muscidæ. Many are their species. To distinguish them from one another by rearing them in order to obtain the perfect stage would involve a great expenditure of time to little profit. We will describe them by the general name of maggots.

The Life of the Fly

To see them at work, I select, as the field
of exploitation, the Satanic boletus (*Boletus
Satanas*, LENZ.), one of the largest mushrooms
that I can gather in my neighbourhood. It
has a dirty-white cap; the mouths of the tubes
are a bright orange-red; the stem swells into
a bulb with a delicate net-work of carmine
veins. I divide a perfectly sound specimen
into equal parts and place these in two deep
plates, put side by side. One of the halves is
left as it is: it will act as a control, a term of
comparison. The other half receives on the
pores of its under-surface a couple of dozen
maggots taken from a second boletus in full
process of decomposition.

The dissolving action of the grub asserts
itself on the very day whereon these preparations are made. The under-surface, originally
a bright red, turns brown and runs in every
direction into a mass of dark stalactites. Soon,
the flesh of the cap is attacked and, in a few
days, becomes a gruel similar to liquid asphalt.
It is almost as fluid as water. In this broth the
maggots wallow, wriggling their bodies and,
from time to time, sticking the breathing holes
in their sterns above the water. It is an exact
repetition of what the liquefiers of meat, the
grubs of the Grey Flesh-fly and the Bluebottle,

Insects and Mushrooms

have lately shown us. As for the second half of the Boletus, the half which I did not colonize with vermin, it remains compact, the same as it was at the start, except that its appearance is a little withered by evaporation. The fluidity, therefore, is really and truly the work of the grubs and of them alone.

Does this liquefaction imply an easy change? One would think so at first, on seeing how quickly it is performed by the action of the grubs. Moreover, certain mushrooms, the coprini, liquefy spontaneously and turn into a black fluid. One of them bears the expressive name of the inky mushroom (*Coprinus atramentarius*, BULL.), and dissolves into ink of its own accord. The conversion, in certain cases, is singularly rapid. One day, I was drawing one of our prettiest coprini (*Coprinus sterquilinus*, FRIES), which comes out of a little purse or volva. My work was barely done, a couple of hours after gathering the fresh mushroom, when the model had disappeared, leaving nothing but a pool of ink upon the table. Had I procrastinated ever so little, I should not have had time to finish and I should have lost a rare and interesting find.

This does not mean that the other mushrooms, especially the boleti, are of ephemeral

duration and lacking in consistency. I made
the attempt with the edible boletus (*Boletus
edulis*, BULL.), the famous *cèpe* of our
kitchens, so highly esteemed for its flavour.
I was wondering whether it would not be possible to obtain from it a sort of Liebig's extract of fungus, which would be useful in cooking. With this purpose, I had some of these
mushrooms cut into small pieces and boiled,
on the one hand, in plain water and, on the
other, in water with bicarbonate of soda
added. The treatment lasted two whole days.
The flesh of the boletus was indomitable. To
attack it, I should have had to employ violent
drugs, which were inadmissible in view of the
result to be attained.

What prolonged boiling and the aid of bicarbonate of soda leave almost intact the Fly's
grubs quickly turn into fluid, even as the Flesh-
worms fluidify hard-boiled white of egg. This
is done in each instance without violence, probably by means of a special pepsin, which is not
the same in both cases. The liquefier of meat
has its own brand; the liquefier of the boletus
has another sort. The plate, then, is filled
with a dark, running gruel, not unlike tar in
appearance. If we allow evaporation free
course, the broth sets into a hard, easily

crumbled slab, something like toffee. Caught
in this matrix, grubs and pupæ perish, incapable of freeing themselves. Analytical
chemistry has proved fatal to them. The conditions are quite different when the attack is
delivered on the surface of the ground. Gradually absorbed by the soil, the excess of liquid
disappears, leaving the colonists free. In my
dishes, it collects indefinitely, killing the inhabitants when it dries up into a solid layer.

The purple boletus (*Boletus purpureus,*
FRIES), when subjected to the action of the
maggots, gives the same result as the Satanic
boletus, namely, a black gruel. Note that both
mushrooms turn blue if broken and especially
if crushed. With the edible boletus, whose
flesh invariably remains white when cut, the
product of its liquefaction by the vermin is a
very pale brown. With the *oronge,* or imperial mushroom, the result is a broth which
the eye would take for a thin apricot-jam.
Tests made with sundry other mushrooms con-
firm the rule: all, when attacked by the maggot, turn into a more or less fluid mess, which
varies in colour.

Why do the two boleti with the red tubes,
the purple boletus and the Satanic boletus,
change into a dark gruel? I have an inkling

of the reason. Both of them turn blue, with
an admixture of green. A third species, the
bluish boletus (*Boletus cyanescens*, BULL.,
var. *lacteus*, LÉVEILLÉ), possess remarkable
colour- sensitiveness. Bruise it ever so lightly,
no matter where, on the cap, the stem, the
tubes of the under-surface: forthwith, the
wounded part, originally a pure white, is
tinted a beautiful blue. Place this boletus in
an atmosphere of carbonic acid gas. We can
now knock it, crush it, reduce it to pulp; and
the blue no longer shows. But extract a fragment from the crushed mass: immediately, at
the first contact with the air, the matter turns
a most glorious blue. It reminds us of a process employed in dyeing. The indigo of commerce, steeped in water containing lime and
sulphate of iron, or copperas, is deprived of a
part of its oxygen; it loses its colour and becomes soluble in water, as it was in the original
indigo-plant, before the treatment which the
plant underwent. A colourless liquid results.
Expose a drop of this liquid to the air.
Straightway, oxidization works upon the product: the indigo is reformed, insoluble and
blue.

This is exactly what we see in the boleti that
turn blue so readily. Could they, in fact, con-

tain soluble, colourless indigo? One would
say so, if certain properties did not give
grounds for doubt. When subjected to prolonged exposure to the air, the boleti that are
apt to turn blue, particularly the most remarkable, *Boletus cyanescens,* lose their colour, instead of retaining the deep blue which would
be a sign of real indigo. Be this as it may,
these mushrooms contain a colouring-prin-
ciple which is very liable to change under the
influence of the air. Why should we not regard it as the cause of the black tint when the
maggots have liquefied the boleti which turn
blue? The others, those with the white flesh,
the edible boletus, for instance, do not assume
this asphalty appearance once they are lique-
fied by the grubs.

All the boleti that change to blue when
broken have a bad reputation; the books treat
them as dangerous, or at least open to suspicion. The name of Satanic awarded to one
of them is an ample proof of our fears. The
caterpillar and the maggot are of another
opinion: they greedily devour what we hold in
dread. Now here is a strange thing: those
passionate devotees of *Boletus Satanas* absolutely refuse certain mushrooms which we find
delightful eating, including the most cele-

brated of all, the *oronge*, the imperial mushroom, which the Romans of the empire, past-
masters in gluttony, called the food of the
gods, *cibus deorum*, the agaric of the Cæsars,
Agaricus cæsareus. It is the most elegant of
all our mushrooms. When it prepares to make
its appearance by lifting the fissured earth,
it is a handsome ovoid formed by the outer
wrapper, the volva. Then this purse gently
tears and the jagged opening partly reveals a
globular object of a magnificent orange. Take
a hen's egg, boil it, remove the shell: what
remains will be the imperial mushroom in its
purse. Remove a part of the white at the top,
uncovering a little of the yolk. Then you
have the nascent imperial. The likeness is
perfect. And so the people of my part, struck
by the resemblance, call this mushroom *lou
rousset d' ioù*, or, in other words, yolk-of-egg.
Soon, the cap emerges entirely and spreads
into a disk softer than satin to the touch and
richer to the eye than all the fruit of the Hes-
perides. Appearing amid the pink heather,
it is an entrancing object.

Well, this gorgeous agaric (*Amanita
cæsarea*, scop.), this food of the gods the maggot absolutely refuses. My frequent examinations have never shown me an imperial at-

tacked by the grubs in the field. It needs imprisonment in a jar and the absence of other victuals to provoke the attempt; and even then the treacle hardly seems to suit them. After the liquefaction, the grubs try to make off, showing that the fare is not to their liking. The Mollusc also, the Arion, is anything but an ardent consumer. Passing close to an imperial mushroom and finding nothing better, he stops and takes a bite, without lingering. If, therefore, we required the evidence of the insect, or even of the Slug, to know which mushrooms are good to eat, we should refuse the best of them all. Though respected by the vermin, the glorious imperial is nevertheless ruined not by larvæ, but by a parasitic fungus, the *Mycogone rosea,* which spreads in a purply stain and turns it into a putrid mass. This is the only despoiler that I know it to possess.

A second amanita, the sheathed amanita (*Amanita vaginata,* BULL.), prettily streaked on the edges of the cap, is of an exquisite flavour, almost equal to the imperial. It is called *lou pichot gris,* the greyling, in these parts, because of its colouring, which is usually an ashen grey. Neither the maggot nor the even more enterprising Moth ever touches it. They likewise refuse the mottled amanita

(*Amanita pantherina,* D. C.), the vernal amanita (*Amanita verna,* FRIES) and the lemon-yellow amanita · (*Amanita citrina,* SCHAEFF.), all three of which are poisonous. In short, whether it be to us a delicious dish or a deadly poison, no amanita is accepted by the grubs. The Arion alone sometimes bites at it. The cause of the refusal escapes us. It were vain, speaking of the mottled amanita, for instance, to allege as a reason the presence of an alkaloid fatal to the grubs, for we should have to ask ourselves why the imperial, the amanita of the Cæsars, which is wholly free from poison, is rejected no less uncompromisingly than the venomous species. Could it perhaps be lack of relish, a deficiency of seasoning for stimulating the appetite? In point of fact, when eaten raw, the amanitæ have no particular flavour.

What shall we learn from the sharper-flavoured mushrooms? Here, in the pinewoods, is the woolly milk-mushroom (*Lactarius torminosus,* SCHAEFF.), turned in at the edges and wrapped in a curly fleece. Its taste is biting, worse than Cayenne pepper. *Torminosus* means colic-producing. The name is very suitable. Unless he possessed a stomach built for the purpose, the man who touched such food as this would have a singularly bad time before him. Well, that stomach the vermin possess: they revel in the pungency of the woolly milk-mushroom even as the Spurge-caterpillar browses with delight on the loathsome leaves of the euphorbiæ. As for us, we might as well, in either case, eat live coals.

Is a condiment of this kind necessary to the grubs? Not at all. Here, in the same pine-woods, is the "delicious" milk-mushroom (*Lactarius deliciosus,* LIN.), a glorious orange-red crater, adorned with concentric zones. If bruised, it assumes a verdigris hue, possibly a variant of the indigo tint peculiar to the blue-turning boleti.. From its flesh laid bare by being broken or cut ooze blood-red drops, a well-defined characteristic peculiar to this milk-mushroom. Here the violent spices of the woolly milk-mushroom disappear; the flesh has a pleasant taste when eaten raw. No matter: the vermin devour the mild milk-mushroom with the same zest with which they devour the horribly peppered one. To them the delicate and the strong, the insipid and the peppery are all alike.

The epithet 'delicious' applied to the mushroom whose wound weeps tears of blood is highly exaggerated. It is edible, no doubt,

417

but it is coarse eating and difficult to digest. My household refuses it for cooking purposes. We prefer to put it to soak in vinegar and afterwards to use it as we might use pickled gherkins. The real value of this mushroom is largely overrated thanks to a too-laudatory epithet.

Is a certain degree of consistency required, to suit the grubs: something midway between the softness of the amanitæ and the firmness of the milk-mushrooms? Let us begin by questioning the olive-tree agaric or luminous mushroom (*Pleurotus phosphoreus,* BATT.), a magnificent mushroom coloured jujube-red. Its popular name is not particularly appropriate. True, it frequently grows at the base of old olive-trees, but I also pick it at the foot of the box, the holm-oak, the plum-tree, the cypress, the almond-tree, the Guelder-rose and other trees and shrubs. It seems fairly indifferent to the nature of the support. A more remarkable feature distinguishes it from all the other European mushrooms: it is phosphorescent. On the lower surface and there only, it sheds a soft, white gleam, similar to that of the Glowworm. It lights up to celebrate its nuptials and the emission of its spores. There is no question of chemist's phosphorus here. This is a slow combustion, a sort of more active respiration than usual. The luminous emission is extinguished in the unbreathable gases, nitrogen and carbonic acid; it continues in aerated water; it ceases in water deprived of its air by boiling. It is exceedingly faint, however, so much so that it is not perceptible except in the deepest darkness. At night and even by day, if the eyes have been prepared for it by a preliminary wait in the darkness of a cellar, this agaric is a wonderful sight, looking indeed like a piece of the full moon.

Now what do the vermin do? Are they drawn by this beacon? In no wise: maggots, Caterpillars and Slugs never touch the resplendent mushroom. Let us not be too quick to explain this refusal by the noxious properties of the olive-tree agaric, which is said to be extremely poisonous. Here, in fact, on the pebbly ground of the wastelands, is the eryngo agaric (*Pleurotus eryngii*, D. C.), which has the same consistency as the other. It is the *berigoulo* of the Provençaux, one of the most highly-esteemed mushrooms. Well, the vermin will have none of it: what is a treat to us is detestable to them.

It is superfluous to continue this method of investigation: the reply would be everywhere

the same. The insect, which feeds on one
sort of mushroom and refuses others, cannot
tell us anything about the kinds that are good
or bad for us. Its stomach is not ours. It
pronounces excellent what we find poisonous;
it pronounces poisonous what we think excellent. That being so, when we are lacking in
the botanical knowledge which most of us have
neither time nor inclination to acquire, what
course are we to take? The course is extremely simple.

During the thirty years and more that I
have lived at Sérignan, I have never heard of
one case of mushroom-poisoning, even the
mildest, in the village; and yet there are plenty
of mushrooms eaten here, especially in autumn.
Not a family.but, when on a walk in the mountains, gathers a precious addition to its modest
alimentary resources. What do these people
gather? A little of everything. Often, when
rambling in the neighbouring woods, I inspect
the baskets of the mushroom-pickers, who are
delighted for me to look. I see things fit to
make mycological experts stand aghast. I
often find the purple boletus, which is classed
among the dangerous varieties. I made the
remark one day. The man carrying the basket
stared at me in astonishment:

Insects and Mushrooms

'That a poison! The wolf's bread!'[1] he said, patting the plump boletus with his hand. 'What an idea! It's beef-marrow, sir, regular beef-marrow!'

He smiled at my apprehensions and went away with a poor opinion of my knowledge in the matter of mushrooms.

In the baskets aforesaid, I find the ringed agaric (*Armillaria mellea*, FRIES), which is stigmatized as *valde venenatus* by Persoon,[2] an expert on the subject. It is even the mushroom most frequently made use of, because of its being so plentiful, especially at the foot of the mulberry-trees. I find the Satanic boletus, that dangerous tempter; the belted milk-mushroom (*Lactarius zonarius*, BULL.), whose burning flavour rivals the pepper of its woolly kinsman; the smooth-headed amanita (*Amanita leiocophala*, D. C.), a magnificent white dome rising out of an ample volva and fringed at the edges with floury relics resembling flakes of casein. Its poisonous smell and soapy after-

---

[1] The boleti are known hereabouts by the generic name of *pan de loup*, or wolf's bread. The people use them indiscriminately for cooking-purposes, after removing the tubes on the under side, which are easily separated from the rest of the mushroom.—*Author's Note.*

[2] Christiaan Hendrik Persoon (1770-1836), a Dutch naturalist, author of various works on fungology.—*Translator's Note.*

taste should lead to suspicion of this ivory dome; but nobody seems to mind them.

How, with such careless picking, are accidents avoided? In my village and for a long way around, the rule is to blanch the mushrooms, that is to say, to bring them to the boil in water with a little salt in it. A few rinsings in cold water conclude the treatment. They are then prepared in whatever manner one pleases. In this way, what might at first be dangerous becomes harmless, because the preliminary boiling and rinsing have removed the noxious elements.

My personal experience confirms the efficacy of this rustic method. At home, we very often make use of the ringed agaric, which is reputed extremely dangerous. When rendered wholesome by the ordeal of boiling water, it becomes a dish of which I have naught but good to say. Then again the smooth-headed amanita frequently appears upon my table, after being duly boiled: if it were not first treated in this fashion, it would be hardly safe. I have tried the blue-turning boleti, especially the purple boletus and the Satanic. They answered very well to the eulogistic term of beef-marrow applied to them by the mushroom-picker who scouted my prudent counsels. I have sometimes employed the mottled amanita, so ill-famed in the books, without disastrous result. One of my friends, a doctor, to whom I communicated my ideas about the boiling-water treatment, thought that he would make the experiment on his own account. He chose the lemon-yellow amanita, which has as bad a reputation as the mottled variety, and ate it at supper. Everything went off without the slightest inconvenience. Another, a blind friend, in whose company I was one day to taste the Cossus of the Roman epicures, treated himself to the olive-tree agaric, said to be so formidable. The dish was, if not excellent, at least harmless.

It results from these facts that a good preliminary boiling is the best safeguard against accidents arising from mushrooms. If the insect, devouring one species and refusing another, cannot guide us in any way, at least rustic wisdom, the fruit of long experience, prescribes a rule of conduct which is both simple and efficacious. You are tempted by a basketful of mushrooms, but you do not feel very sure as to their good or evil properties. Then have them blanched, well and thoroughly blanched. When it leaves the purgatory of the stewpan, the doubtful mushroom can be eaten without fear.

But this, you will tell me, is a system of cookery fit for savages: the treatment with boiling water will reduce the mushrooms to a mash; it will take away all their flavour and all their succulence. That is a complete mistake. The mushroom stands the ordeal exceedingly well. I have described my failure to subdue the *cèpes* when I was trying to obtain an extract from them. Prolonged boiling, with the aid of bicarbonate of soda, so far from reducing them to a mess, left them very nearly intact. The other mushrooms whose size entitles them to culinary consideration offer the same degree of resistance. In the second place, there is no loss of succulence and hardly any of flavour. Moreover, they become much more digestible, which is a most important condition in a dish generally so heavy for the stomach. For this reason, it is the custom, in my family, to treat them one and all with boiling water, including even the glorious imperial.

I am a Philistine, it is true, a barbarian caring little for the refinements of cookery. I am not thinking of the epicure, but of the frugal man, the husbandman especially. I should

Insects and Mushrooms

consider myself amply repaid for my persistent observations if I succeeded in popularizing, however little, the wise Provençal recipe for mushrooms, an excellent food that makes a pleasant change from the dish of beans or potatoes, when we can overcome the difficulty of distinguishing between the harmless and the dangerous.

CHAPTER XIX

A MEMORABLE LESSON

I TAKE leave of the mushrooms with regret: there would be so many other questions to solve concerning them! Why do the maggots eat the Satanic boletus and scorn the imperial mushroom? How is it that they find delicious what we find poisonous and why is it that what seems exquisite to our taste is loathsome to theirs? Can there be special compounds in mushrooms, alkaloids, apparently, which vary according to the botanical genus? Would it be possible to isolate them and study their properties fully? Who knows whether medical science could not employ them in relieving our ailments, even as it employs quinine, morphia and other alkaloids? One might enquire into the cause of the liquefaction of the coprini, which is spontaneous, and that of the boleti, which is brought about by the maggots. Do both cases come within the same category? Does the coprinus digest itself by virtue of a pepsin similar to the maggots'? One would like to discover the oxidi-

A Memorable Lesson

zable substance that gives the luminous mushroom its soft, white light, which is like the
beams of the full moon. It would be interesting to know whether certain boletiturn blue
owing to the presence of an indigo which is
more liable to change than dyers' indigo and
whether the green of the so-called delicious
milk-mushroom when bruised is due to a like
cause.

All these patient chemical investigations
would tempt me, if the rudimentary equipment
of my laboratory and especially the irrevo-
cable flight of age-worn hopes permitted it.
The day has passed for it now; there is no
time left to me. No matter: let us talk chemistry once more, for a little while; and, for
want of something better, let us revive old
memories. If the historian, now and again,
takes a small place in the story of his animals,
the reader will kindly excuse him: old age is
prone to these reminiscences, the bloom of
later days.

I have received, in all, two lessons of a scientific character in the course of my life: one
in anatomy and one in chemistry. I owe the
first to the learned naturalist Moquin-Tandon,
who, on our return from a botanizing expedition to Monte Renoso, in Corsica, showed

me the structure of a Snail in a plate filled
with water. It was short and fruitful. From
that moment, I was initiated. Henceforth, I
was to wield the scalpel and decently to explore an animal's interior without any other
guidance from a master. The second lesson,
that of chemistry, was less fortunate. I will
tell you what happened.

In my normal school, the scientific teaching
was on an exceedingly modest scale, consisting
mainly of arithmetic and odds and ends of
geometry. Physics was hardly touched. We
were taught a little meteorology, in a summary
fashion: a word or two about a red moon, a
white frost, dew, snow and wind; and, with
this smattering of rustic physics, we were considered to know enough of the subject to dis-
cuss the weather with the farmer and the
ploughman.

Of natural history, absolutely nothing. No
one thought of telling us anything about flowers and trees, which give such zest to one's
aimless rambles, nor about insects, with their
curious habits, nor about stones, so instructive
with their fossil records. That entrancing
glance through the windows of the world was
refused us. Grammar was allowed to strangle
life.

A Memorable Lesson

Chemistry was never mentioned either: that goes without saying. I knew the word, however. My casual reading, only half-understood for want of practical demonstration, had taught me that chemistry is concerned with the shuffle of matter, uniting or separating the various elements. But what a strange idea I formed of this branch of study! To me it smacked of sorcery, of alchemy and its search for the philosopher's stone. To my mind, every chemist, when at work, should have had a magic wand in his hand and the wizard's pointed, star-studded cap on his head.

An important personage who sometimes visited the school, in his capacity as an honorary lecturer, was not the man to rid me of those foolish notions. He taught physics and chemistry at the grammar-school. Twice a week, from eight to nine o'clock in the evening, he held a free public class in an enormous building adjacent to our school-house. This was the former Church of Saint-Martial, which has to-day become a Protestant meeting-house.

It was a wizard's cave certainly, just as I had pictured it. At the top of the steeple, a rusty weathercock creaked mournfully; in the dusk, great Bats flew all around the edifice or dived down the throats of the gargoyles; at night, Owls hooted upon the copings of the leads. It was inside, under the immensities of the vault, that my chemist used to perform. What infernal mixtures did he compound? Should I ever know?

It is the day for his visit. He comes to see us with no pointed cap: in ordinary garb, in fact, with nothing very queer about him. He bursts into our schoolroom like a hurricane. His red face is half-buried in the enormous stiff collar that digs into his ears. A few wisps of red hair adorn his temples; the top of his head shines like an old ivory ball. In a dictatorial voice and with wooden gestures, he questions two or three of the boys; after a moment's bullying, he turns on his heel and goes off in a whirlwind as he came. No, this is not the man, a capital fellow at heart, to inspire me with a pleasant idea of the things which he teaches.

Two windows of his laboratory look out upon the garden of the school. One can just lean on them; and I often come and peep in, trying to make out, in my poor brain, what chemistry can really be. Unfortunately, the room into which my eyes penetrate is not the sanctuary but a mere outhouse where the

A Memorable Lesson

learned implements and crockery are washed.
Leaden pipes with taps run down the walls;
wooden vats occupy the corners. Sometimes,
those vats bubble, heated by a spray of steam.
A reddish powder, which looks like brick-dust,
is boiling in them. I learn that the simmering
stuff is a dyer's root, known as madder, which
will be converted into a purer and more concentrated product. This is the master's pet
study.

What I saw from the two windows was not
enough for me. I wanted to see farther, into
the very class-room. My wish was satisfied.
It was the end of the scholastic year. A stage
ahead in the regular work, I had just obtained
my certificate. I was free. A few weeks remain before the holidays. Shall I go and
spend them out of doors, in all the gaiety of
my eighteen summers? No, I will spend them
at the school which, for two years past, has
provided me with an untroubled roof and my
daily crust. I will wait until a post is found
for me. Employ my willing service as you
think fit, do with me what you will: as long
as I can study, I am indifferent to the rest.

The principal of the school, the soul of
kindness, has grasped my passion for knowledge. He encourages me in my determina-

tion; he proposes to make me renew my acquaintance with Horace and Virgil, so long since forgotten. He knows Latin, he does; he will rekindle the dead spark by making me translate a few passages. He does more: he lends me an *Imitation* with parallel texts in Latin and Greek. With the first text, which I am almost able to read, I will puzzle out the second and thus increase the small vocabulary which I acquired in the days when I was translating Æsop's Fables. It will be all the better for my future studies. What luck! Board and lodging, ancient poetry, the classical languages, all the good things at once!

I did better still. Our science-master—the real, not the honorary one—who came twice a week to discourse of the rule of three and the properties of the triangle, had the brilliant idea of letting us celebrate the end of the school-year with a feast of learning. He promised to show us oxygen. As a colleague of the chemist in the grammar-school, he obtained leave to take us to the famous laboratory and there to handle the object of his lesson under our very eyes. Oxygen, yes, oxygen, the all-consuming gas; that was what we were to see on the morrow. I could not sleep all night for thinking of it.

A Memorable Lesson

Thursday afternoon came at last. As soon as the chemistry-lesson is over, we were to go for a walk to Les Angles, the pretty village over yonder, perched on a steep rock. We were therefore in our Sunday best, our out-of-doors clothes: black frock-coats and tall hats. The whole school was there, some thirty of us, in the charge of an usher, who knew as little as we did of the things which we were about to see. We crossed the threshold of the laboratory, not without excitement. I entered a great nave with a Gothic roof, an old, bare church through which one's voice echoed, into which the light penetrated discreetly through stained-glass windows set in ribs and rosettes of stone. At the back were huge raised benches, with room for an audience of many hundreds; at the other end, where the choir once was, stood an enormous chimney-mantel; in the middle was a large, massive table, corroded by the chemicals. At one end of this table was a tarred tub, lined inside with lead and filled with water. This, I at once learnt, was the pneumatic trough, the vessel in which the gases were collected.

The professor begins the experiment. He takes a sort of large, long glass bulb, bent abruptly in the region of the neck. This, he

informs us, is a retort. He pours into it, from
a screw of paper, some black stuff that looks
like powdered charcoal. This is manganese
dioxide, the master tells us. It contains in
abundance, in a condensed state and retained
by combination with the metal, the gas which
we propose to obtain. An oily-looking liquid,
sulphuric acid, an excessively powerful agent,
will set it at liberty. Thus filled, the retort is
placed on a lighted stove. A glass tube brings
it into communication with a bell-jar full of
water on the shelf of the pneumatic trough.
Those are all the preparations. What will be
the result? We must wait for the action of
heat.

My fellow-pupils gather eagerly round the
apparatus, cannot come close enough to it.
Some of them play the part of the fly on the
wheel and glory in contributing to the success
of the experiment. They straighten the retort, which is leaning to one side; they blow
with their mouths on the coals in the stove.
I do not care for these familiarities with the
unknown. The good-natured master raises no
objection; but I have never been able to en-
dure the thronging of a crowd of gapers, who
are very busy with their elbows and force their
way to the front row to see whatever is hap-

pening, even though it be merely a couple of
mongrels fighting. Let us withdraw and leave
these officious ones to themselves. There is
so much to see here, while the oxygen is being
prepared. Let us make the most of the occasion and take a look round the chemist's
arsenal.

Under the spacious chimney-mantel is a
collection of queer stoves, bound round with
bands of sheet-iron. There are long and short
ones, high and low ones, all pierced with little
windows that are closed with a terra-cotta
shutter. This one, a sort of little tower, is
formed of several parts placed one above the
other and each supplied with big round
handles to hold them by when you take the
monument to pieces. A dome, with an iron
chimney, tops the whole edifice, which must
be capable of producing a very hell-fire to
roast a stone of no significance. Another, a
squat one, stretches out like a curved spine.
It has a round hole at either end; and a thick
porcelain tube sticks out from each. It is impossible to conceive the purpose which such instruments as these can serve. The seekers of
the philosopher's stone must have had many
like them. They are torturers' engines, tearing the metals' secrets from them.

The Life of the Fly

The glass things are arranged on shelves. I see retorts of different sizes, all with necks bent at a sudden angle. In addition to their long beak, some of them have a narrow little tube coming out of their bulb. Look, youngster, and do not try to guess the object of these curious vessels. I see glasses with feet to them, funnel-shaped and deep; I stand amazed at strange-looking bottles with two or three mouths to each, at phials swelling into a balloon with a long, narrow tube. What an odd array of implements! And here are glass cupboards with a host of bottles and jars, filled with all manner of chemicals. The labels apprise me of their contents: molybdate of ammonia, chloride of antimony, permanganate of potash and ever so many other strange terms. Never, in all my reading, have I met with such repellent language.

Suddenly, bang! And there is running and stamping and shouting and cries of pain! What has happened? I rush up from the back of the room. The retort has burst, squirting its boiling vitriol in every direction. The wall opposite is all stained with it. Most of my fellow-pupils have been more or less struck. One poor youth has had the splashes full in his face, right into his eyes. He is yelling like a madman. With the help of a friend who has come off better than the others, I drag him outside by main force, take him to the sink, which fortunately is close at hand, and hold his face under the tap. This swift ablution serves its purpose. The horrible pain begins to be allayed, so much so that the sufferer recovers his senses and is able to continue the washing-process for himself.

My prompt aid certainly saved his sight. A week later, with the help of the doctor's lotions, all danger was over. How lucky it was that I took it into my head to keep some way off! My isolation, as I stood looking into the glass case of chemicals, left me all my presence of mind, all my readiness of resource. What are the others doing, those who got splashed through standing too near the chemical bomb? I return to the lecture-hall. It is not a cheerful spectacle. The master has come off badly: his shirt-front, waistcoat and trousers are covered with smears, which are all smouldering and burning into holes. He hurriedly divests himself of a portion of his dangerous raiment. Those of us who possess the smartest clothes lend him something to put on so that he can go home decently.

The Life of the Fly

One of the tall, funnel-shaped glasses which I was admiring just now is standing, full of ammonia, on the table. All, coughing and snivelling, dip their handkerchiefs into it and rub the moist rag over their hats and coats. In this way, the red stains left by the horrible compound are made to disappear. A drop of ink will presently restore the colour completely.

And the oxygen? There was no more question, I need hardly say, of that. The feast of learning was over. Never mind: the disastrous lesson was a mighty event for me. I had been inside the chemist's laboratory; I had had a glimpse of those wonderful jars and tubes. In teaching, what matters most is not the thing taught, whether well or badly grasped: it is the stimulus given to the pupil's latent aptitudes; it is the fulminate awakening the slumbering explosives. One day, I shall obtain on my own account that oxygen which ill-luck has denied me; one day, without a master, I shall yet learn chemistry.

Yes, I shall learn this chemistry, which started so disastrously. And how? By teaching it. I do not recommend that method to anybody. Happy the man who is guided by a master's word and example! He has a smooth

A Memorable Lesson

and easy road before him, lying straight ahead. The other follows a rugged path, in which his feet often stumble; he goes groping into the unknown and loses his way. To recover the right road, if want of success have not discouraged him, he can rely only on perseverance, the sole compass of the poor. Such was my fate. I taught myself by teaching others, by passing on to them the modicum of seed that had ripened on the barren moor cleared, from day to day, by my patient ploughshare.

A few months after the incident of the vitriol-bomb, I was sent to Carpentras to take charge of junior classes at the college there. The first year was a difficult one, swamped as I was by the excessive number of pupils, a set of duffers kept out of the more advanced classes and all at different stages in spelling and grammar. Next year, my school is divided into two; I have an assistant. A weeding-out takes place in my crowd of scatterbrains. I keep the older, the more intelligent ones; the others are to have a term in the preparatory division. From that day forward, things are different. Curriculum there is none. In those happy times, the master's personality counted for something; there was no such thing as the scholastic piston working with the regularity of a machine. It was left for me to act as I thought fit. Well, what should I do to make the school earn its title of 'upper primary'?

Why, of course! Among other things, I shall do some chemistry! My reading has taught me that it does no harm to know a little chemistry, if you would make your furrows yield a good return. Many of my pupils come from the country; they will go back to it to improve their land. Let us show them what the soil is made of and what the plant feeds on. Others will follow industrial careers; they will become tanners, metal-founders, distillers; they will sell cakes of soap and kegs of anchovies. Let us show them pickling, soap-making, stills, tannin and metals. Of course, I know nothing about these things, but I shall learn, all the more so as I shall have to teach them to the boys; and your schoolboy is a little demon for jeering at the master's hesitation.

As it happens, the college boasts a small laboratory, containing just what is strictly indispensable: a receiver, a dozen glass balloons, a few tubes and a niggardly assortment of chemicals. That will do, if I can have the

run of it. But the laboratory is a sanctum
reserved for the use of the sixth form. No
one sets foot in it except the professor and his
pupils preparing for their degree. For me,
the outsider, to enter that tabernacle with
my band of young imps would be most un-
seemly; the rightful occupant would never
think of allowing it. I feel it myself: elementary teaching dare not aspire to such
familiarity with the higher culture. Very
well, we will not go there, so long as they will
lend me the things.

I confide my plan to the principal, the su-
preme dispenser of those riches. He is a
classics man, knows hardly anything of science, at that time held in no great esteem, and
he does not quite understand the object of
my request. I humbly insist and exert my
powers of persuasion. I discreetly emphasize
the real point of the matter. My group of
pupils is a numerous one. It takes more
meals at the school-house—the real concern
of a principal—than any other section of the
college. This group must be encouraged,
lured on, increased if possible. The prospect
of disposing of a few more platefuls of soup
wins the battle for me; my request is granted.
Poor science! All that diplomacy to gain

your entrance among the despised ones, who have not been nourished on Cicero and Demosthenes!

I am authorized to move, once a week, the material required for my ambitious plans. From the first floor, the sacred dwelling of the scientific things, I shall take them down to a sort of cellar where I give my lessons. The troublesome part is the pneumatic trough. It has to be emptied before it is carried downstairs and to be filled again afterwards. A day-scholar, a zealous acolyte, hurries over his dinner and comes to lend me a hand an hour or two before the class begins. We effect the move between us.

What I am after is oxygen, the gas which I once saw fail so lamentably. I thought it all out at my leisure, with the help of a book. I will do this, I will do that, I will go to work in this or the other fashion. Above all, we will run no risks, perhaps of blinding ourselves; for it is once more a question of heating manganese dioxide with sulphuric acid. I am filled with misgivings at the recollection of my old school-fellow yelling like mad. Who cares? Let us try for all that: fortune favours the brave! Besides, we will make one prudent condition, from which I shall never depart: no one but myself shall come near the table. If an accident happen, I shall be the only one to suffer; and, in my opinion, it is worth a burn or two to make acquaintance with oxygen.

Two o'clock strikes; and my pupils enter the class-room. I purposely exaggerate the likelihood of danger. They are all to stay on their benches and not stir. This is agreed. I have plenty of elbow-room. There is no one by me, except my acolyte, standing by my side, ready to help me when the time comes. The others look on in profound silence, reverent towards the unknown.

Soon the gaseous bubbles come "gloo-glooing" through the water in the bell-jar. Can it be my gas? My heart beats with excitement. Can I have succeeded without any trouble at the first attempt? We will see. A candle blown out that moment and still retaining a red tip to its wick is lowered by a wire into a small test-jar filled with my product. Capital! The candle lights with a little explosion and burns with extraordinary brilliancy. It is oxygen right enough.

The moment is a solemn one. My audience is astounded and so am I, but more at my own success than at the relighted candle. A puff

of vainglory rises to my brow; I feel the fire
of enthusiasm run through my veins. But I
say nothing of these inner sensations. Before
the boys' eyes, the master must appear an old
hand at the things he teaches. What would
the young rascals think of me if I allowed
them to suspect my surprise, if they knew that
I myself am beholding the marvellous subject of my demonstration for the first time in
my life? I should lose their confidence, I
should sink to the level of a mere pupil.

Sursum corda! Let us go on as if chemistry were a familiar thing to me. It is the
turn of the steel ribbon, an old watch-spring
rolled corkscrew-fashion and furnished with
a bit of tinder. With this simple lighted bait,
the steel should take fire in a jar filled with
my gas. And it does burn; it becomes a
splendid firework, with cracklings and a blaze
of sparks and a cloud of rust that tarnishes the
jar. From the end of the fiery coil a red drop
breaks off at intervals, shoots quivering
through the layer of water left at the bottom
of the vessel and embeds itself in the glass
which has suddenly grown soft. This metallic
tear, with its indomitable heat, makes every
one of us shudder. All stamp and cheer and
applaud. The timid ones place their hands

A Memorable Lesson

before their faces and dare not look except through their fingers. My audience exults; and I myself triumph. Ha, my friends, isn't it grand, this chemistry!

All of us have red-letter days in our lives. Some, the practical men, have been successful in business; they have made money and hold their heads high in consequence. Others, the thinkers, have gained ideas; they have opened a new account in the ledger of nature and they silently taste the hallowed joys of truth. One of my great days was that of my first acquaintance with oxygen. On that day, when my class was over and all the materials put back in their place, I felt myself grow several inches taller. An untrained workman, I had shown, with complete success, that which was unknown to me a couple of hours before. No accident whatever, not even the least stain of acid.

It is, therefore, not so difficult nor so dangerous as the pitiful finish of the Saint-Martial lesson might have led me to believe. With a vigilant eye and a little prudence, I shall be able to continue. The prospect is enchanting.

And so, in due season, comes hydrogen, carefully contemplated in my reading, seen

The Life of the Fly

and reseen with the eye of the mind before being seen with the eyes of the body. I delight my little rascals by making the hydrogen-flame sing in a glass tube, which trickles with the drops of water resulting from the combustion; I make them jump with the explosions of the thunderous mixture. Later, I show them, with the same invariable success, the splendours of phosphorus, the violent powers of chlorine, the loathsome smells of sulphur, the metamorphoses of carbon and so on. In short, in a series of lessons, the principal non-metallic elements and their compounds are passed in review during the course of the year.

The thing was bruited abroad. Fresh pupils came to me, attracted by the marvels of the school. Additional places were laid in the dining-hall; and the principal, who was more interested in the profits on his beans and bacon than in chemistry, congratulated me on this accession of boarders. I was fairly started. Time and an indomitable will would do the rest.

CHAPTER XX

INDUSTRIAL CHEMISTRY

EVERYTHING happens sooner or later. When, through the low windows over-looking the garden of the school, my eye glanced at the laboratory, where the madder-vats were steaming; when, in the sanctuary it-self, I was present, by way of a first and last chemistry-lesson, at the explosion of the retort of sulphuric acid that nearly disfigured every one of us, I was far indeed from suspecting the part which I was destined to play under that same vaulted roof. Had a prophet foretold that I should one day succeed the master, never would I have believed him. Time works these surprises for us.

Stones would have theirs too, if anything were able to astonish them. The Saint-Mar-tial building was originally a church; it is a protestant place of worship now. Men used to pray there in Latin; to-day they pray in French. In the intervening period, it was for some years in the service of science, the noble orison that dispels the darkness. What has the future in store for it? Like many another in the ringing city, to use Rabelais' epithet,[1] will it become a home for the fuller's teazles, a warehouse for scrap-iron, a carrier's stable? Who knows? Stones have their destinies no less unexpected than ours.

When I took possession of it as a laboratory for the municipal course of lectures, the nave remained as it was at the time of my former short and disastrous visit. To the right, on the wall, a number of black stains struck the eye. It was as though a madman's hand, armed with the ink-pot, had smashed its fragile projectile at that spot. I recognized the stains at once. They were the marks of the corrosive which the retort had splashed at our heads. Since those days of long ago, no one had thought fit to hide them under a coat of whitewash. So much the better: they will serve me as excellent counsellors. Always before my eyes, at every lesson, they will speak to me incessantly of prudence.

[1] The allusion is to the many churches and chapels at Avignon and to *Pantagruel*, Book v, chap. i: 'Our pilot told us that it was the Ringing Island; and indeed we heard a kind of a confused and often repeated noise . . . not unlike the sound of great, middle-sized and little bells, rung all at once, as it is customary at Paris, Tours, Gergeau, Nantes and elsewhere on high holidays; and, the nearer we came to the land, the louder we heard that jangling.'—*Translator's Note*.

Industrial Chemistry

For all its attractions, however, chemistry did not make me forget a long-cherished plan well-suited to my tastes, that of teaching natural history at a university. Now, one day, at the grammar-school, I had a visit from a chief-inspector which was not of an encouraging nature. My colleagues used to call him the Crocodile. Perhaps he had given them a rough time in the course of his inspections. For all his boorish ways, he was an excellent man at heart. I owe him a piece of advice which greatly influenced my future studies.

That day, he suddenly appeared, alone, in the schoolroom, where I was taking a class in geometrical drawing. I must explain that, at this time, to eke out my ridiculous salary and, at all costs, to provide a living for myself and my large family, I was a mighty pluralist, both inside the college and out. At the college in particular, after two hours of physics, chemistry or natural history, came, without respite, another two hours' lesson, in which I taught the boys how to make a projection in descriptive geometry, how to draw a geodetic plane, a curve of any kind whose law of generation is known to us. This was called graphics.

The sudden irruption of the dread personage causes me no great flurry. Twelve o'clock strikes, the pupils go out and we are left alone. I know him to be a geometrician. The transcendental curve, perfectly drawn, may work upon his gentler mood. I happen to have in my portfolio the very thing to please him. Fortune serves me well, in this special circumstance. Among my boys, there is one who, though a regular dunce at everything else, is a first-rate hand with the square, the compass and the drawing-pen: a deft-fingered numskull, in short.

With the aid of a system of tangents of which I first showed him the rule and the method of construction, my artist has obtained the ordinary cycloid, followed by the interior and the exterior epicycloid and, lastly, the same curves both lengthened and shortened. His drawings are admirable Spider's webs, encircling the cunning curve in their net. The draughtsmanship is so accurate that it is easy to deduce from it beautiful theorems, which would be very laborious to work out by the calculus.

I submit the geometrical masterpieces to my chief-inspector, who is himself said to be smitten with geometry. I modestly describe the method of construction, I call his attention to the fine deductions which the drawing enables

Industrial Chemistry

one to make. It is labour lost: he gives but a
heedless glance at my sheets and flings each on
the table as I hand it to him.

'Alas!' said I to myself. 'There is a storm
brewing; the cycloid won't save you; it's your
turn for a bite from the Crocodile!'

Not a bit of it. Behold the bugbear growing genial. He sits down on a bench, with one
leg here, another there, invites me to take a
seat by his side and, in a moment, we are discussing graphics. Then, bluntly:

'Have you any money?' he asks.

Astounded at this strange question, I answer with a smile.

'Don't be afraid,' he says. 'Confide in me.
I'm asking you in your own interest. Have
you any capital?'

'I have no reason to be ashamed of my
poverty, monsieur l'inspecteur général. I
frankly admit, I possess nothing; my means
are limited to my modest salary.'

A frown greets my answer; and I hear,
spoken in an undertone, as though my con-
fessor were talking to himself:

'That's sad, that's really very sad.'

Astonished to find my penury treated as sad,
I ask for an explanation: I was not accustomed
to this solicitude on the part of my superiors.

The Life of the Fly

'Why, yes, it's a great pity,' continues the man reputed so terrible. 'I have read your articles in the *Annales des sciences naturelles*. You have an observant mind, a taste for research, a lively style and a ready pen. You would have made a capital university-professor.'

'But that's just what I'm aiming at!'

'Give up the idea.'

'Haven't I the necessary attainment?'

'Yes, you have; but you have no capital.'

The great obstacle stands revealed to me: woe to the poor in pocket! University-teaching demands a private income. Be as ordinary, as commonplace as you please, but, above all, possess the coin that lets you cut a dash. That is the main thing; the rest is a secondary condition.

And the worthy man tells me what poverty in a frock-coat means. Though less of a pauper than I, he has known the mortification of it; he describes it to me, excitedly, in all its bitterness. I listen to him with an aching heart; I see the refuge which was to shelter my future crumbling before my eyes:

'You have done me a great service, sir,' I answered. 'You put an end to my hesitation. For the moment, I give up my plan. I will first see if it is possible to earn the small fortune which I shall need if I am to teach in a decent manner.'

Thereupon we exchanged a friendly grip of the hand and parted. I never saw him again. His fatherly arguments had soon convinced me: I was prepared to hear the blunt truth. A few months earlier, I had received my nomination as an assistant-lecturer in zoology at the university of Poitiers. They offered me a ridiculous salary. After paying the costs of moving, I should have had hardly three francs a day left; and, on this income, I had to keep my family, numbering seven in all. I hastened to decline the very great honour.

No, science ought not to practise these jests. If we humble persons are of use to her, she should at least enable us to live. If she can't do that, then let her leave us to break stones on the highway. Oh, yes, I was prepared for the truth when that honest fellow talked to me of frock-coated poverty! I am telling the story of a not very distant past. Since then, things have improved considerably; but, when the pear was properly ripened, I was no longer of an age to pick it.

And what was I to do now, to overcome the difficulty mentioned by my inspector and confirmed by my personal experience? I would take up industrial chemistry. The municipal lectures at Saint-Martial placed a spacious and fairly well-equipped laboratory at my disposal. Why not make the most of it?

The chief manufacture of Avignon was madder. The farmer supplied the raw material to the factories, where it was turned into purer and more concentrated products. My predecessor had gone in for it and done well by it, so people said. I would follow in his footsteps and use the vats and furnaces, the expensive plant which I had inherited. So to work.

What should I set myself to produce? I proposed to extract the colouring-substance, alizarin, to separate it from the other matters found with it in the root, to obtain it in the pure state and in a form that allowed of the direct printing of the stuffs, a much quicker and more artistic method than the old dyeing-process.

Nothing could be simpler than this problem, once the solution was known; but how tremendously obscure while it had still to be solved! I dare not call to mind all the imagination and patience spent upon endless endeavours which nothing, not even the madness of them, discouraged. What mighty meditations in the sombre church! What glowing dreams, soon to be followed by sore disappointment, when experiment spoke the last word and upset the scaffolding of my plans. Stubborn as the slave of old amassing a peculium for his enfranchisement, I used to reply to the check of yesterday by the fresh attempt of to-morrow, often as faulty as the others, sometimes the richer by an improvement, and I went on indefatigably, for I too cherished the indomitable ambition to set myself free.

Should I succeed? Perhaps so. I at last had a satisfactory answer. I obtained, in a cheap and practical fashion, the pure colouring-matter, concentrated in a small volume and excellent for both printing and dyeing. One of my friends took up my process on a large scale in his works; a few calico-factories adopted the produce and expressed themselves delighted with it. The future smiled at last; a pink rift opened in my grey sky. I should possess the modest fortune without which I must deny myself the pleasure of teaching in a university. Freed of the torturing anxiety about my daily bread, I should be able to live at ease among my insects.

In the midst of the joys of seeing these

The Life of the Fly

problems solved by chemistry, yet another ray
of sunshine was reserved for me, adding its
gladness to that of my success. Let us go
back a couple of years. The chief-inspectors
visited our grammar-school. These person-
ages travel in pairs: one attends to literature,
the other to science. When the inspection was
over and the books checked, the staff was summoned to the principal's drawing-room, to receive the parting admonitions of the two lumi-
naries. The man of science began. I should
be sadly put to it to remember what he said.
It was cold professional prose, made up of
soulless words which the hearer forgot once
the speaker's back was turned, words merely
boring to both. I had heard enough of these
chilly sermons in my time; one more of them
could not hope to make an impression on me.

The inspector in literature spoke next. At
the first words which he uttered, I said to myself:

'Oho! This is a very different business!'

The speech was alive and vigorous and
imageful; indifferent to scholastic commonplaces, the ideas soared, hovering gently in the
serene heights of a kindly philosophy. This
time, I listened with pleasure; I even felt
stirred. Here was no official homily: it was

full of impassioned zeal, of words that carried you with them, uttered by an honest man accomplished in the art of speaking, an orator in the true sense of the word. In all my school-experience, I had never had such a treat.

When the meeting broke up, my heart beat faster than usual:

'What a pity,' I thought, 'that my side, the science side, cannot bring me into contact, some day, with that inspector! It seems to me that we should become great friends.'

I enquired his name of my colleagues, who were always better-informed than I. They told me it was Victor Duruy.

Well, one day, two years later, as I was looking after my Saint-Martial laboratory in the midst of the steam from my vats, with my hands the colour of boiled lobster-claws from constant dipping in the indelible red of my dyes, there walked in, unexpectedly, a person whose features straightway seemed familiar. I was right, it was the very man, the chief-inspector whose speech had once stirred me. M. Duruy was now minister of public instruction. He was styled, 'Your excellency;' and this style, usually an empty formula, was well-deserved in the present case, for our new min-

ister excelled in his exalted functions. We all
held him in high esteem. He was the workers' minister, the man for the humble toiler.

'I want to spend my last half-hour at
Avignon with you,' said my visitor, with a
smile. 'That will be a relief from the official
bowing and scraping.'

Overcome by the honour paid me, I apolo-
gized for my costume—I was in my shirt-
sleeves—and especially for my lobster-claws,
which I had tried, for a moment, to hide behind my back.

'You have nothing to apologize for. I
came to see the worker. The working-man
never looks better than in his overall, with the
marks of his trade on him. Let us have a
talk. What are you doing just now?'

I explained, in a few words, the object of
my researches; I showed my product; I executed under the minister's eyes a little attempt
at printing in madder-red. The success of the
experiment and the simplicity of my apparatus, in which an evaporating-dish, maintained at boiling-point under a glass funnel,
took the place of a steam-chamber, caused him
some surprise.

'I will help you,' he said. 'What do you
want for your laboratory?'

Industrial Chemistry

'Why, nothing, monsieur le ministre, no-thing! With a little application, the plant I have is ample.'

'What, nothing! You are unique there! The others overwhelm me with requests; their laboratories are never well enough supplied. And you, poor as you are, refuse my offers!'

'No, there is one thing which I will accept.'

'What is that?'

'The signal honour of shaking you by the hand.'

'There you are, my friend, with all my heart. But that's not enough. What else do you want?'

'The Paris Jardin des Plantes[1] is under your control. Should a crocodile die, let them keep the hide for me. I will stuff it with straw and hang it from the ceiling. Thus adorned, my workshop will rival the wizard's cave.'

The minister cast his eyes round the nave and glanced up at the Gothic vault:

'Yes, it would look very well.' And he gave a laugh at my sally. 'I now know you as a chemist,' he continued. 'I knew you already as a naturalist and a writer. I have

[1] The Zoological and Botanical Gardens on the left bank of the Seine.—*Translator's Note.*

459

heard about your little animals. I am sorry
that I shall have to leave without seeing them.
They must wait for another occasion. My
train will be starting presently. Walk with me
to the station, will you? We shall be alone
and we can chat a bit more on the way.'

We strolled along, discussing entomology
and madder. My shyness had disappeared.
The self-sufficiency of a fool would have left
me dumb; the fine frankness of a lofty mind
put me at my ease. I told him of my experiments in natural history, of my plans for a
professorship, of my fight with harsh fate, my
hopes and fears. He encouraged me, spoke to
me of a better future. We reached the station and walked up and down outside, talking
away delightfully.

A poor old woman passed, all in rags, her
back bent by age and years of work in the
fields. She furtively put out her hand for
alms. Duruy felt in his waistcoat, found a
two-franc piece and placed it in the out-
stretched hand; I wanted to add a couple of
sous as my contribution, but my pockets were
empty, as usual. I went to the beggar-woman
and whispered in her ear:

'Do you know who gave you that? It's
the emperor's minister.'

Industrial Chemistry

The poor woman started; and her astounded eyes wandered from the open-handed swell to the piece of silver and from the piece of silver to the open-handed swell. What a surprise! What a windfall!

'*Que lou bon Dièu ié done longo vido e santa, pecaïre!*' she said, in her cracked voice.

And, curtseying and nodding, she withdrew, still staring at the coin in the palm of her hand.

'What did she say?' asked Duruy.

'She wished you long life and health.'

'And *pecaïre?*'

'*Pecaïre* is a poem in itself: it sums up all the gentler passions.'

And I myself mentally repeated the artless vow. The man who stops so kindly when a beggar puts out her hand has something better in his soul than the mere qualities that go to make a minister.

We entered the station, still alone, as promised, and I quite without misgivings. Had I but foreseen what was going to happen, how I should have hastened to take my leave! Little by little, a group formed in front of us. It was too late to fly; I had to screw up my courage. Came the general of division and his officers, came the prefect and his secretary, the mayor and his deputy, the school-inspector and the pick of the staff. The minister faced the ceremonial semicircle. I stood next to him. A crowd on one side, we two on the other. Followed the regulation spinal contortions, the empty obeisances which my dear Duruy had come to my laboratory to forget. When bowing to St. Roch,[1] in his corner niche, the worshipper at the same time salutes the saint's humble companion. I was something like St. Roch's dog in the presence of those honours which did not concern me. I stood and looked on, with my awful red hands concealed behind my back, under the broad brim of my felt hat.

After the official compliments had been exchanged, the conversation began to languish; and the minister seized my right hand and gently drew it from the mysterious recesses of my wide-awake.

'Why don't you show those gentlemen your hands?' he said. 'Most people would be proud of them.'

I vainly protested with a jerk of the elbow. I had to comply and I displayed my lobster-claws.

[1] St. Roch (1295-1327) is always represented in his statues with the dog that saved his life by discovering him in the solitude where, after curing the plague-stricken Italians, he had hidden himself lest he should communicate the pestilence to others.—*Translator's Note.*

Industrial Chemistry

'Workman's hands,' said the prefect's secretary. 'Regular workman's hands.'

The general, almost scandalized at seeing me in such distinguished company, added:

'Hands of a dyer and cleaner.'

'Yes, workman's hands,' retorted the minister, 'and I wish you many like them. Believe me, they will do much to help the chief industry of your city. Skilled as they are in chemical work, they are equally capable of wielding the pen, the pencil, the scalpel and the lens. As you here seem unaware of it, I am delighted to inform you.'

This time, I should have liked the ground to open and swallow me up. Fortunately, the bell rang for the train to start. I said goodbye to the minister and, hurriedly taking to flight, left him laughing at the trick which he had played me.

The incident was noised about, could not help being so, for the peristyle of a railway-station keeps no secrets. I then learnt to what annoyances the shadow of the great exposes us. I was looked upon as an influential person, having the favour of the gods at my disposal. Place-hunters and canvassers tormented me. One wanted a license to sell tobacco and stamps, another a scholarship for his son, an-

other an increase of his pension. I had only
to ask and I should obtain, said they.

O simple people, what an illusion was
yours! You could not have hit upon a worse
intermediary. I figuring as a postulant! I
have many faults, I admit, but that is certainly
not one of them. I got rid of the importunate
people as best I could, though they were utterly unable to fathom my reserve. What
would they have said had they known of the
minister's offers with regard to my laboratory
and my jesting reply, in which I asked for a
crocodile-skin to hang from my ceiling! They
would have taken me for an idiot.

Six months elapsed; and I received a letter
summoning me to call upon the minister at
his office. I suspected a proposal to promote
me to a more important grammar-school and
wrote begging that I might be left where I
was, among my vats and my insects. A second
letter arrived, more pressing than the first and
signed by the minister's own hand. This letter said:

'Come at once, or I shall send my gendarmes to fetch you.'

There was no way out of it. Twenty-four
hours later, I was in M. Duruy's room. He
welcomed me with exquisite cordiality, gave

me his hand and, taking up a number of the *Moniteur:*

'Read that,' he said. 'You refused my chemical apparatus; but you won't refuse this.'

I looked at the line to which his finger pointed. I read my name in the list of the Legion of Honour. Quite stupid with surprise, I stammered the first words of thanks that entered my head.

'Come here,' said he, 'and let me give you the accolade. I will be your sponsor. You will like the ceremony all the better if it is held in private, between you and me: I know you!'

He pinned the red ribbon to my coat, kissed me on both cheeks, made me telegraph the great event to my family. What a morning, spent with that good man!

I well know the vanity of decorative ribbonry and tinware, especially when, as too often happens, intrigue degrades the honour conferred; but, coming as it did, that bit of ribbon is precious to me. It is a relic, not an object for show. I keep it religiously in a drawer.

There was a parcel of big books on the table, a collection of the reports on the pro-

The Life of the Fly

gress of science drawn up for the International
Exhibition of 1867, which had just closed.

'Those books are for you,' continued the
minister. 'Take them with you. You can
look through them at your leisure: they may
interest you. There is something about your
insects in them. You're to have this too: it
will pay for your journey. The trip which I
made you take must not be at your own expense. If there is anything over, spend it on
your laboratory.'

And he handed me a roll of twelve hundred
francs. In vain I refused, remarking that my
journey was not so burdensome as all that; besides, his embrace and his bit of ribbon were
of inestimable value compared with my dis-
bursements. He insisted:

'Take it,' he said, 'or I shall be very
angry. There's something else: you must
come to the emperor's with me to-morrow, to
the reception of the learned societies.'

Seeing me greatly perplexed and as though
demoralized by the prospect of an imperial in-
terview:

'Don't try to escape me,' he said, 'or look
out for the gendarmes of my letter! You saw
the fellows in the bearskin caps on your way
up. Mind you don't fall into their hands. In

466

any case, lest you should be tempted to run away, we will go to the Tuileries together, in my carriage.'

Things happened as he wished. The next day, in the minister's company, I was ushered into a little drawing-room at the Tuileries by chamberlains in knee-breeches and silver-buckled shoes. They were queer people to look at. Their uniforms and their stiff gait gave them the appearance, in my eyes, of Beetles who, by way of wing-cases, wore a great, gold-laced dress-coat, with a key in the small of the back. There were already a score of persons from all parts waiting in the room. These included geographical explorers, botanists, geologists, antiquaries, archæologists, collectors of prehistoric flints, in short, the usual representatives of provincial scientific life.

The emperor entered, very simply dressed, with no parade about him beyond a wide, red, watered-silk ribbon across his chest. No sign of majesty, an ordinary man, round and plump, with a large moustache and a pair of half-closed, drowsy eyelids. He moved from one to the other, talking to each of us for a moment as the minister mentioned our names and the nature of our occupations. He

showed a fair amount of information as he
changed his subject from the ice-floes of Spitz-
bergen to the dunes of Gascony, from a Car-
lovingian charter to the flora of the Sahara,
from the progress in beetroot-growing to
Cæsar's trenches before Alesia. When my
turn came, he questioned me upon the hyper-
metamorphosis of the Meloidæ,[1] my last essay in entomology. I answered as best I
could, floundering a little in the proper mode
of address, mixing up the everyday *monsieur*
with *sire,* a word whose use was so entirely
new to me. I passed through the dread straits
and others succeeded me. My five minutes'
conversation with an imperial majesty was,
they tell me, a most distinguished honour.
I am quite ready to believe them, but I never
had a desire to repeat it.

The reception came to an end, bows were
exchanged and we were dismissed. A
luncheon awaited us at the minister's house.
I sat on his right, not a little embarrassed by
the privilege; on his left was a physiologist of
great renown. Like the others, I spoke of all
manner of things, including even Avignon
Bridge. Duruy's son, sitting opposite me,
chaffed me pleasantly about the famous bridge

[1] A family of Beetles, including the Oil-beetle and the
Spanish Fly.—*Translator's Note.*

Industrial Chemistry

on which everybody dances;[1] he smiled at my impatience to get back to the thyme-scented hills and the grey olive-yards rich in Grass-hoppers.

'What!' said his father. 'Won't you visit our museums, our collections? There are some very interesting things there.'

'I know, monsieur le ministre, but I shall find better things, things more to my taste, in the incomparable museum of the fields.'

'Then what do you propose to do?'

'I propose to go back to-morrow.'

I did go back, I had had enough of Paris: never had I felt such tortures of loneliness as in that immense whirl of humanity. To get away, to get away was my one idea.

Once home among my family, I felt a mighty load off my mind and a great joy in my heart, where rang a peal of bells pro-claiming the delights of my approaching emancipation. Little by little, the factory that was to set me free rose skywards, full of promises. Yes, I should possess the modest income which would crown my ambition by allowing me to descant on animals and plants in a university chair.

[1] The old, partly-demolished bridge at Avignon, which figures in the well-known French catch:
'Sur le pont d'Avignon,
Tout le monde y danse en rond.'
—*Translator's Note.*

The Life of the Fly

'Well, no,' said Fate, 'you shall not acquire the freedman's peculium; you shall remain a slave, dragging your chain behind you; your peal of bells rings false!'

Hardly was the factory in full swing when a piece of news was bruited, at first a vague rumour, an echo of probabilities rather than certainties, and then a positive statement leaving no room for doubt. Chemistry had obtained the madder-dye by artificial means; thanks to a laboratory-concoction, it was utterly overthrowing the agriculture and industries of my district. This result, while destroying my work and my hopes, did not surprise me unduly. I myself had toyed with the problem of artificial alizarin and I knew enough about it to foresee that, in no very distant future, the work of the chemist's retort would take the place of the work of the fields.

It was finished; my hopes were dashed to the ground. What to do next? Let us change our lever and begin to roll Sisyphus' stone once more. Let us try to draw from the ink-pot what the madder-vat declines to yield. *Laboremus!*

INDEX

A

Acorn-shell, 27
Æsculapean Snake, 231
Æsop, 432
d'Alembert, Jean Baptiste le Rond, 315
Amazon-ant, 22
Amber-snail, 187
Ammonite, 175, 176
Ammophila, 11, 22, 34, 87
Andrena, 20
Anisotoma cinnamomea, 404
Annelid, 26
Ant (*see also* Amazon-ant), 217, 218, 222
Anthidium, 9, 67, 89
Anthophora, 19, 67, 70-71, 88, 96
Anthophora personata (*see* Anthophora)
Anthrax, 28-63, 72, 78-110
Anthrax sinuata, A. trifasciata (*see* Anthrax)
Arion (*see* Slug)

B

Bacon-beetle (*see* Dermestes)
Baker, W. S. Graff, 9
Barnacle, 27
Barn-owl (*see* Owl)
Bee, 11, 16, 18, 26, 47, 51, 92, 112, 152, 254, 371
Bee-fly (*see* Bombylius)

Beetle, 51, 154, 159, 468
Bembex, 21, 26
Bison Oritis, 111
Black-bellied Tarantula (*see* Narbonne Lycosa)
Black-eared Chat, 21
Blackbird, 326-327
Black Psen, 50
Blistering Beetle (*see* Spanish Fly)
Bluebottle, 233, 255, 316-363, 408
Blue Osmia (*see* Osmia)
Bolboceras gallicus, 404-406
Bombylius, 91
Brullé, Gaspard August, 75
Bulimus, 188
Bumble-bee Fly, 252-275
Buprestis, 219
Burying-beetle (*see* Necrophorus)
Butterfly, 51, 393

C

Caddis-worm, 162, 182-211, 352
Cæsar, Caius Julius, 468
Cahen, Edward, 9
Calicurgus (*see* Pompilus)
Calliphora vomitoria (*see* Bluebottle)
Cantharis (*see* Spanish Fly)

Index

Capricorn, 374, 376, 383
Carabus, 376
Carbuncle-fly, 382
Carcan (*see* Cicada)
Carrion-beetle (*see* Silpha)
Cat, 397
Caterpillar (*see* also Looper), 22, 34, 119, 403, 406, 419
Centipede, 349
Cerambyx (*see* Capricorn)
Cerceris, 13, 86
Cerceris bupresticida, C. tuberculata (*see* Cerceris)
Cetonia, 23, 46, 219, 266, 374, 376-377, 383, 387
Chaffinch, 317
Chalcidid, 31, 364-369, 384-385
Chalicodoma, *Chalicodoma muraria, C. sicula* (*see* Mason-bee)
Chat (*see* Black-eared Chat)
Chicken, 217
Cicada, 14, 94, 152
Cicadella, 21
Cicero, 442
Cigale (*see* Cicada)
Cinnamon-coloured Anisotoma (*see* Anisotoma *cinnamomea*)
Cirriped, 27
Clausilium, 188
Cleonus, 86
Clod-hopper (*see* Wheatear)
Cockchafer, 50, 152, 374, 376, 387
Copris (*see* Lunary Copris, Spanish C.)

Cossus, 423
Crayfish, 144
Cricket, 87, 213, 346
Crow, 152, 229-230, 395-396
Cul-blanc (*see* Wheatear)
Cuttlefish, 40
Cuvier, Georges Léopold Chrétien Frédéric Dagobert Baron, 85

D

Darwin, Charles Robert, 116
Dasypoda, 19
Death-watch, 308
Demosthenes, 442
Dermestes, 218, 356-357
Dermestes vulpinus (*see* Dermestes)
Diadem Anthidium (*see* Anthidium)
Dog, 238
Dolphin, 163
Dragon-fly, 163
Dufour, Léon, 29*n*
Dung-beetle, 111, 122, 161, 230, 375, 380
Duruy, Jean Victor, 457-470
Dytiscus (*see* Waterbeetle)

E

Earth-worm, 163
Ephialtes mediator, 50
Ephippigera, 21
Eucera, 19
Eumenes, 25
Eyed Lizard, 21

Index

F

Fabre, Antoine, the author's father, 119, 126, 131, 145, 149, 165, 174

Fabre, Mme. Antoine, *née* Salgues, the author's mother, 149, 164, 174-175

Fabre, Jean, the author's great-great-grandfather, 119*n*

Fabre, Mme. Jean, *née* Desmazes, the author's great-great-grandmother, 119*n*

Fabre, Pierre, the author's great-grandfather, 119*n*

Fabre, Mme. Pierre, *née* Fages, the author's great-grandmother, 119*n*

Fabre, Pierre Jean, the author's grandfather, 119-122, 126-127, 130, 134

Fabre, Mme. Pierre Jean, *née* Poujade, the author's grandmother, 119-125, 128, 134

Fages, Pierre, the author's great-great-great-grandfather, 119*n*

Fages, Mme. Pierre, *née* Baumelou, the author's great-great-great-grandmother, 119*n*

Favier, the author's factotum, 94

Fallow-chat (*see* Wheatear)

Flesh-fly (*see* Grey Flesh-fly)

Fox, 78

Frog, 23-24, 144, 162, 169, 347, 349-350

Froghopper (*see* Cicadella)

G

Geometrid Moth, Caterpillar of the (*see* Looper)

Giant Scarites, 202

Glass-snail, 188

Gnat, 63

Grasshopper, 14, 122, 129, 469

Great Peacock Moth, 366-367, 377

Greenbottle, 212-232, 233-235, 237, 238, 240, 248, 255, 364

Greenfinch, 23

Grey Flesh-fly, 219, 233-251, 255, 331-336, 339, 365-366, 368-369, 383-386, 408, 410, 413

Ground-beetle (*see also* Carabus, Procrustes), 153, 383

H

Hairy-footed Anthophora (*see* Anthophora)

Halictus, 20

Helix, 188

Hen, 395

Hoplia, 144, 170, 175-176

Horace, 294, 398, 432

Horse, 385

House-fly, 261

Hugo, Victor Marie, 302

Hydrometra (*see* Pond-skater)

Hylotoma, 259, 270

473

Index

I

Ichneumon-fly (*see Ephialtes mediator*)

J

Jelly-fish, 27
Jussieu, Bernard, 85

K

Kepler, Johann, 310, 312
Kitten, 217

L

La Fontaine, Jean, 149-150
Lagrange, Joseph Louis Comte, 85
Lamb, 38
Lamellicorn, 23
Languedocian Scorpion, 370-377, 383
Languedocian Sphex, 21
Laplace, Pierre Simon Marquis de, 85
Large White Butterfly, 130
Lark, 325-326
Leaf-beetle, 219
Leaf-cutter, Leaf-cutting Bee (*see* Megachile)
Leech, 163
Leibnitz, Gottfried Wilhelm Baron von, 85
Leptocerus, 204
Leucospis, 31, 50, 64-66, 68, 72, 105
Leucospis gigas (*see* Leucospis)
Limnæa, 163, 187
Limnophilus flavicornis, 182

Linnet

Linnet, 152, 317-325, 328
Lion, 78, 385
Lizard (*see also* Eyed Lizard), 216
Locust, 87-88, 145, 312
Looper, 87, 102, 104
Lucilia cadaverina, L. Caesar (*see* Greenbottle)
Lucilia cuprea, 219
Lunary Copris, 111-112

M

Macrocera, 19
Mademoiselle Mori, Author of, 11*n*
Mammoth, 63
Mantis (*see* Praying Mantis)
Masked Anthophora (*see* Anthophora)
Mason-bee, 19, 21, 25, 28-31, 34-35, 38-62, 64-67, 69, 71, 76, 78-80
Mastodon, 63
Measuring-worm (*see* Looper)
Megachile, 19
Megatherium, 63
Melecta, 88
Miall, Bernard, 14*n*
Midge, 63-64
Midwife Toad, 24
Milesia fulminans, 259
Minnow, 395
Minotaurus typhoeus, 111
Mitchell, Dr. Peter Chalmers, 7
Mole, 213-217, 220, 222, 231, 238
Monodontomerus cupreus, 63-77

Index

Moquin-Tandon, Horace Bénédict Alfred, 157-159, 427
Mosquito, 163
Moth (*see also* Caterpillar), 128, 336-337, 406, 407
Motteux (*see* Wheatear)
Mussel, 305
Myodites, 50

N

Nanny-goat, 152
Napoleon III, the Emperor, 466-468
Narbonne Lycosa, 22
Natterjack, 23
Necrophorus, 217
Newt, 162
Newton, Sir Isaac, 85, 283
Nightingale, 253-255
Notonecta (*see* Water-boatman)

O

Oil-beetle, 28, 86, 88, 468
Oritis (*see* Bison Oritis)
Oryctes, 23, 374
Oryctes nasicornis (*see* Oryctes)
Osmia, 19-20, 28, 47, 71, 75, 81, 88-90, 92
Osmia cyanea, O. tricornis (*see* Osmia)
Owl, 23, 252, 351-356
Oxyporus rufus, 403-404, 406

P

Paludina, 187
Partridge, 326, 336, 387
Pelopaeus, 25
Persoon, Christiaan Hendrik, 421
Philanthus apivorus, 87
Phryganea (*see* Caddis-worm)
Physa, 163, 187
Pig, 342, 345
Pine Cockchafer, 153
Pisidium, 187
Planorbis, 163, 187, 205, 212, 215
Plato, 294
Plover, 326
Polistes, 25, 258-259
Pompilus, 22
Pond-skater, 162, 212
Pond-snail, 163, 187
Porto-canèu, Porto-fais (*see* Caddis-worm)
Poujade, Antoine, the author's great-grandfather, 119*n*
Poujade, Mme. Antoine, *née* Azémar, the author's great-grandmother, 119*n*
Praying Mantis, 87, 350
Procrustes, 376
Psen (*see* Black Psen)
Pupa, 188

R

Rabbit, 217
Rabelais, François, 448

Index

Réaumur, René Antoine Ferchault de, 237
Red Admiral, 130
Redi, Francesco, 348, 350
Requien, Esprit, 156-157
Rhinoceros Beetle (see Oryctes)
Ringed Calicurgus (see Pompilus)
River-snail, 163
Rodwell, Miss Frances, 9
Rose-chafer (see Cetonia)
Rove-beetle (see Staphylinus)

S

Sacred Beetle, 21, 112, 144, 170, 230, 374-377, 380, 383
Salamander, 394
Salgues, the author's grandfather, 118
Salgues, Mme., the author's grandmother, 118
Saperda, 259, 270-273
Saperda scalaria (see Saperda)
Saprinus, 218, 364-365
Sarcophaga carnaria (see Grey Flesh-fly)
Sarcophaga haemorrhoidalis, 235
Sardine, 349
Saxicola (see Wheatear)
Scarab (see Sacred Beetle)
Scolia, 23, 46
Scolopendra, 350
Scops, 23
Scorpion (see Languedocian Scorpion)
Screech-owl (see Owl)
Sea-anemone, 27, 240

Sericostoma, 204
Serin-finch, 23
Sheep, 342, 345
Silk-worm, 235
Silky Ammophila (see Ammophila)
Silpha, 218
Sisyphus, 112
Sitaris, 88, 103, 107
Skate, 349
Slug, 350, 403, 407, 415-416, 419
Snail, 144, 159, 428
Snake (see also Æsculapian Snake), 213, 216, 223-224, 226, 231, 235, 238
Snipe, 326, 336
Spallanzani, Abbé Lazaro, 229
Spanish Copris, 112
Spanish Fly, 373, 468
Sparrow, 23, 317
Sphex (see also Languedocian Sphex, White-banded S.), 11, 13, 87
Spider, 22, 276, 316, 450
Sponge, 27
Spurge-caterpillar, 417
Staphylinid (see Oxyporus rufus)
Staphylinus, 218
Star-fish, 27
Stelis, 66-68, 70
Stizus, 21
Stizus ruficornis (see Stizus)
Stone-chat (see Wheatear)

T

Tachytus, 88
Tadpole, 162, 166, 169

Index

Tailor-bee (*see* Anthidium)

Tarantula (*see* Narbonne Lycosa)

Teal, 326

Teixeira de Mattos, Alexander, 20*n*

Theocritus, 294

Three-horned Osmia (*see* Osmia)

Thrush, 326, 336

Tree-frog, 24, 350

Triplax russica, 404, 406

Toad, 23-24, 216

Tournefort, Joseph Pitton de, 398

Turtle-dove, 152

V

Virgil, 82, 152, 278, 294, 432

Vitrina (*see* Glass-snail)

Volucella, *Volucella zonaria* (*see* Bumble-bee Fly)

W

Warbler, 23, 253

Wasp, 11, 16, 18, 25, 26, 41, 46-47, 112, 153, 234, 252-275, 371

Water-beetle, 162, 201-203, 212

Water-boatman, 168, 212

Water-scorpion, 163

Wheatear, 312, 390-393

Whin-chat (*see* Wheatear)

Whirligig, 162, 212

White-banded Sphex, 24

White-tail (*see* Wheatear)

Whiting, 349

Wood-louse, 309

Z

Zoophyte, 27